Geobiology: Objectives, Concepts, Perspectives

Geobiology: Objectives, Concepts, Perspectives

Edited by

N. Noffke

Old Dominion University, Ocean, Earth & Atmospheric Sciences
Norfolk, VA 23529, USA

Reprinted from Palaeogeography, Palaeoclimatology, Palaeoecology,
Volume 219, Nos. 1-2

2005

ELSEVIER

Amsterdam – Boston – Heidelberg – London – New York – Oxford – Paris
San Diego – San Francisco – Singapore – Sydney – Tokyo

ELSEVIER B.V.
Radarweg 29
P.O. Box 211, 1000 AE
Amsterdam, The Netherlands

ELSEVIER Inc.
525 B Street, Suite 1900
San Diego, CA 92101-4495
USA

ELSEVIER Ltd
The Boulevard, Langford Lane
Kidlington, Oxford OX5 1GB
UK

ELSEVIER Ltd
84 Theobalds Road
London WC1X 8RR
UK

First edition 2005

Library of Congress Cataloging in Publication Data
A catalog record is available from the Library of Congress.

British Library Cataloguing in Publication Data
A catalogue record is available from the British Library.

ISBN: 0 444 52019 8

Transferred to digital print 2007
Printed and bound by CPI Antony Rowe, Eastbourne

Working together to grow
libraries in developing countries

www.elsevier.com | www.bookaid.org | www.sabre.org

ELSEVIER   BOOK AID International   Sabre Foundation

CONTENTS

Available online at www.sciencedirect.com

Palaeogeography, Palaeoclimatology, Palaeoecology 219 (2005) 1–3

PALAEO

www.elsevier.com/locate/palaeo

Introduction

Geobiology–a holistic scientific discipline

In less than a decade, a new scientific discipline, "geobiology", has been defined. Rapidly, geobiological studies are published, geobiological research institutions are established, and journals are founded that deal exclusively with this new, pulsing research branch. What is geobiology, and what makes it so fascinating?

Geobiology understands Earth as a system, and life as part of it. In space and time, life influences Earth's development, and Earth's changing environments moulds life. Although geobiology appears new and yet to unfold, its objectives have been focus of research already of the earliest geologists. James Hutton discussed in 1788 the interaction of organisms with geological processes, and described how life affected the history of Earth (Knoll, 2003). Later generations of scientists proved him right. Today, geobiology experiences an enormous expansion and popularity. Why?

One reason lies in the development of science itself. During the past centuries, the increase of knowledge about nature's laws, their development, and their various applications lead to the definition of more and more scientific branches. This diversification was supported by progress in technology. Today, we understand Earth as a holistic system, and we are challenged by the interwoven complexity of nature. In response to this new understanding, and equipped with our highly developed technological abilities, we demand a more appropriate scientific approach: a complex, but holistic system Earth can be understood only by a holistic, interdisciplinary concept of research. The tool of choice is not the diversification,

but the fusion of the disciplines of life sciences. Geobiology is one example of this modern trend.

Geobiology mirrors not only a new development in sciences, but also responses to the new challenges and demands imposed by today's world and society. The three domains of geobiological research are (i) to understand environmental problems of global scale, and to predict unforeseen damages in the future; (ii) to reconstruct the history of our planet, analyzing causes and consequences of life–environment interactions during the joint evolution of life and Earth; and (iii) to explore extraterrestrial worlds by studying analogue environments on Earth. But, how do geobiologists approach this great variety of research topics? How does the concept of a holistic scientific discipline look like?

In order to understand the complex interactions of life in system Earth, geobiologists must determine coupled biological and geological processes across different spatial and temporal scales. That is, they investigate macro- and micro-scale changes in the present (the today's Earth and its biota), but also in the fossil record. The fossil record is provided by million year old rocks, which constitute an archive of ancient life in former and unfamiliar environments. In response to this twofold approach, geobiological studies comprise a specific, dual concept that mirrors the nature of the parent disciplines biology and geology (Noffke, 2002). How can geobiology been seen in the context of its parent disciplines?

Biology and geology are basic research disciplines. Traditionally, their objectives and methodological approaches have different foci and rarely overlap.

0031-0182/$ - see front matter © 2004 Elsevier B.V. All rights reserved.
doi:10.1016/j.palaeo.2004.10.010

Biology predominantly studies present life forms and their interactions with their environments. In consequence, most of the methods of investigations have direct access in the field, and in laboratory experiments. Only short-termed processes can be considered, and biologists seldom have access to data bases that cover more than the last 100 years. Biology does structure and describe nature today, but does not detect or observe long-term processes.

Geoscientists focus on ancient life forms, the reconstruction of the strange environments of Earth's past, and on processes that last millions of years. But in comparison to biology, the geological methods of investigations can only access fossils, sedimentary structures, or minerals that are witnesses of ancient processes manifested in rock. Large-scale geological processes may take very long, so that commonly we cannot observe them during our life time. One example would be the formation of orogens. But the study of the fossil record allows us to conclude on such slow processes, to decipher their causes, and finally to understand, what effects they have on today's nature and the future. The disadvantage of the geological methods is that they require subjective interpretation. Why? The reason is that the fossil record is a history of gaps. Long periods of the Earth history are not recorded simply because rock successions are incomplete. For example, the oldest portion of Earth's history, the Archean eon, comprises 2000 Ma of time, but is represented only by rock successions of a few kilometers thickness. In addition, taphonomy, that is the change of an organism to a lithified fossil, is very selective, and only parts of organisms, or only parts of ancient ecological communities become preserved. And finally, the diagenetic and tectonic alteration of rocks in course of time also overprints information. In all those cases, the biological understanding of life processes of the modern system Earth, help to fill in the gaps of the geological past. For example, we study metabolic cycles of bacterial communities to reconstruct former life conditions of ancient environments not documented in lithologies.

Clearly, we can observe the traditional gap between the objectives and concepts of the parent disciplines. Geobiology bridges this traditional gap. How? The question of geobiologists is that of the interface of life and Earth's environments. This question frames each hypothesis, which originate either from a more biological or from a more geological root (Fig. 1). To answer this question, geobiology combines both direct and indirect methodologies of the parent disciplines, and employs the multifold technologies provided by the pool of their subdisciplines such as microbiology, biogeochemistry, geomicrobiology, astrobiology, paleontology, mineralogy, and many others. For geobiological studies, it is fundamental, that results of investigations in the laboratory or in the modern environment are compared with studies on rocks, and vice versa. This dualism in objective, and methodological approach is the core of geobiology, mirroring its parentage. The twofold research concept permits the synthesis of data, and the fusion of a theoretical

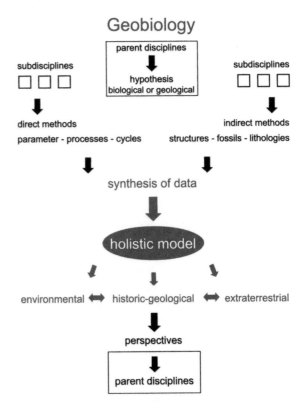

Fig. 1. Geobiology investigates life in the context of system Earth, past and present. The research discipline reflects the trend in modern sciences to study system Earth by fusion of disciplines, not by their diversification. Geobiology bridges the traditional gap between its parent disciplines biology and geology. Its dual concept of approach and methodology creates a holistic model of system Earth, which gives insight into past, present, and future changes. Environmental, historic-geological, and extraterrestrial aspects are topics of geobiological research. Further explanation in text.

model of Earth's environment and life (Fig. 1). This holistic model does not document a present and rigid status. Rather, it reflects flexibilities or weaknesses of the living system Earth, and documents triggering and feed-back effects in global changes over time. The holistic model of system Earth serves to simulations with different parameters. Our understanding of past and present joins causes and consequences, and enables us to predict the future. But then, do we still need the traditional research disciplines?

The core disciplines of biology and geology (including their subdisciplines) are not replaced by geobiology, but continue to exist and function as sources of basic knowledge. Rather, the parental research disciplines get support from a new methodological approach that provides exciting perspectives in environmental, historic-geological, and extraterrestrial research (Fig. 1). In consequence, the term geobiology reflects earlier fruitful combinations of geology with other disciplines, such as physics (geophysics), chemistry (geochemistry), or mathematics (geomathematics). Geobiology touches various subdisciplines of geology and biology in many ways (Olszewski, 2001). This new scientific discipline will serve biogeochemists, paleontologists, biomineralogists, astrobiologists, geomicrobiologists, and many others as a forum for critical discussion and as an opportunity to determine future directions in research. Indeed, geobiology can be understood as a holistic science.

Acknowledgement

The production of such a volume involves many more people than those, whose names appear here. I wish to acknowledge the reviewers, the Editor-in-Chief, and the Editorial Office at Elsevier for their contributions to the realization of this Special Issue.

References

Knoll, A.H., 2003. The geological consequences of evolution. Geobiology 1 (1), 3–14.

Noffke, N., 2002. The concept of geobiological studies: the example of bacterially generated structures in physical sedimentary systems. Palaios 17 (6), 1–2.

Olszewski, T., 2001. Geobiology: a golden opportunity and a call to action. Palaios 16 (6), 1–2.

Nora Noffke
Department of Ocean,
Earth and Atmospheric Sciences,
Old Dominion University,
Norfolk, VA 23529, USA
E-mail address: nnoffke@odu.edu.

Available online at www.sciencedirect.com

Palaeogeography, Palaeoclimatology, Palaeoecology 219 (2005) 5–21

www.elsevier.com/locate/palaeo

Geobiology and the fossil record: eukaryotes, microbes, and their interactions

David J. Bottjer*

Department of Earth Sciences, University of Southern California, Los Angeles, CA 90089-0740, United States

Received 7 July 2003; accepted 29 October 2004

Abstract

Geobiology attempts to understand the interactions between Earth and the life which has evolved on it. It is an all-encompassing field that embraces both living systems on Earth as well as an understanding of the history of these systems since life first evolved. The fossil record and associated geological and geochemical data provide a major avenue towards understanding the evolution of geobiological systems.

Among the many new and exciting directions of research for geobiology, a particularly fertile field is represented by studies of the various interactions between eukaryotes and microbes that can be detected through examination of the fossil record. For example, such interactions include the various roles which microbes play in taphonomy and preservation of eukaryotes as fossils, particularly cases of exceptional fossil preservation. Similarly, recognition of microbially induced sedimentary structures (MISS) in siliciclastic sediments has been a subject of significant recent interest. Documentation of MISS has led to a greater understanding of the role of microbial mats in providing a distinctive structure to marine subtidal seafloors, which has fostered studies of benthic eukaryote adaptations to such mat-dominated environments. The continued evolution of bioturbation in the Cambrian led to elimination of microbial mats on shallow subtidal seafloors, causing evolutionary and ecological changes in eukaryotes adapted to living on and in seafloors structured by such mats. Studies of these changes, termed the "Cambrian substrate revolution", have been documented for a variety of echinoderms, molluscs, and trace fossils, and are some of the first to illuminate ecological interactions between eukaryotes and microbes through study of the fossil record.

This new awareness which geobiology represents has begun to produce a whole host of approaches that causes researchers in the laboratory of the molecular or microbiologist to interact with those from the laboratory of the geochemist as they are led on a field trip by the palaeontologist exploring some fundamental issue in the history of life. Such interactions are not only important for considerations of life on Earth, but provide a framework for the search for life that may have once existed on other planetary bodies, such as Mars. Data from deep time and hence the fossil record plays a central role in such research, and thus geobiology opens up vast new research opportunities for palaeontology.

Keywords: Palaeontology; Palaeoecology; Taphonomy; Microbe; Eukaryote; Precambrian

---

* Tel.: +1 213 740 6100; fax: +1 213 740 8801.
  E-mail address: dbottjer@usc.edu.

0031-0182/$ - see front matter © 2004 Elsevier B.V. All rights reserved.
doi:10.1016/j.palaeo.2004.10.011

1. Introduction

Palaeontology, the study of fossils, is the offspring of geology and biology, a hybrid that differs from biology because its data (fossils) have turned to stone, and from geology because it asks biological questions as well as geological ones. The ebb and flow of palaeontology has varied over the centuries from emphasis on answering geological questions, such as the development of the geological time scale, to answering biological questions, such as elucidating macroevolutionary processes.

With a history of serious study ranging back at least into the seventeenth century (Rudwick, 1985), palaeontology is arguably the founding discipline of geobiology. It has focused primarily on the history of eukaryotic life on Earth and the evolutionary processes that have shaped that history through deep time. In particular, the fossil and stratigraphic records encode much of the history and evidence for long-term evolutionary processes that have shaped the Earth's current biosphere. A large part of what we know about the fossil record has been learned through intensive efforts in the search for natural resources including development of the relative geological time scale. Most of this biostratigraphic effort has been done through study of fossil eukaryotes, which has thus primarily been limited to the Phanerozoic, where there is a fossil record of organisms with shells and bones (e.g., Berry, 1987).

Palaeobiology developed in the 1970s and 1980s as Phanerozoic eukaryotic palaeontologists began to increasingly ask biological questions of the fossil record (e.g., Schopf, 1972; Valentine, 1985). During this time, Precambrian palaeobiology also blossomed as a basic research interest. Here, there was a new emphasis on studying fossil microbes and their sedimentary structures, the stromatolites. With an emphasis on attaining an understanding of the early evolution of life, Precambrian palaeobiology developed almost as a separate culture from Phanerozoic eukaryotic palaeobiology (e.g., Schopf and Klein, 1992; Bengtson, 1994; Schopf, 1999; Knoll, 2003).

During the 1970s and 1980s rapid progress was also made in a variety of geochemical approaches, perhaps most prominently stable isotope studies (e.g., Hoefs, 1997). These studies have focused on understanding climate and environmental changes in the younger parts of Earth history, through analyses of various components of the fossil and stratigraphic record (e.g., Corfield, 1995). Slowly but steadily, these approaches have been extended back in time, deep into the Precambrian, and have provided a unique source of data on environments and evolution of life. Such geochemical approaches have been extremely useful in understanding the nature of Phanerozoic and Precambrian oceans and atmospheres, and geochemical data is crucial to determining the conditions of such remarkable phenomena as Snowball Earth (e.g., Hallam and Wignall, 1997; Knoll, 2003).

Palaeontology has also experienced a burst of research interest towards study of modern environments and organisms, in order to understand palaeoecology (e.g., Schäfer, 1972). Originally, this was largely driven by the search for natural resources, particularly petroleum. However, this research direction widened as eukaryotic palaeoecologists sought to understand the effects of bioturbation upon sediments and sedimentary rocks and the meaning of trace fossils (e.g., Frey, 1975; Bromley, 1996). Related studies, commonly by Precambrian palaeobiologists, sought to understand the role that microbes play in generating sedimentary structures such as stromatolites (e.g., Golubic, 1991). Study on modern environments has also focused on the processes that lead to preservation of mineralized skeletons, as part of the broad field of taphonomy (e.g., Behrensmeyer and Hill, 1980; Weigelt, 1989; Allison and Briggs, 1991a; Martin, 1999; Di Renzi et al., 2002), with the ultimate goal of understanding how the fossil record has been generated. Recently, these research fields have developed to generate questions on how environment affects evolution, under the broad purview of evolutionary palaeoecology (e.g., Allmon and Bottjer, 2001).

The discovery of extraterrestrial iridium in a bed at the Cretaceous–Tertiary boundary by the Alvarezes in the early 1980s founded the modern study of mass extinctions (e.g., Alvarez, 1997). From the flood of research that ensued, we have learned that widespread environmental stress can occur during a very short time interval through extraterrestrial bolide impact (e.g., Koeberl and MacLeod, 2002). An appreciation of mass extinctions has contributed greatly to the understanding of evolution and the history of life. The study of the fossil record of mass extinctions and associated data by Phanerozoic eukaryotic palaeobiologists and Earth scientists in general has strongly increased awareness of the extraterrestrial influence of life on Earth.

Studies on the fossil record of Ediacaran life and the Cambrian explosion have pushed Phanerozoic eukaryotic palaeontologists and sedimentologists back into the Precambrian, and forced them to think more about microbes (e.g., Seilacher, 1999; Bottjer, 2002). The Ediacara fossils, generally of late Neoproterozoic age, are among the most remarkable fossil biotas known from the stratigraphic record. This stems from the fact that this biota is thought to include fossils of some of the earliest larger organisms, whose nature has been much debated. Because Ediacara organisms typically lived on seafloors covered with microbial mats, they were largely adapted to an environment structured by microbes (e.g., Seilacher, 1999; Seilacher et al., 2003). Similarly, as discussed later, their unusual mode of preservation was largely controlled by microbial processes (Gehling, 1999).

As exemplified by this direction in studies of Ediacara fossils, it is an increased awareness by the broad palaeontological community of the role of microbes in the long history of life (e.g., Hagadorn et al., 1999), as well as in mediating biological and geological processes (e.g., Banfield and Nealson, 1997), that has caused much of the recent upsurge of interest in geobiology, particularly geomicrobiology. The conclusion that extraterrestrial impacts can have a strong influence on life has also primed palaeontologists to appreciate processes outside of planet Earth. In particular, renewed interest in the possibility of life on other planets, and the likelihood that if it exists it would be most like microbial life, is reflected in the expansion of astrobiology (e.g., Goldsmith and Owen, 2001). Such interest has been matched by a great interest among molecular and developmental biologists towards understanding the evolutionary origin of animals, represented in the new field of "evolutionary–developmental biology", known as "evo–devo" (e.g., Arthur, 1997; Wilkins, 2002). Questions on the origin of animals logically lead biologists to the Ediacara fossils, and the older metazoan fossils of the Doushantuo Formation in China. This injection of geomicrobiology, astrobiology, and evo–devo into the agenda of palaeontology

has caused much ferment and promises to continue to produce significant advances in our understanding of the history of life and the evolutionary processes which have shaped it. To illustrate this bright future for geobiology and the fossil record, a sampling of some of the promising research avenues that have opened up as part of these new interdisciplinary interactions follows, with an emphasis upon eukaryote–microbe interactions.

2. Microbes and exceptional preservation of fossil eukaryotes

Largely due to interest in how soft-bodied eukaryotes are preserved as fossils, much recent work in palaeontology has focused on how soft tissues are preserved, as part of the science of taphonomy. Taphonomy is a fundamental branch of palaeontology and aims to understand the processes behind the preservation of all fossils ranging from the most common bones and shells to the rare, soft-body remains found in many Lagerstätten (e.g., Donovan, 1991; Allison and Briggs, 1991a; Kidwell and Behrensmeyer, 1993). As such, taphonomy involves the study of the ecological, biogeochemical, and sedimentary processes that occur in the environment before and after burial of organisms. Thus, taphonomy–and the study of the formation of Lagerstätten–is necessarily a science which integrates biology, chemistry, and geology (e.g., Bottjer et al., 2002a).

Under typical taphonomic conditions in marine settings where oxygen is present, scavenging and microbial decay rapidly remove soft tissue from mineralized skeletal elements such as shells and bones. These elements are also subject to scattering by carnivores and scavengers, degradation by agents such as boring microorganisms, chemical dissolution, and physical erosion by waves and currents. Thus, biological remains are typically destroyed before they can be buried by sediment.

However, a small proportion of organic remains do become buried below the seafloor. If the sediment pore waters are undersaturated in dissolved calcium carbonate (the mineral of which most shells are made of), or calcium phosphate (which bones are made of), chemical dissolution will occur. If they are not dissolved, continued deposition of sediment can bury organic remains to the point where they are no longer in the "taphonomically active zone" (TAZ) (Davies et al., 1989; Walker and Goldstein, 1999), and become immune to re-exposure by erosion and damage by organisms which burrow through the sediment surface. It is these biological, chemical, and sedimentary processes that almost all fossils must pass through in order to become preserved.

Very rarely, these processes of decay and destruction are inhibited, producing a conservation Lagerstätte with exceptional preservation of biological remains (e.g., Bottjer et al., 2002b). If, for example, a carcass is buried before it can be disturbed, the quality of preservation may be very high. One mechanism for this is "obrution" (e.g., Seilacher et al., 1985; Brett et al., 1997), where the organism (either alive or recently dead) is smothered and immediately buried by a pulse of sedimentation, such as a storm or turbidite. Once the remains are buried below the TAZ, they are protected from further disturbance. Another way to decrease the chance of being disturbed is to introduce the organism into an environment where dissolved oxygen concentrations are so low that scavengers and aerobic microorganisms are not present; conservation Lagerstätten resulting from this process are termed "stagnation deposits" (Seilacher et al., 1985).

Due to the lack of disturbance under both obrution and stagnation conditions, skeletal articulations will be maintained but soft tissues will commonly still not be preserved (e.g., Allison and Briggs, 1991b,c). So how is eukaryote soft-tissue preservation accomplished in marine environments? Usually it is through the presence of unique geochemical conditions, commonly mediated by the presence of microbial activity, which leads either to the preservation of original organic material or to early precipitation of pyrite, calcium carbonate, or phosphate, which produces a mineral replicate of the soft tissues (e.g., Allison and Briggs, 1991b,c).

2.1. Microbial mats and preservation of Neoproterozoic eukaryotes

These processes for preservation of soft tissues in marine environments, outlined above, are thought to apply primarily to Phanerozoic settings. However, once one steps back into the Precambrian, seafloor conditions are different, producing some different pathways for preservation of soft eukaryote tissues, also involving microbes. For example, although many studies on taphonomy of the Ediacara biota (Fig. 1a) have concentrated on proposed degradational properties of Ediacara soft tissues, several recent studies have focused on the effects of microbial mats on preservation of these fossils (e.g., Gehling, 1986, 1999; Narbonne and Dalrymple, 1992; Narbonne et al., 1997). Evidence is beginning to develop that Neoproterozoic shallow siliciclastic seafloors were typically covered with microbial mats (e.g., Pflüger and Sarkar, 1996; Hagadorn and Bottjer, 1997, 1999; Grazhdankin, 2004). Gehling (1999) has proposed that when event beds covered siliciclastic seafloors on which Ediacara organisms lived, that the smothered microbial mats inhibited the vertical movememt of pore fluids, hence promoting rapid cementation of a sole veneer in the overlying sand (Fig. 2). In this way, the microbial mats may have acted as "death masks" for buried Ediacara organisms (Gehling, 1999).

Similarly, much work is currently being done on the phosphatized microfossils of the Neoproterozoic Doushantuo Formation in China, which are older than the Ediacara fauna and represent fossils of the earliest known animals on Earth (e.g., Li et al., 1998; Xiao et al., 1998; Barfod et al., 2002; Chen et al., 2002, 2004) (Fig. 1b). A review of previous work on the Doushantuo as well as other phosphorites (e.g., Lucas and Prevot, 1991; Siegmund, 1997; Tang et al., 1997; Li et al., 1998; Xiao et al., 1998; Trappe, 1998; Xiao and Knoll, 1999) suggests an environmental model for phosphatization of Doushantuo animal eggs and embryos that also involves microbial mats. For Doushantuo stratigraphic intervals with phosphatized microfossils, deposition would have included both carbonate and siliciclastic sediments, with the relatively shallow seafloors covered by microbial mats as well as dense populations of algae and microscopic animals (infaunal burrowing organisms had not yet evolved). Occasionally, a storm passed through the Doushantuo sea, mobilizing seafloor sediment and the organisms living on it. This mixture of sediment and organisms was transported and deposited elsewhere on the seafloor as a storm layer. While organic material in the storm layer

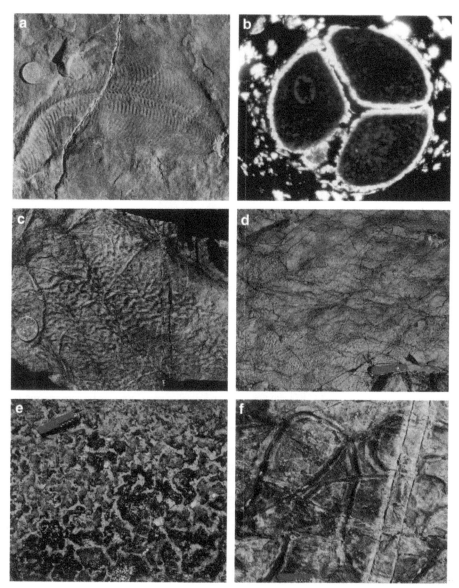

Fig. 1. Photographs exemplifying the geobiology of eukaryotes, microbes, and their interactions. (a) Fossil of the large soft-bodied organism *Pteridinium simplex*, a frondose member of the Ediacara biota, in a sandstone bed of the terminal Neoproterozoic Nama Group in Namibia. (b) View from petrographic thin section showing x-section of phosphatized 4-cell stage embryo from the Neoproterozoic Doushantuo Formation (Weng'an Phosphorite), Guizhou Province, southwest China; embryo is ~400 μm across. (c) Bedding plane surface from Lower Cambrian Harkless Formation (from near Cedar Flats, White-Inyo Mountains, California) showing wrinkle structures with associated horizontal trace fossils. (d) Bedding plane surface from Lower Cambrian Harkless Formation (from near Cedar Flats, White-Inyo Mountains, California) showing extensive but patchy occurrence of wrinkle structures. (e) Modern cryptobiotic soil from Canyonlands National Park, Utah. (f) Bedding plane surface from Lower Cambrian Poleta Formation in Deep Springs Valley of White-Inyo Mountains, California showing the large looping and meandering trace fossil *Taphrhelminthopsis*.

began to degrade, a tough relatively impervious microbial mat would have formed quickly on the surface of the storm layer. Continued degradation of organic matter in the storm layer would have then led to pore water anoxia and caused a sharp increase in the PO4 content of pore waters. Sealing

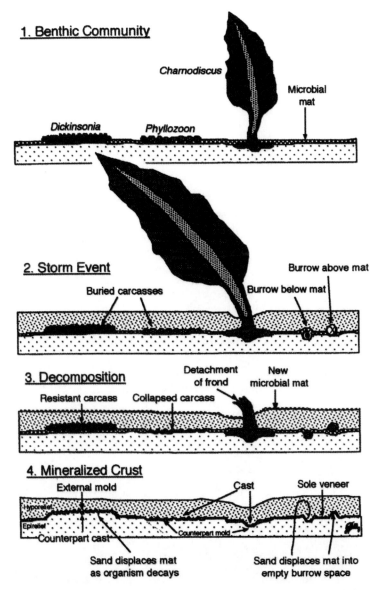

Fig. 2. A schematic model for Ediacara taphonomy. (1) Benthic community of Ediacara biota on a microbial mat; prostrate *Dickinsonia* and *Phylozoon*; upright frond, *Charniodiscus*. (2) Storm event buries Ediacara organisms and mat; frond survives with buried holdfast; infaunal burrows made above mat and below mat. (3) Decomposition of organisms begins with rapid collapse of *Phylozoon* and slow decay of *Dickinsonia*; storm detaches frond, but new mat prevents erosion. Bacterial reduction of iron in the sole veneer. (4) Mineralized crust forms as a death mask in the sole veneer, after complete decay of organic material, the underlying sand forms epirelief counterpart casts and molds. From Bottjer et al. (2002c), modified from Gehling (1999).

by the microbial mat on the surface of the storm layer would have allowed very little to no exchange between pore waters and overlying seawater. Within a matter of days, PO4 concentrations in the storm layer pore waters would have reached supersaturation, and phosphatization of soft tissues, including microscopic animal eggs and embryos, would have occurred. Remobilization and transport of sediments, including phosphatized grains such as eggs, embryos (Fig. 1b) and other accompanying biota, may have occurred numerous times before final deposition. Thus, as for the Ediacara fossils, the presence of a microbial mat on the seafloor may have been the key to enabling preservation of these tiny eukaryotic remains.

As these two examples from the Neoproterozoic show, palaeontologists have developed sophisticated models for eukaryote preservation that infer the key role of microbes and microbial mats in preserving soft tissues, but these models have been developed primarily from observational and geochemical studies (e.g., Briggs and Kear, 1993). The study of how both Precambrian and Phanerozoic fossil Lagerstätten are preserved is thus ripe for experiments where the specific roles of microbes in preservation of soft tissues can be examined in detail. Thus, the next stage of experimental approaches in taphonomy will involve interactions with geomicrobiologists, so that the behavior of microbes in processes of eukaryote fossil preservation can be fully elucidated.

3. Microbial fossils and microbially induced sedimentary structures

In 1996, a report on the possible existence of ancient life on Mars, from the Martian Allan Hills (ALH84001) meteorite reopened the question of whether there was or is life on Mars (McKay et al., 1996). This report reinvigorated scientists to appreciate the possibility that life may exist elsewhere in the solar system, and coupled with studies of mass extinctions reinforced the idea that life and the influences on it is a topic bigger than the Earth. Because the evidence for Martian life in ALH84001 was thought to be microbial, this also brought Precambrian palaeobiologists, and their expertise in studying microbial fossils, to the fore (e.g., Schopf, 1999). These events helped to bring together geomicrobiologists and earth scientists, which has been a strong influence in developing the fields of geobiology and astrobiology. In particular, such efforts have spawned the concept of the "biosignature", which is any evidence for the current or past existence of life, on Earth or any other planetary body (e.g., Conrad and Nealson, 2001). Since palaeontologists are trained to detect biosignatures of ancient life on Earth, they have been particularly well suited in contributing to the search for biosignatures that can be used to search for life on other planets, such as Mars.

3.1. Renewed search for microbially induced sedimentary structures on land and sea

Microbial communities are thought to have flourished in the marine biosphere, particularly during the Archean and Proterozoic. In marine settings, stromatolites provide ample evidence about microbial activity in Archean, Proterozoic, and early Phanerozoic carbonate-dominated facies (e.g., Awramik, 1991). However, in contrast, very limited information has been available for microbial activities in coeval marine siliciclastic-dominated facies.

With the increased emphasis on searching for evidence of microbial activity, there has been a renewed search for sedimentary structures, built or mediated by microbes. Microbes and microbial mats on and in subaqueous surface sediments are typically known for the sedimentary structures that they produce in mineral-precipitating environments (e.g., MacIntyre et al., 1996; Fouke et al., 2000; MacIntyre et al., 2000). The best examples of such structures are the stromatolites from carbonate sedimentary environments. However, it turns out that in subaqueous sedimentary surface environments with little to no mineral precipitation, such as those made of siliciclastic sediments, microbes and microbial mats also mediate the production of sedimentary structures. These structures are much more subtle than the traditional stromatolites, which can range to meters in height. Overall, they have been termed "microbially induced sedimentary structures", or MISS (Noffke et al., 2001).

For example, MISS include wrinkle structures, which have dimensions in the millimeter size range, and which have only recently received extensive attention (e.g., Hagadorn and Bottjer, 1997, 1999; Noffke et al., 2002; Pruss et al., 2004). Wrinkle structures appear on bedding surfaces as irregular quasi-polygonal ridges with dimensions on the millimeter scale, that can be recognized with the naked eye (Hagadorn and Bottjer, 1997, 1999) (Figs. 1c,d and 3). For much of the 20th century, wrinkle structures were recognized in ancient sedimentary sequences, but their genesis was controversial. It has only been in the last five years that a consensus has built that such structures are formed through the mediation of microbial mats (e.g., Hagadorn and Bottjer, 1997, 1999; Noffke et al., 2001, 2002). Studies of ancient

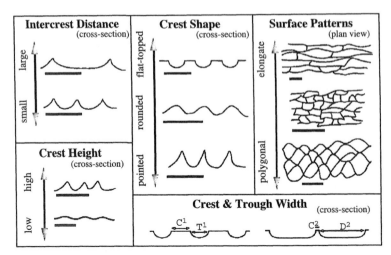

Fig. 3. Schematic representation of morphologic characteristics used to describe wrinkle structures. Surface patterns depicted as lines show orientations of crests or ridges, separated by troughs. All scale bars are 0.5 cm. From Hagadorn and Bottjer (1999).

wrinkle structures have included analyses of morphology as well as microbial filaments that were part of the original mat. These unusual structures can be readily recognized on bedding planes and they have been proposed as potentially important biosignatures for siliciclastic sediments and sedimentary rocks on planetary bodies that have or had standing bodies of water (Bottjer, 2001; Bottjer et al., 2002a). This is analogous to the role that stromatolites can play as biosignatures in subaqueous carbonate environments.

On Earth, wrinkle structures are commonly found in sedimentary rocks deposited in intertidal to deep-sea marine settings of Proterozoic–Cambrian age (e.g., Hagadorn and Bottjer, 1999). In post-Cambrian time, wrinkle structures have typically been restricted to supratidal, intertidal, and deep-sea environments, and this restriction has been attributed to the increase in levels of bioturbation and consequent mixed-layer development after the Cambrian (e.g., Hagadorn and Bottjer, 1999). An exception to this Phanerozoic pattern is the reappearance of shallow subtidal wrinkle structures during the aftermaths of mass extinction periods, such as occurred in the Early Triassic after the end-Permian mass extinction (Pruss et al., 2004).

Do such structures form on land in dry environments as well? A recent report by Prave (2002) demonstrated the likely existence of microbially mediated sedimentary structures in Proterozoic environments that formed on land surfaces which underwent relatively prolonged periods of subaerial exposure. This study by Prave (2002) spotlights the previously unrecognized potential of microbially mediated sedimentary structures as potential biosignatures from dry environments on Earth and other planetary bodies.

An example of such an unusual microbially mediated desert sedimentary structure is cryptobiotic soil, also known as cryptogamic crusts or soil algal crusts (Fig. 1e). Such soils produce a surface crust in widespread arid and semiarid regions of the world, with cyanobacteria as the dominant organism (e.g., Johansen, 1993). These soils have never before been studied in a geological context as biogenic sedimentary structures that could be preserved in ancient sedimentary sequences. However, preliminary investigations show that cryptobiotic soils produce recognizable surface patterns that in many ways are similar to wrinkle structures (e.g., Johansen, 1993) (Fig. 1e).

Cryptobiotic soil is a desert soil that at first inspection appears to be abiotic. However, closer examination reveals that it forms a surface crust which includes extensive microorganisms, particularly filamentous cyanobacteria, but also lichens and mosses (e.g., Johansen, 1993). The sticky sheath material of the cyanobacteria helps to bind the soil particles together, leading to development of the crust (Johansen, 1993). Both live and dead sheaths perform this binding function.

In the deserts of the southwestern US, cryptobiotic soil crusts are found extensively. Due to the binding activity of the cyanobacterial sheaths upon soil particles, cryptobiotic soil has received extensive attention by land managers as a biotic feature that promotes soil stability and retards erosion (e.g., Johansen, 1993). Recently, Dott (2003), expanding on the work of Prave (2002) and others, has proposed that cryptobiotic crusts were extensively developed on Precambrian land surfaces. However, no studies to date have focused on determination of cryptobiotic soil as a sedimentary structure that might be recognized in ancient sedimentary sequences. If criteria can be developed for their recognition on modern desert surfaces as well as in ancient sedimentary sequences, cryptobiotic soils may come to serve as important biosignatures for relatively dry planets that have or had microbial life.

4. Early evolution of animals and the fossil record

4.1. A convergence of disciplines

Work on fossil biotas just before and at the beginning of the Phanerozoic has outlined the empirical pattern of appearance of fossils of animals with mineralized skeletons during this time. This geologically sudden appearance in the fossil record of a diverse array of organisms has been termed the "Cambrian explosion". Much debate exists on whether this represents a primary flowering of metazoan phyla, or the evolution of mineralized skeletons by these animals. Recognition of the Cambrian explosion has been of great interest to all those interested in how animals evolved on Earth (e.g., Gould, 1989; Bengtson, 1994; Knoll, 2003).

Advances in developmental and molecular biology have also led to new insights on how eukaryotic animals may have evolved, with much insight gained through understanding of the *Hox* genes and genomic regulatory systems, which has provided a first glimpse of how different types of animals may have evolved (e.g., Davidson et al., 1995; Knoll and Carroll, 1999; Davidson, 2001). Such evo–devo studies have since been coupled with new molecular clock approaches that provide another pathway to understanding the time of the evolutionary origin of the major groups of animals (e.g., Wray et al., 1996; Peterson et al., 2004). These three research strategies provide the fundamen-

tal basis towards making new and significant progress on the evolutionary origin of animals. They have provided a fruitful basis for interaction with palaeontologists who study, for example, Precambrian fossils of the earliest animals, such as those from the Doushantuo Formation, or palaeoecologists who examine the ecological settings in which these early animals evolved.

4.2. Doushantuo fossils

Molecular biology can tell one how modern organisms are built, and molecular clocks can indicate when these organisms have their evolutionary roots. But these approaches, which involve casting back into time, are vastly strengthened if actual evidence of early life, through fossils, can be found. One of the exceptional discoveries of the past few years has been the recognition of fossils from the Neoproterozoic Doushantuo phosphorites in southwestern China that represent sponges and cnidarians, as well as early bilaterians (e.g., Xiao et al., 1998; Li et al., 1998; Chen et al., 2000, 2002, 2004) (Fig. 1b). There is much to be learned from the Doushantuo fossils, particularly when one has the possibility of actually studying some of the earliest animals right in front of you (e.g., Chen et al., 2004).

The Doushantuo Formation is a fossil Lagerstätten that exhibits a variety of modes of preservation, other than in phosphorites (as previously discussed). For example, black shales of the Doushantuo contain compressions of benthic macroalgae that attached as seaweeds to the seafloor (e.g., Knoll, 2003). Cherts in the Doushantuo Formation also contain abundant spores of algae (e.g., Knoll, 2003). These were some of the first fossils to be discovered in the Doushantuo.

However, it has been the phosphorites that have yielded fossils of the greatest importance. In 1998, a stunning discovery was made by two research groups (Li et al., 1998; Xiao et al., 1998), when the presence of phosphatized animal eggs and embryos, as well as microscopic adult sponges, was reported. The embryos are most probably of sponges or cnidarians (e.g., Knoll, 2003). A variety of microscopic adult cnidarian forms have since been identified (e.g., Chen et al., 2002), and fossil embryos of bilaterians, as well as adults, have also been found (Chen et al., 2000, 2004).

Work on the Doushantuo has only just started. A recent study has served to identify this fossil deposit as possibly as old as 600 million years (Barfod et al., 2002), making these fossils the oldest animals yet found on Earth. Thus, their continued study will be a high priority for geobiologists interested in how animals first evolved (e.g., Chen et al., 2002, 2004).

5. Microbe–eukaryote interactions in ancient benthic environments

5.1. The Cambrian substrate revolution

As already discussed, much of the studied fossil record has been that of eukaryotes, with the microbial fossil record studied primarily by Precambrian paleobiologists. With the convergence of knowledge described above, we are beginning to integrate the fossil record of microbes and eukaryotes, and to determine how interactions between microbes and eukaryotes in ecological settings have influenced the evolution of life.

As shown through studies of stromatolites and wrinkle structures, microbial mats commonly occurred on late Neoproterozoic shallow seafloors. With the advent of vertical bioturbation that marks the Cambrian (Droser et al., 1999), seafloor conditions changed so that mats became less common in subtidal environments (Hagadorn and Bottjer, 1997, 1999; Bottjer et al., 2000; Dornbos et al., 2004). However, with a mosaic of mat-covered and bioturbated seafloors, it is likely that many Cambrian benthic metazoans also were largely adapted for life on mat-covered surfaces. The skeletonized representatives of the Cambrian explosion constitute the "Cambrian Evolutionary Fauna" of Sepkoski (1981). This Cambrian Fauna had its peak during the late Middle and early Late Cambrian, but subsequently faded away in benthic environments (Sepkoski, 1981, 1990). If many of the organisms of the Cambrian Evolutionary Fauna had particular adaptations for living in a benthic world dominated by seafloors covered with microbial mats, then as such mat-covered benthic settings became increasingly rare due to increased bioturbation (e.g., Droser and Bottjer, 1988, 1989; Bottjer et al., 2000; Dornbos et al., 2004), organisms of the Cambrian Fauna unable to adapt to these

changes might have declined, thus leading to their observed steady decrease after the Cambrian (Sepkoski, 1981, 1990). Such a faunal pattern is part of a larger phenomenon that has been termed the "Cambrian substrate revolution" (Bottjer et al., 2000). The Cambrian substrate revolution involved both evolutionary and ecological changes occurring at different time scales, including extinction, adaptation, and environmental restriction, that were caused by the trend towards increasing bioturbation of seafloor sediments in the early Paleozoic (Bottjer et al., 2000).

Through analogy with the development of agriculture and its resulting effects upon soils, Seilacher and Pfluger (1994) have termed this evolution of increasing bioturbation the "agronomic revolution". Late Neoproterozoic seafloors, typically characterized by well-developed microbial mats (e.g., Gehling, 1986, 1996, 1999; Schieber, 1986; Hagadorn and Bottjer, 1997, 1999), had poor development of sediment mixing by vertically oriented burrowing (e.g., Droser et al., 1999; McIlroy and Logan, 1999). Sediment layers on the seafloor thus had relatively low water content and were characterized by a sharp water–sediment interface. Work on carbonates (e.g., Awramik, 1991) and more recently on siliciclastics (e.g., Hagadorn and Bottjer, 1997, 1999) and phosphorites (Dornbos et al., 2004) has shown that in the Cambrian, shallow marine environments characterized by seafloors covered with microbial mats became increasingly scarce, largely due to increasing vertically oriented bioturbation (e.g., Seilacher, 1999). This change to a more "Phanerozoic-style" seafloor resulted in relatively greater water content of seafloor sediment and a blurry water–sediment interface, which led to the first appearance of a mixed layer (e.g., Bottjer et al., 2000). Mixed layers constitute the soupy upper few centimeters of the substrate that are homogenized by bioturbation and are characteristic of later Phanerozoic fine-grained substrates (e.g., Ekdale et al., 1984). With near elimination of microbial mats in shallow marine environments, microbial or mat-related food sources in sediment changed from being well-layered to having a more homogeneously diffuse distribution in the sediment layers on the seafloor. Thus, this agronomic revolution led to the soft-sediment substrates that we commonly see in shallow carbonate and siliciclastic marine environments today (e.g., Bottjer et al., 2000).

Palaeontologists have long been interested in the morphological features evolved by organisms that live on soft sediment seafloors (e.g., Thayer, 1975). Until recently, such adaptations could only be adequately assessed for later Phanerozoic benthic organisms, due to an incomplete understanding of late Neoproterozoic and Cambrian paleobiology and paleoenvironments. New data from the Neoproterozoic–Phanerozoic transition has allowed palaeobiologists to begin to address the adaptive morphology of these early animals. Environments of the Neoproterozoic–Phanerozoic transition were different from those today, requiring the use of non-uniformitarian approaches to analyze the palaeobiology and palaeoecology of animals living at this time (e.g., Bottjer, 1998). Seilacher (e.g., 1999) has postulated that lifestyles of organisms that lived on late Neoproterozoic sediments characterized by microbial mats, or "matgrounds", would include: (1) "mat encrusters", which were permanently attached to the mat; (2) "mat scratchers", which grazed the surface of the mat without destroying it; (3) "mat stickers", which were suspension feeders that were partially embedded in the mat, and comprise a subset of adaptations resulting in organisms broadly termed "sediment stickers"; and (4) "undermat miners", which burrowed underneath the mat and fed on decomposing mat material.

The presence of metazoan fossils perhaps as old as 600 million years in the Doushantuo Formation (Barfod et al., 2002) suggests that there was an early stage of evolution for most benthic metazoan groups before they evolved mineralized skeletons (e.g., Fortey et al., 1996, 1997). This early stage of evolution for benthic organisms was within the environmental context of a "Neoproterozoic-style" minimally bioturbated seafloor covered with microbial mats. Thus, how did this late Neoproterozoic–Phanerozoic transition to more Phanerozoic-style seafloor conditions affect the evolution, dispersal, and palaeoenvironmental distribution of metazoans which were adapted to these Neoproterozoic seafloor sediments? Were there animals and perhaps entire communities that were adapted to these seafloor conditions, in the manner proposed by Seilacher (1999)?

We cannot yet fully answer these questions. However, mounting evidence suggests that many evolutionary and ecological changes which took place during this time interval were due to the transition in substrate style from the late Neoproterozoic marine environments and lifestyles described by Seilacher (e.g., 1999), to the bioturbated sedimentary environments and morphological adaptations documented for later Phanerozoic benthic organisms (e.g., Thayer, 1975). Early suspension-feeding echinoderms provide an example of the effects of this change in substrate character (Dornbos and Bottjer, 2000, 2001).

5.2. Echinoderms and the Cambrian substrate revolution

Evolution of Cambrian suspension-feeding echinoderms that had an immobile, or sessile, lifestyle provides strong evidence for the short-term impact of the Cambrian substrate revolution. For example, the unusual Early Cambrian helicoplacoid echinoderms were well-adapted for survival on Neoproterozoic-style substrates. These small (1–5 cm) suspension-feeding echinoderms (Fig. 4) lived as sediment stickers on a substrate that underwent only low to moderate levels of horizontally directed bioturbation and did not have a mixed layer (Dornbos and Bottjer, 2000). Helicoplacoids lacked typical Phanerozoic soft-substrate adaptations, such as the ability to attach to available hard substrates or presence of a rootlike holdfast. Significant increase in depth and intensity of bioturbation in shallow-water muds and sands through the Cambrian (e.g., Droser, 1987) destroyed the stable substrates that these small echinoderms required and likely led to their extinction (Dornbos and Bottjer, 2000) (Fig. 4).

In contrast, edrioasteroids and eocrinoids, the other groups of undisputed Cambrian sessile suspension-feeding echinoderms, were both able to adapt to the change in substrates created by increased bioturbation (Bottjer et al., 2000). The earliest edrioasteroids lived unattached on the seafloor during the Early and Middle Cambrian, but by the Late Cambrian, edrioasteroids lived attached to available hard substrates (e.g., Sprinkle and Guensburg, 1995) (Fig. 4). Similarly, several Early and Middle Cambrian eocrinoids were stemless and lived unattached on the seafloor (Ubaghs,

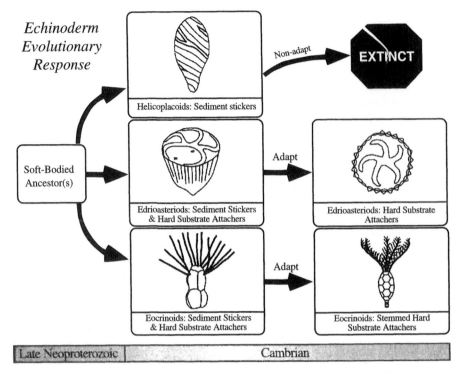

Fig. 4. Evolutionary response of Cambrian sessile suspension-feeding echinoderms as part of the Cambrian substrate revolution. Arrows do not represent a direct evolutionary relationship between the specific echinoderms shown, but imply a general evolutionary trend through the Cambrian within each of the groups examined, with these echinoderms serving as individual examples of different adaptive strategies. The helicoplacoid drawing is modified from Paul and Smith (1984); specimen is 3 cm in height. For edrioasteroids, *Camptostroma* (left) is modified from Paul and Smith (1984); specimen shown is 5 cm in height. Edrioasteroid on right is a schematic of a typical attaching edrioasteroid, modified from Paul and Smith (1984); size is 5 cm in width. For eocrinoids, *Lichenoides* (left) is modified from Ubaghs (1967); specimen is approximately 2.5 cm in height. Eocrinoid on right is *Tatonkacystis*, modified from Sumrall et al. (1997); specimen is approximately 5 cm in height. Geological time not to scale and boxes do not represent the precise age range of the echinoderms they contain. From Bottjer et al. (2000).

1967; Sprinkle, 1992) (Fig. 4). By the Late Cambrian, however, eocrinoids had evolved stems and also lived attached to available hard substrates (Fig. 4). Thus, by attaching to hard substrates or by developing stems, each of these Cambrian echinoderm groups avoided the detrimental effects of increased substrate instability caused by increasing bioturbation (Fig. 4), and they survived into the post-Cambrian Paleozoic (Bottjer et al., 2000). The remaining undisputed Cambrian echinoderms were all mobile deposit- or suspension-feeders (e.g., Sprinkle, 1992). Their mobility likely exacerbated the substrate changes occurring during this time, and, because they could adjust their position relative to the sediment–water interface, they would have been relatively immune to the effects of this change in substrate character.

5.3. Additional implications of the Cambrian substrate revolution

Evidence from trace fossils in the Neoproterozoic and Early Cambrian that consists of radular scratch marks upon the seafloor in neritic palaeoenvironments are likely indicators that early molluscs crawled over the mat-bound seafloor and grazed the sediment surface for microbes (e.g., Bottjer et al., 2000; Dornbos et al., 2004). Such trace fossils are usually only found after the Early Cambrian as scratch marks on hard substrates. The most primitive molluscs that have a fossil record, the monoplacophorans and the polyplacophorans, are in the modern typically restricted to hard substrate environments and the deep sea (e.g., Bottjer et al., 2000). Thus, although little currently is known about the ecology of soft-bodied late Neoproterozoic and Cambrian ancestors of poly-placophorans and monoplacophorans, they may have lived on soft as well as hard substrates in shallow marine environments and grazed microbial mats which covered the seafloor, a lifestyle which today is typically restricted to hard substrates and the deep sea.

Because adaptations to these mat-covered and more coherent Neoproterozoic-style soft substrates required different morphologies and behaviors than soupier Phanerozoic-style soft substrates, the Cambrian explosion is also characterized by a unique variety of bedding-parallel trace fossils. For example, large meandering trace fossils such as *Plagiogmus* and *Taphrhelminthopsis* (Fig. 1f) were common in Early Cambrian shallow marine environments, yet were likely made by soft-bodied metazoans for which we have no body fossil record (McIlroy and Heys, 1997; Hagadorn et al., 2000). Similarly, several other meandering trace fossils as well as those exhibiting a network pattern, including *Helminthoida* and *Paleodictyon*, also occur in Cambrian strata deposited in shallow marine environments (Crimes and Fedonkin, 1994). A number of these Cambrian trace fossil genera, as well as ichnogenera with similar morphologies, are found only in deep-sea strata after the Cambrian, and thus are united by a similar paleoenvironmental history of onshore–offshore retreat (Bottjer et al., 1988; Crimes and Fedonkin, 1994; Hagadorn et al., 2000). This pattern of post-Cambrian restriction to the deep sea by bedding-parallel trace fossils is mirrored by the record of microbial structures produced in siliciclastic sediments (Hagadorn and Bottjer, 1999). Thus, as for grazing molluscs, the environmental restriction shown by trace fossils is likely also an effect of the Cambrian substrate revolution, caused by the broad increase in vertically directed bioturbation, and consequent decrease in development of microbial mats, in shallow marine soft substrate environments (Bottjer et al., 2000).

Further analysis of the Cambrian substrate revolution may contribute to a better understanding of broader evolutionary phenomena. The Cambrian is characterized by a wide variety of metazoans, reflected in both body and trace fossils, many of which have morphologies that appear strange to the modern eye (e.g., Gould, 1989). Perhaps the co-occurrence during the Cambrian of benthic metazoans adapted more to Neoproterozoic-style soft substrates, with those more adapted to Phanerozoic-style substrates, contributed significantly to the high morphological disparity exhibited by animals of the Cambrian explosion.

6. Conclusions

The fossil record and associated geological and geochemical data provide a major avenue towards understanding the evolution of geobiological systems. Among the many new and exciting directions of research for geobiology, a particularly fertile field is represented by studies of the numerous interactions between eukaryotes and microbes that can be detected through examination of the fossil record. For example, such interactions include the various roles which microbes play in taphonomy and preservation of eukaryotes as fossils, particularly cases of exceptional fossil preservation. Similarly, recognition of microbially induced sedimentary structures (MISS) in siliciclastic sediments has been a subject of significant recent interest. Documentation of MISS has led to a greater understanding of the role of microbial mats in providing a distinctive structure to marine subtidal seafloors, which has fostered studies of benthic eukaryote adaptations to such mat-dominated environments. The continued evolution of bioturbation through the Cambrian led to elimination of microbial mats on subtidal seafloors, causing evolutionary and ecological changes in eukaryotes adapted to living on and in seafloors structured by such mats. Studies of these changes, termed the "Cambrian substrate revolution", have been documented for a variety of echinoderms, molluscs, and trace fossils, and are some of the first to illuminate ecological interactions between eukaryotes and microbes through study of the fossil record.

Eukaryote palaeontologists have a long tradition of studying modern biological systems in order to make sense of the fossil bones, shells and biogenic sedimentary structures that they find in the fossil record. Precambrian palaeobiologists have also provided much of the impetus for studies on modern microbes and the sedimentary structures that they produce. Such research has provided much of the foundation for the studies from the fossil record described herein on interactions between eukaryotes and microbes.

Similarly, there are many research programs in several other rapidly growing disciplines of geobiology which focus on modern materials and processes. This can be seen in the efforts of geomicrobiologists to understand how microbes affect, for example, rates of weathering, and other processes that heretofore had been treated primarily as physical and chemical phenomena (e.g., Banfield and Nealson, 1997). And, much of astrobiology has been concerned with modern processes towards finding current life on other planets. But, a significant number of these research programs are at some point forced to consider that evidence from deep time is necessary when trying to search for and understand the development of biological systems. This trend is likely to continue, and it ensures a bright future for studies in geobiology that include information from the fossil record. Although we are a species that tends to focus on short-term outcomes, the increasing realization of the complexity of natural systems compels us to take the long view.

Acknowledgements

Thanks are made to P.G. Conrad, J.-Y. Chen, F. Corsetti, E.H. Davidson, S.Q. Dornbos, M.L. Droser, A.G. Fischer, J.P. Grotzinger, J.W. Hagadorn, K.H. Nealson, C.-W. Li, N. Noffke, P. Oliveri, J.W. Schopf, and C.M. Tang for stimulating discussion that led to the development of this paper.

References

Allison, P.A., Briggs, D.E.G., 1991a. Taphonomy: Releasing the Data Locked in the Fossil Record. Plenum Press, New York (560 pp.).

Allison, P.A., Briggs, D.E.G., 1991b. The taphonomy of soft-bodied animals. In: Donovan, S.K. (Ed.), The Processes of Fossilization. Columbia University Press, New York, pp. 120–140.

Allison, P.A., Briggs, D.E.G., 1991c. Taphonomy of nonmineralized tissues. In: Allison, P.A., Briggs, D.E.G. (Eds.), Taphonomy: Releasing the Data Locked in the Fossil Record. Plenum Press, New York, pp. 25–70.

Allmon, W.D., Bottjer, D.J. (Eds.), 2001. Evolutionary Paleoecology: The Ecological Context of Macroevolutionary Change. Columbia University Press, New York (357 pp.).

Alvarez, W., 1997. T. rex and the Crater of Doom. Vintage Books, New York (185 pp.).

Arthur, W., 1997. The Origin of Animal Body Plans: A Study in Evolutionary Developmental Biology. Cambridge University Press (338 pp.).

Awramik, S.M., 1991. Archaean and Proterozoic stromatolites. In: Riding, R. (Ed.), Calcareous Algae and Stromatolites. Springer-Verlag, Berlin, pp. 289–304.

Banfield, J.F., Nealson, K.H. (Eds.), 1997. Geomicrobiology: Interactions Between Microbes and Minerals, Reviews in Mineralogy, vol. 35. The Mineralogical Society of America, Washington, DC (448 pp.).

Barfod, G.H., Albarede, F., Knoll, A.H., Xiao, S., Baker, J., Frei, R., 2002. New Lu–Hf and Pb–Pb age constraints on the earliest animal fossils. Earth Planet. Sci. Lett. 201, 203–212.

Behrensmeyer, A.K., Hill, A.P. (Eds.), 1980. Fossils in the Making. University of Chicago Press (338 pp.).

Bengtson, S. (Ed.), 1994. Early Life on Earth. Columbia University Press, New York (630 pp.).

Berry, W.B.N., 1987. Growth of a Prehistoric Time Scale. Blackwell Scientific Publications, New York (158 pp.).

Bottjer, D.J., 1998. Phanerozoic non-actualistic paleoecology. Geobios 30, 885–893.

Bottjer, D.J., 2001. Prospecting for life on Mars: Using the 500–600 million-year-old Earth record of microbial siliciclastic structures as an analogue. In: Eleventh Annual V.M. Goldschmidt Conference, Abstract #3383. LPI Contribution No. 1088, Lunar and Planetary Institute, Houston (CD-ROM).

Bottjer, D.J., 2002. Enigmatic Ediacara fossils: ancestors or aliens? In: Bottjer, D.J., Etter, W., Hagadorn, J.W., Tang, C.M. (Eds.), Exceptional Fossil Preservation: A Unique View on the Evolution of Marine Life. Columbia University Press, New York, pp. 11–33.

Bottjer, D.J., Droser, M.L., Jablonski, D., 1988. Palaeoenvironmental trends in the history of trace fossils. Nature 333, 252–255.

Bottjer, D.J., Hagadorn, J.W., Dornbos, S.Q., 2000. The Cambrian substrate revolution. GSA Today 10 (9), 1–7.

Bottjer, D.J., Dornbos, S.Q., Corsetti, F.A., Hagadorn, J.W., 2002a. Microbial mats and siliciclastic substrates: implications for understanding life on Earth and the search for life on Mars. Abstracts, 16th international sedimentological congress. Rand Afrikaans University, Johannesburg, South Africa, p. 36.

Bottjer, D.J., Etter, W., Hagadorn, J.W., Tang, C.M., 2002b. Fossil-Lagerstätten: jewels of the fossil record. In: Bottjer, D.J., Etter, W., Hagadorn, J.W., Tang, C.M. (Eds.), Exceptional Fossil Preservation: A Unique View on the Evolution of Marine Life. Columbia University Press, New York, pp. 1–10.

Bottjer, D.J., Etter, W., Hagadorn, J.W., Tang, C.M. (Eds.), 2002c. Exceptional fossil preservation: a unique view on the evolution of marine life. Columbia University Press, New York 403 pp.

Brett, C.E., Baird, G.C., Speyer, S.E., 1997. Fossil Lagerstätten: stratigraphic record of paleontological and taphonomic events. In: Brett, C.E., Baird, G.C. (Eds.), Paleontological Events: Stratigraphic, Ecological, and Evolutionary Implications. Columbia University Press, New York, pp. 3–40.

Briggs, D.E.G., Kear, A.J., 1993. Fossilization of soft-tissues in the laboratory. Science 259, 1439–1442.

Bromley, R.G., 1996. Trace Fossils: Biology, Taphonomy and Applications. Chapman and Hall, London (361 pp.).

Chen, J.-Y., Oliveri, P., Li, C.-W., Zhou, G.-Q., Gao, F., Hagadorn, J.W., Peterson, K.J., Davidson, E.H., 2000. Precambrian animal diversity: putative phosphatized embryos from the Doushantuo Formation of China. Proc. Natl. Acad. Sci. 97, 4457–4462.

Chen, J.-Y., Oliveri, P., Gao, F., Dornbos, S.Q., Li, C.-W., Bottjer, D.J., Davidson, E.H., 2002. Precambrian animal life: probable developmental and adult cnidarian forms from southwest China. Dev. Biol. 248, 182–196.

Chen, J.-Y., Bottjer, D.J., Oliveri, P., Dornbos, S.Q., Gao, F., Ruffins, S., Li, C.-W., Davidson, E.H., 2004. Small bilaterian fossils from 40–55 my before the Cambrian. Science 305, 218–222.

Conrad, P.G., Nealson, K.H., 2001. A non-earthcentric approach to life detection. Astrobiology 1, 15–24.

Corfield, R.M., 1995. An introduction to the techniques, limitations and landmarks of carbonate oxygen isotope palaeothermometry. In: Bosence, D.W.J., Allison, P.A. (Eds.), Marine Palaeoenvironmental Analysis from Fossils, Geol. Soc., Spec. Publ., vol. 183. The Geological Society, London, pp. 27–42.

Crimes, T.P., Fedonkin, M.A., 1994. Evolution and dispersal of deepsea traces. Palaios 9, 74–83.

Davidson, E.H., 2001. Genomic Regulatory Systems: Development and Evolution. Academic Press, San Diego (261 pp.).

Davidson, E.H., Peterson, K.J., Cameron, R.A., 1995. Origin of bilaterian body plans: evolution of developmental regulatory mechanisms. Science 270, 1319–1325.

Davies, D.J., Powell, E.N., Stanton Jr., R.J., 1989. Taphonomic signature as a function of environmental process: shells and shell beds in a hurricane-influenced inlet on the Texas coast. Palaeogeogr. Palaeoclimatol. Palaeoecol. 72, 317–356.

Di Renzi, M., Pardo Alonso, M.V., Belinchon, M., Penalver, E., Montoya, P., Marquez-Aliaga, A. (Eds.), 2002. Current Topics on Taphonomy and Fossilization. Ajuntament de Valencia 544 pp.

Donovan, S.K. (Ed.), 1991. The Processes of Fossilization. Columbia University Press, New York 303 pp.

Dornbos, S.Q., Bottjer, D.J., 2000. Evolutionary paleoecology of the earliest echinoderms: helicoplacoids and the Cambrian substrate revolution. Geology 28, 839–842.

Dornbos, S.Q., Bottjer, D.J., 2001. Taphonomy and environmental distribution of helicoplacoid echinoderms. Palaios 16, 197–204.

Dornbos, S.Q., Bottjer, D.J., Chen, J.-Y., 2004. Evidence for seafloor microbial mats and associated metazoan lifestyles in Lower Cambrian phosphorites of southwest China. Lethaia 37, 1–11.

Dott Jr., R.H., 2003. The importance of eolian abrasion in supermature quartz sandstones and the paradox of weathering on vegetation-free landscapes. J. Geol. 111, 387–405.

Droser, M.L., 1987. Trends in extent and depth of bioturbation in Great Basin Precambrian–Ordovician strata, California, Nevada, and Utah: Unpublished PhD thesis, University of Southern California, Los Angeles 365 pp.

Droser, M.L., Bottjer, D.J., 1988. Trends in extent and depth of bioturbation in Cambrian carbonate marine environments, western United States. Geology 16, 233–236.

Droser, M.L., Bottjer, D.J., 1989. Ordovician increase in extent and depth of bioturbation: implications for understanding early Paleozoic ecospace utilization. Geology 17, 850–852.

Droser, M.L., Gehling, J.G., Jensen, S., 1999. When the worm turned: concordance of Early Cambrian ichnofabric and trace-fossil record in siliciclastic rocks of South Australia. Geology 27, 625–628.

Ekdale, A.A., Muller, L.N., Novak, M.T., 1984. Quantitative ichnology of modern pelagic deposits in the abyssal Atlantic. Palaeogeogr. Palaeoclimatol. Palaeoecol. 45, 189–223.

Fortey, R.A., Briggs, D.E.G., Wills, M.A., 1996. The Cambrian evolutionary "explosion": decoupling cladogenesis from morphological disparity. Biol. J. Linn. Soc. 57, 13–33.

Fortey, R.A., Briggs, D.E.G., Wills, M.A., 1997. The Cambrian evolutionary "explosion" recalibrated. BioEssays 19, 429–434.

Fouke, B.W., Farer, J.D., Des Marais, D.J., Pratt, L., Sturchio, N.C., Burns, P.C., Discipulo, M.K., 2000. Depositional facies and aqueous-solid geochemistry of travertine-depositing hot springs (Angel Terrace, Mammoth Hot Springs, Yellowstone National Park, U.S.A.). J. Sediment. Res. 70, 565–586.

Frey, R.W. (Ed.), 1975. The Study of Trace Fossils. Springer-Verlag, New York 562 pp.

Gehling, J.G., 1986. Algal binding of siliciclastic sediments: a mechanism in the preservation of Ediacaran fossils. 12th International Sedimentology Congress, Abstracts, p. 117.

Gehling, J.G., 1996. Taphonomy of the Terminal Proterozoic Ediacaran Biota, South Australia. Unpublished PhD thesis, University of California, Los Angeles 222 pp.

Gehling, J.G., 1999. Microbial mats in terminal Proterozoic siliciclastics. Ediacaran death masks. Palaios 14, 40–57.

Goldsmith, D., Owen, T., 2001. The Search for Life in the Universe, 3rd ed. University Science Books, Sausalito, California (573 pp.).

Golubic, S., 1991. Modern stromatolites: a review. In: Riding, R. (Ed.), Calcareous Algae and Stromatolites. Springer-Verlag, Berlin, pp. 541–561.

Gould, S.J., 1989. Wonderful Life. W.W. Norton and Company, New York (347 pp.).

Grazhdankin, D., 2004. Patterns of distribution in the Ediacaran biotas: facies versus biogeography and evolution. Paleobiology 30, 203–221.

Hagadorn, J.W., Bottjer, D.J., 1997. Wrinkle structures: microbially mediated sedimentary structures common in subtidal siliciclastic settings at the Proterozoic–Phanerozoic transition. Geology 25, 1047–1050.

Hagadorn, J.W., Bottjer, D.J., 1999. Restriction of a late Neoproterozoic biotope: suspect-microbial structures and trace fossils at the Vendian–Cambrian transition. Palaios 14, 73–85.

Hagadorn, J.W., Pflüger, F., Bottjer, D.J. (Eds.), 1999. Unexplored Microbial Worlds, Palaios, vol. 14, pp. 1–93.

Hagadorn, J.W., Schellenberg, S.A., Bottjer, D.J., 2000. Paleoecology of a large Early Cambrian bioturbator. Lethaia 33, 142–156.

Hallam, A., Wignall, P.B., 1997. Mass Extinctions and their Aftermath. Oxford University Press (321 pp.).

Hoefs, J., 1997. Stable Isotope Geochemistry. Springer-Verlag, Berlin (201 pp.).

Johansen, J.R., 1993. Cryptogamic crusts of semiarid and arid lands of North America. J. Phycol. 29, 140–147.

Kidwell, S.M., Behrensmeyer, A.K. (Eds.), 1993. Taphonomic Approaches to Time Resolution in Fossil Assemblages, Short Courses in Paleontology, vol. 6. The Paleontological Society, University of Tennessee, Knoxville 302 pp.

Knoll, A.H., 2003. Life on a Young Planet. Princeton University Press (277 pp.).

Knoll, A.H., Carroll, S.B., 1999. Early animal evolution: emerging views from comparative biology and geology. Science 284, 2129–2137.

Koeberl, C., MacLeod, K.G. (Eds.), 2002. Catastrophic Events and Mass Extinctions: Impacts and Beyond, Spec. Pap.-Geol. Soc. Am., vol. 356 (746 pp.).

Li, C.-W., Chen, J.-Y., Hua, T.E., 1998. Precambrian sponges with cellular structures. Science 279, 879–882.

Lucas, J., Prevot, L.E., 1991. Phosphates and fossil preservation. In: Allison, P.A., Briggs, D.E.G. (Eds.), Taphonomy: Releasing the Data Locked in the Fossil Record. Plenum, New York 560 pp.

MacIntyre, I.G., Reid, R.P., Steneck, R.S., 1996. Growth history of stromatolites in a Holocene fringing reef, Stocking Island, Bahamas. J. Sediment. Res. 66, 142–231.

MacIntyre, I.G., Prufert-Bebout, L., Redi, R.P., 2000. The role of endolithic cyanobacteria in the formation of lithified laminae in Bahamian stromatolites. Sedimentology 47, 915–921.

Martin, R.E., 1999. Taphonomy: A Process Approach. Cambridge University Press (508 pp.).

McIlroy, D., Heys, G.R., 1997. Palaeobiological significance of *Plagiogmus arcuatus* from the Lower Cambrian of central Australia. Alcheringa 21, 161–178.

McIlroy, D., Logan, G.A., 1999. The impact of bioturbation on infaunal ecology and evolution during the Proterozoic–Cambrian transition. Palaios 14, 58–72.

McKay, D.S., Gibson Jr., E.K., Thomas-Keprta, K.L., Vali, H., Romaneck, C.S., Clemett, S.J., Cillier, X.D.F., Maechling, C.R., Zare, R.N., 1996. Search for past life on Mars: possible relic biogenic activity in martian meteorite ALH84001. Science 273, 924–930.

Narbonne, G.M., Dalrymple, R.W., 1992. Taphonomy and ecology of deep-water Ediacaran organisms from Northwestern Canada. In: Lidgard, S., Crane, P.R. (Eds.), Fifth North American Paleontological Convention Abstracts with Programs, Paleontological Society Special Publication, vol. 6, p. 219.

Narbonne, G.M., Dalrymple, R.W., MacNaughton, R.B., 1997. Deep-water microbialites and Ediacara-type fossils from Northwestern Canada. Abstr. Programs-Geol. Soc. Am. 29, A-193.

Noffke, N., Gerdes, G., Klenke, Th., Krumbein, W.E., 2001. Microbially induced sedimentary structures—a new category within the classification of primary sedimentary structures. J. Sediment. Res. 71, 649–656.

Noffke, N., Knoll, A.H., Grotzinger, J.P., 2002. Sedimentary controls on the formation and preservation of microbial mats in siliciclastic deposits: a case study from the Upper Neoproterozoic Nama Group, Namibia. Palaios 17, 533–544.

Paul, C.R.C., Smith, A.B., 1984. The early radiation and phylogeny of echinoderms. Biol. Rev. Camb. Philos. Soc. 59, 443–481.

Peterson, K.J., Lyons, J.B., Nowak, K.S., Takacs, C.M., Wargo, M.J., McPeek, M.A., 2004. Estimating metazoan divergence times with a molecular clock. Proc. Natl. Acad. Sci. 101, 6536–6541.

Pflüger, F., Sarkar, S., 1996. Precambrian bedding planes—bound to remain. Abstr. Programs-Geol. Soc. Am. 28, 491.

Prave, A.R., 2002. Life on land in the Proterozoic: evidence from the Torridonian rocks of northwest Scotland. Geology 30, 811–814.

Pruss, S., Fraiser, M., Bottjer, D.J., 2004. Proliferation of Early Triassic wrinkle structures: implications for environmental stress following the end-Permian mass extinction. Geology 32, 461–464.

Rudwick, M.J.S., 1985. The Meaning of Fossils, 2nd ed. University of Chicago Press (287 pp.).

Schäfer, W., 1972. Ecology and Palaeoecology of Marine Environments. University of Chicago Press (568 pp.).

Schieber, J., 1986. The possible role of benthic microbial mats during the formation of carbonaceous shales in shallow Mid-Proterozoic basins. Sedimentology 33, 521–536.

Schopf, T.J.M. (Ed.), 1972. Models in Paleobiology. Freeman, Cooper and Co., San Francisco (250 pp.).

Schopf, J.W., 1999. Cradle of Life. Princeton University Press (367 pp.).

Schopf, J.W., Klein, C., 1992. The Proterozoic Biosphere. Cambridge University Press (1348 pp.).

Seilacher, A., 1999. Biomat-related lifestyles in the Precambrian. Palaios 14, 86–93.

Seilacher, A., Pfluger, F., 1994. From biomats to benthic agriculture: a biohistoric revolution. In: Krumbein, W.E., et al. (Eds.), Biostabilization of Sediments. Bibliotheks- und Informationssystem der Universität Oldenburg, pp. 97–105.

Seilacher, A., Reif, W.E., Westphal, F., 1985. Sedimentological, ecological and temporal patterns of fossil Lagerstätten. Philos. Trans. R. Soc. Lond., B 311, 5–23.

Seilacher, A., Grazhdankin, D., Legouta, A., 2003. Ediacaran biota: the dawn of animal life in the shadow of giant protists. Paleontol. Res. 7, 43–54.

Sepkoski Jr., J.J., 1981. A factor analytic description of the Phanerozoic marine fossil record. Paleobiology 7, 36–53.

Sepkoski Jr., J.J., 1990. Evolutionary faunas. In: Briggs, D.E.G., Crowther, P.R. (Eds.), Palaeobiology. A Synthesis. Blackwell, Oxford, pp. 37–41.

Siegmund, H., 1997. Microfacies, geochemistry, and genetic aspects of lowermost Cambrian phosphorites of South China. In: Chen, J.-Y., et al. (Eds.), The Cambrian Explosion and the Fossil Record, Bulletin of the National Museum of Natural Science, Taiwan, vol. 10, pp. 143–159.

Sprinkle, J., 1992. Radiation of echinodermata. In: Lipps, J.J., Signor, P.W. (Eds.), Origin and Early Evolution of the Metazoa. Plenum Press, New York, pp. 375–398.

Sprinkle, J., Guensburg, T.E., 1995. Origin of echinoderms in the Paleozoic Evolutionary Fauna: the role of substrates. Palaios 10, 437–453.

Sumrall, C.D., Sprinkle, J., Guensburg, T.E., 1997. Systematics and paleoecology of Late Cambrian echinoderms from the western United States. J. Paleontol. 71, 1091–1108.

Tang, T., Shkolnik, E.L., Xue, Y., Yu, C., 1997. Determination of the conditions of formation of Sinian and Early Cambrian granular phosphorites of the Yangtze region, China. In: Chen, J.-Y., et al. (Eds.), The Cambrian Explosion and the Fossil Record, Bulletin of the National Museum of Natural Science, Taiwan, vol. 10, pp. 117–132.

Thayer, C.W., 1975. Morphological adaptations of benthic invertebrates on soft substrata. J. Mar. Res. 33, 177–189.

Trappe, J., 1998. Phanerozoic phosphorite depositional systems: a dynamic model for a sedimentary resource system. Lecture Notes in Earth Science, vol. 76. Springer Verlag, Berlin, 316 pp.

Ubaghs, G., 1967. Eocrinoidea. In: Moore, R.C. (Ed.), Treatise on Invertebrate Paleontology, Part S, Echinodermata, vol. 1, pp. 445–495.

Valentine, J.W., 1985. Phanerozoic Diversity Patterns: Profiles in Macroevolution. Princeton University Press (441 pp.).

Walker, S.E., Goldstein, S.T., 1999. Taphonomic tiering: experimental field taphonomy of molluscs and foraminifera above and below the sediment–water interface. Palaeogeogr. Palaeoclimatol. Palaeoecol. 149, 227–244.

Weigelt, J., 1989. Recent Vertebrate Carcasses and their Paleobiological Implications. University of Chicago Press (188 pp.).

Wilkins, A.S., 2002. The Evolution of Developmental Pathways. Sinauer Associates, Sunderland, Massachusetts (603 pp.).

Wray, G.A., Levinton, J.S., Shapiro, L.H., 1996. Molecular evidence for deep pre-Cambrian divergences among metazoan phyla. Science 274, 568–573.

Xiao, S., Knoll, A.H., 1999. Fossil preservation in the Neoproterozoic Doushantuo phosphorite Lagerstätte, South China. Lethaia 32, 219–240.

Xiao, S., Zhang, Y., Knoll, A.H., 1998. Three-dimensional preservation of algae and animal embryos in a Neoproterozoic phosphorite. Nature 391, 553–558.

Available online at www.sciencedirect.com

Palaeogeography, Palaeoclimatology, Palaeoecology 219 (2005) 23–33

ELSEVIER

www.elsevier.com/locate/palaeo

Geobiology and paleobiogeography: tracking the coevolution of the Earth and its biota

Bruce S. Lieberman*

Department of Geology, 120 Lindley Hall, 1475 Jayhawk Blvd., University of Kansas, Lawrence, KS 66045, United States

Received 13 February 2003; accepted 29 October 2004

Abstract

Paleobiogeographic research is an important area of geobiology that involves the study of the coevolution of the Earth and its biota by considering how tectonic and climatic changes have affected the evolution and distribution of organisms. The intellectual heritage of the discipline stretches back well before Darwin. Phylogenetic approaches to paleobiogeography have played an important part in the expansion of the field, and recent analyses have incorporated Geographic Information Systems (GIS). Each of these approaches, by enhancing the precision of paleobiogeography, helps make the discipline more relevant to geobiology.

The interaction occurring between the geological and biological sciences in paleobiogeography is apparent in several areas. First, there is the emergence of new techniques such as the ability to analyze ancient DNA sequences. Also, paleobiogeographers and biogeographers have realized that geo-dispersal, a biogeographic process first identified through studies of the fossil record, can powerfully influence the evolution and distribution of biotas. Finally, biogeographers have recognized that paleontological incompleteness and extinction constrain our ability to reconstruct biogeographic patterns in the fossil record and the extant biota, respectively. Each of these developments suggests that further growth in paleobiogeography will involve important interactions between studies involving the fossil record and the extant biota, and this, along with the discipline's commitment to studying how tectonic and climatic changes have influenced evolution, reaffirms the validity of the synthetic field that is geobiology.
© 2004 Elsevier B.V. All rights reserved.

Keywords: Paleobiogeography; Geobiology; Phylogenetics; Extinction; Geographic Information Systems (GIS)

1. Introduction

Geobiology is that unifying discipline that seeks to span and link the geological and biological sciences. Paleobiogeography is an important research area within geobiology because it is aimed at tracing the

* Tel.: +1 785 864 2741; fax: +1 785 864 5276.
  E-mail address: blieber@ku.edu.

0031-0182/$ - see front matter © 2004 Elsevier B.V. All rights reserved.
doi:10.1016/j.palaeo.2004.10.012

coevolution of the Earth and its biota. Its neontological cousin, biogeography, has the same aims but generally omits data from the fossil record. Paleobiogeography and biogeography focus on where and why groups of organisms are distributed over the face of the Earth. The Earth system is important to the biological system because it controls, through a variety of processes, many aspects of organismal distribution and evolution. A major research area within paleobiogeography is relating evolutionary patterns in clades of organisms to tectonic and climatic changes. This area of research has a long intellectual heritage extending back well before Darwin's (1859) publication of *On the Origin of Species* (Mayr, 1982; Browne, 1983; Lieberman, 2000), and a brief reprise of this heritage helps illustrate the future prospects for synthesis between the geo- and biosciences. Paleobiogeography also contributed crucial evidence to the development of theories of continental drift and plate tectonics (Hallam, 1981).

More recently, the discipline has shown renewed interest and growth for several reasons, including the use of phylogenetic approaches and Geographic Information Systems (GIS) to analyze paleobiogeographic patterns; these techniques enhance the quantitative nature of the discipline, increasing precision in testing hypotheses. Other exciting and important recent advances include the application of ancient DNA to paleobiogeography; the extension of techniques to the fossil record that were once restricted to the modern biota further blurs the line between geology and biology. Also important is that although some biologists have claimed that phylogenetic biogeography should be based solely on the search for congruent patterns of speciation linked to vicariance, paleontologists have recognized that there is another type of congruent biogeographic process that can be studied: geo-dispersal. Vicariance is the process whereby geographic barriers form within the range of one or more species due to geological or climatic changes, and the formation of these barriers causes populations to become geographically isolated; the populations then diverge and eventually speciate. It is a type of allopatric speciation. Geo-dispersal is the process whereby geographic barriers fall, again due to geological or climatic changes, and then several species can subsequently and congruently

expand their geographic ranges (Lieberman and Eldredge, 1996; Lieberman, 1997). If geo-dispersal is not considered in biogeographic studies, the resulting biogeographic patterns may be incomplete or inaccurate. Finally, paleobiogeographers have come to recognize that extinction is an important mechanism which effects a biologists' ability to study biogeographic patterns when only the modern biota is considered. This means that biologists must more fully utilize paleontological data, again offering another important bridge between the geo- and biosciences.

The increasingly quantitative nature of paleobiogeography, the interdigitation of biogeographic and paleobiogeographic analyses through the touchstone of ancient DNA, documenting the effects extinction has on the modern biota, and the importance of biogeographic processes first uncovered in the fossil record, like geo-dispersal, signals that paleobiogeography is an area of geobiology where the links between the geo- and biosciences continue to grow.

2. The history of biogeography and the move towards a geobiologically centered science

2.1. Early approaches to biogeography

Buffon may have been the first to realize the powerful affect the Earth system had on its biota; for example, he argued "the Earth makes the plants; the Earth and the plants make the animals" (Buffon, 1749–1804 in Mayr, 1982, p. 441). Although operating in a pre-Darwinian framework, many of Buffon's writings indicate that he believed that life showed a pattern akin to evolutionary descent from a common ancestor, and then if different regions have similar species, at one time, they must have been connected (Lieberman, 2000). Buffon's view on the control the Earth exerted on life was refined by Augustin de Candolle (1817, 1820), who recognized a difference between factors that control organismal distribution on the small and large scales. Factors that operate on the small scale include temperature, light, and ecology, while factors at the large scale were related to geology (Nelson, 1978; Browne, 1983). Lyell (1832), building on de Candolle's work, recognized that the geographic ranges of species and clades were dynamic and changed with geological or climatic changes; he argued that in order to understand the geographic distribution of organisms, it was necessary to learn the geological and climatic history of the regions that they occupied. Darwin (1839) offered a series of examples to try to integrate Lyell's (1832) approach to the study of the distribution of organisms.

2.2. Evolutionary biogeography and allopatric speciation

It was Wallace (1855) who was the first to recognize the correlation between geographic distribution and evolutionary relationship (Brooks, 1984; Lieberman, 2000). Wallace (1855) in fact described how a process akin to what is now called vicariance might have produced modern faunal differences in the Galapagos Islands if these now distinct islands were once joined.

In effect, Wallace (1855) was arguing that one way the geological world impinges on the biological world is through the mechanism we now refer to as allopatric speciation. If speciation is allopatric, species can disperse over geographic barriers (that have geological or climatic causes) and become isolated, or geological or climatic changes can cause populations of species to become isolated from one another by creating barriers within formerly continuous ranges (Wiley and Mayden, 1985; Brooks and McLennan, 1991, 2002); the latter is termed vicariance. In either case, the isolated populations diverge and eventually speciate. Mayr (1942) argued cogently that most speciation occurs allopatrically, a view that subsequent studies (e.g., Mayr, 1982; Wiley and Mayden, 1985; Brooks and McLennan, 1991, 2002) continue to confirm. If newly introduced geographic barriers become particularly prominent, they may induce evolutionary divergence among a host of biotic elements confined to the same region (Funk and Brooks, 1990). If it is geological or climatic changes that cause most cases of geographic isolation, and these in turn motivate much of evolutionary change, then the biological organisms are essentially passive, with their evolution driven by the contingent geological changes (Lieberman, 2000). Such a view provides a clear rationale for a synthetic geobiology in both biogeography and paleobiogeography and

suggests that, to borrow from Croizat (1964), the Earth and its biota have coevolved.

Hooker (1853), Murray (1866), and Huxley (1870) were early scientists who emphasized the important role vicariance played in driving evolutionary divergence (Bowler, 1996). Darwin also championed a similar view in the early years of his notebooks (Grinnell, 1974; Mayr, 1976, 1982; Kottler, 1978; Sulloway, 1979; Richardson, 1981). Thus, in the 19th century, there were real possibilities for a synthetic geobiology based on biogeography and paleobiogeography, but this synthesis was not to be. The primary reason involved a subtle shift in Darwin's views on the role geographic isolation plays in motivating divergence and speciation, first in his notebooks (in Barrett et al., 1987) and later in Darwin (1859, 1872), coupled with Darwin's great intellectual influence. Darwin came to focus more on the role competition between species and other biological mechanisms played in promoting divergence, while de-emphasizing geological change as an agent of evolution (Lieberman, 2000). In essence, his later views on speciation more closely match sympatric speciation than allopatric speciation, thereby diminishing the role that geological phenomena like tectonics and climate change play in evolution. Undoubtedly, Darwin (1859) is the most important publication in the history of biology, and also clearly had many positive influences on geology, especially paleontology. However, it is ironic that Darwin's shift away from a geologically centered view of speciation put the brakes on an emerging geobiological synthesis in the fields of biogeography and paleobiogeography, though undoubtedly, it accelerated syntheses in other areas.

Wallace's role in the development of biogeography and geobiology is important, for he not only tied evolutionary change to geological change, helping to create an early synthetic geobiology, but he also was the first to argue that scientists can use the evolutionary histories of faunas to elucidate geological history. For example, "it is evident that, for the complete elucidation of the present state of the fauna of each island and each country, we require a knowledge of its geological history...a knowledge of the fauna and its relation to that of the neighboring countries will often throw great light upon the geology, and enable us to trace out with tolerable certainty its past history" (Wallace, 1857 in Brooks, 1984, p. 163).

Even though, on the whole, allopatric speciation was de-emphasized by evolutionary biologists towards the end of the 19th century and the beginning of the 20th century, eventually, allopatric speciation became more and more accepted. This renewed emphasis on allopatric speciation was championed by Mayr (1942) and subsequently utilized in the punctuated equilibrium hypothesis of Eldredge and Gould (1972). In their hypothesis, Eldredge and Gould (1972) emphasized that species evolve in allopatry (with the role that geology and climate plays in sometimes producing allopatry already described above) and were stable throughout much of their history, with the bulk of evolutionary change occurring at speciation. Mayr (1942) and Eldredge and Gould (1972) validate the role of paleobiogeography as an area of evolutionary research because the fossil record is the place to study how evolution is associated with geological and climatic changes over long periods of time.

3. Geo-dispersal, a paleontological contribution to biogeographic theory

Unfortunately, with the increased acceptance of the importance of allopatric speciation came a dogmatic insistence by some authors (e.g., Croizat et al., 1974; Nelson and Platnick, 1981; Patterson, 1983; Humphries and Parenti, 1986), most of whom studied the extant biota, that vicariance was the only biogeographic process that produced congruent responses in biotas. As already discussed, the emergence of geographic barriers can isolate populations of several different co-occurring species and promote divergence. However, this ignores the opposite side of the coin because just as geological and climatic change can sometimes cause barriers to form, at other times, they may cause barriers to fall, allowing many taxa to simultaneously expand their range. It also ignores a long history of paleontological research which shows that the fossil record is replete with numerous examples of congruent range expansion by independent clades, a pattern that has been termed geo-dispersal by Lieberman and Eldredge (1996) and Lieberman (1997, 2000).

Some prominent examples of geo-dispersal in the paleontological literature include McKenna (1975, 1983), who documented numerous cases of wholesale movement by mammals between Europe and North America and North America and Asia throughout the Cenozoic related to tectonic events and climatic changes. Beard (1998) also recovered more evidence for geo-dispersal by mammals from Asia into North America in the late Paleocene and early Eocene driven by warming events. At a broad scale, Hallam (1992) described numerous cases of geo-dispersal spanning the Phanerozoic in marine invertebrate faunas related to rising and falling sea-level. Lieberman and Eldredge (1996) also documented movements by trilobites between different marine basins in eastern North America during the Middle Devonian related to sea-level rise which joined marine connections between formerly isolated epeiric seas. Finally, Sereno et al. (1996) and Sereno (1997, 1999) found evidence for geo-dispersal by dinosaurian faunas within and across continents during the Early Cretaceous (Fig. 1).

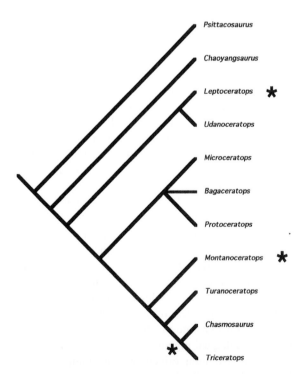

Fig. 1. Phylogeny of ceratopsian dinosaurs, adapted from Sereno (1999), with each "*" marking an inferred dispersal event in the clade.

Studies of the modern biota by Sanmartin et al. (2001) and Conti et al. (2002) have reiterated these results from fossil faunas by their strong support for the existence of geo-dispersal. Not only is geo-dispersal a valid process, but biogeographic methods that only look for vicariance and ignore geo-dispersal will miss important biogeographic patterns and will potentially be inaccurate.

The recognition of the existence of geo-dispersal has a long historical pedigree which extends at least back to paleontological information described by Lyell (1832), suggesting that the initial impetus for the recognition of this process came from the analysis of the fossil record. However, a growing number of biologists have come to recognize the potential biogeographic importance of geo-dispersal (e.g., Riddle, 1996; Ronquist, 1998; Bisconti et al., 2001; Brooks and McLennan, 2002; Conti et al., 2002), although not all of these authors used that term to describe the process. This increasing recognition by biologists of the existence of processes first documented in the fossil record is further evidence for the growing geobiological synthesis between paleobiogeographers and biogeographers.

4. Using paleobiogeography to evaluate the relationship between geology and evolution

4.1. Grand scale analyses

Several authors have tied the intensity of biological evolution to the intensity of geological change. For example, Valentine and Moores (1970, 1972) and Tiffney and Niklas (1990) found an excellent correlation between the position of the Earth's continents and biological diversity: When continents were fragmented and there should have been more opportunities for allopatric speciation, biological diversity is high; when the continents were together with few geographic barriers and less opportunities for allopatric speciation, diversity is low. Times of profound biological diversification, like the Ordovician radiation, also correspond to times of major tectonic change (Miller and Mao, 1995; Owen and Crame, 2002). All of these studies focused on global and regional scale diversity curves and sought to establish correlations between these

and various geological processes. Some studies, such as Wells et al. (1999), also found a link between tectonic activity at mid-ocean ridges and enhanced productivity in the oceans.

4.2. Phylogenetic paleobiogeographic studies

Paleobiogeographic studies have also concentrated more explicitly on the relation between geological change and evolution. For example, there is a long history in paleobiogeography of using quantitative, rigorous approaches to analyze patterns of similarity between faunas from different regions, and how these change through time. Hallam (1977, 1983, 1994) has been one of the pioneers in this area. Phylogenetic approaches to paleobiogeography are also valuable because they allow the relationship between geological change and evolution to be considered in the context of individual speciation events, which are potentially more precise than considering similarities and differences across regional biotas (Lieberman, 2000, 2003a). The increasing emphasis of paleobiogeographic studies on phylogenetic approaches arguably has increased the rigor of the discipline, making this area more likely to appeal to scientists across the geological and biological sciences.

Phylogenetic paleobiogeographic analyses have been used to study the Cambrian radiation: that key episode in the history of life when diverse Metazoan remains appear in the fossil record. Several authors (e.g., Knoll, 1991; Signor and Lipps, 1992; Dalziel, 1997) had suggested that geological changes played an important role in causing the Cambrian radiation, and notably, the Cambrian radiation is preceded by some of the most profound tectonic reorganizations in Earth history. Lieberman (1997, 2002a,b) tested this proposition in greater detail using a phylogenetic paleobiogeographic approach to analyze trilobites, the most diverse and abundant Early Cambrian animals. These analyses used a modified version of Brooks Parsimony Analysis (BPA); BPA is described in detail by Brooks (1985), Wiley (1988), Funk and Brooks (1990), Brooks and McLennan (1991, 2002), and Wiley et al. (1991), while the modified version of BPA is described in detail by Lieberman and Eldredge (1996) and Lieberman (1997, 2000). In essence, the modified version of BPA works by taking the phylogenies of organisms, converting these to area cladograms by substituting the species name with its geographic distribution, and then coding this information into two data matrices: one designed to retrieve congruent episodes of vicariance and one designed to retrieve congruent episodes of geo-dispersal. Each matrix is then analyzed using a parsimony algorithm. The results are expressed as most parsimonious vicariance and geo-dispersal trees which provide information about the relative time that barriers rose between regions, isolating faunas and causing vicariance, or fell joining regions and causing geo-dispersal, respectively.

Lieberman (1997, 2003b,c) found evidence for a tectonic control, especially related to rifting, on paleobiogeographic patterns in the Early Cambrian. Moreover, patterns of faunal relationship between many of the regions considered mirrored the inferred Neoproterozoic and Early Cambrian breakup sequence predicted by other geological techniques. By contrast, cycles of sea-level rise and fall had less of an effect on paleobiogeographic patterns. The patterns of vicariance in the Early Cambrian were more resolved than the patterns of geo-dispersal, which intuitively makes sense, as the Early Cambrian was largely a time of cratonic fragmentation, at least for the regions considered by Lieberman (1997, 2003b,c). Finally, the excessive opportunities for vicariance in trilobites initiated by the tectonic events are associated with high speciation rates, as the Cambrian radiation is a time of high, though not phenomenally high, speciation rates (Lieberman, 2001). These results provide clear evidence that Earth history exerts an important control on evolution during one of the key episodes in the history of life. Was tectonics the sole factor that caused the radiation? Probably not, but tectonic events did leave a fundamental signature in the evolutionary patterns and probably played an important role in influencing, if not precipitating, the Cambrian radiation.

Retrieving the profound signature of Earth history events on biological evolution is a common theme of phylogenetic paleobiogeographic studies. For example, in another study, Lieberman and Eldredge (1996) again applied the modified version of BPA to study phylogenetic paleobiogeographic patterns in trilobites, but this time during a very

different time in Earth history, with a signature very different from the Early Cambrian: the Middle Devonian. Just as the Earth history signatures differed for the two time periods, the corresponding paleobiogeographic and evolutionary signatures differed as well. The Middle Devonian was a time of major cycles of sea-level rise and fall (Hallam, 1992), and the Earth's cratons may have been approaching a Pangea configuration at this time, with abundant episodes of continental amalgamation and collision (Scotese, 1997). The Middle Devonian geo-dispersal tree from trilobites was well resolved and very similar to the resultant vicariance tree. The latter result suggests that the processes controlling vicariance were also producing geo-dispersal, implicating sea-level cycles as important paleobiogeographic factors. Finally, the repeated opportunities for geo-dispersal are associated with reduced speciation rates in Devonian trilobites relative to their Cambrian trilobite kin (Lieberman, 1999). In both the Early Cambrian and Middle Devonian examples, the important role that Earth history plays in motivating evolution speaks eloquently to the validity of the geobiological synthesis.

4.3. Ancient DNA and paleobiogeography

Another area of new methodological developments that may contribute in an important way to the growth of geobiology is the analysis of ancient DNA. This technique has increased the potential for interdigitation between paleobiogeography and biogeography. Existing ancient DNA sequences are rare, and of course in most cases, the DNA of fossil or subfossil remains cannot be isolated; still, this is likely to be an area of future growth. In an innovative study, Haddrath and Baker (2001) obtained complete mitochondrial genomes from two species of moa (large, extinct, flightless birds) from New Zealand that were driven to extinction by humans in the Holocene. They used these sequences, along with sequences from five extant large flightless bird species (this entire group of extinct and extant birds comprises the ratites: a group including the emu, kiwi, moa, ostrich, rhea, and others), in a comparative, phylogenetic framework, to consider the paleobiogeography of Gondwana. Based on the phylogeny derived from the analysis of DNA

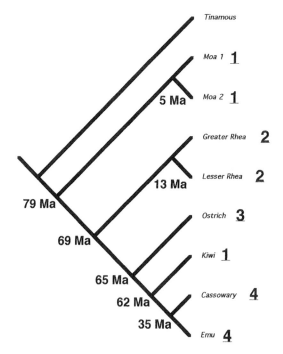

Fig. 2. Phylogeny of extant and extinct ratite birds based on DNA sequences, adapted from Haddrath and Baker (2001). The dates at each node of the phylogeny are the inferred times of lineage divergence; the underlined numbers next to each of the taxon names represent areas of occurrence where: 1=New Zealand; 2=South America; 3=Africa; and 4=Australia. Based on the divergence dates, the kiwi lineage must have dispersed into New Zealand after these islands had become separated from the rest of Gondwana. (Similarly, the arrival of the ostrich lineage in Africa also appears due to dispersal from elsewhere in Gondwana.)

sequences and also on information about divergence times gleaned from the sequences, Haddrath and Baker (2001) found that some of the major splitting events in the ratite lineage were related, via vicariance, to the breakup of Gondwana (Fig. 2). Their results also suggested, however, that New Zealand may have been colonized by ratites (the kiwi lineage) long after those islands had split apart from Australia and the rest of Gondwana.

Another interesting study using ancient DNA is Austin and Arnold (2001), who isolated DNA from now extinct land tortoises that were endemic to the Mascarenes, an island chain east of Malagasy. Their analysis used the phylogeny of the tortoises to adduce paleobiogeographic patterns, and they found that the evolution of these land tortoises was governed by repeated episodes of dispersal from

Mauritius to other islands in the Mascarene island chain.

5. Extinction and the relevance of the fossil record to biogeography

Extinction is a process that can make it difficult to retrieve accurate paleobiogeographic patterns. This is because separated biotas may appear to diverge through time not due to evolutionary change but simply because one or more biotas has been differentially affected by a series of smaller extinction events or by a mass extinction. It seems likely that extinction may explain apparent increases in endemism seen in various dinosaur faunas through time (Sereno, 1999). The effects of extinction are even more profound on biogeographic studies of the extant biota because in a fundamental sense, the modern biota is critically pruned and incomplete (Lieberman, 2000, 2002b).

Phylogenetic biogeographic studies look for biogeographic congruence between clades to uncover biogeographic patterns (Fig. 3). Of course the real world is more complicated than this, and there is often some incongruence between clades, which is why analytical methods are needed to tease apart this type of signal in the data from a variety of sources of noise. However, if we recognize that all modern clades have some deeper history, and furthermore, that extinction occurs in these clades, then incongruence can emerge between the extant representa-

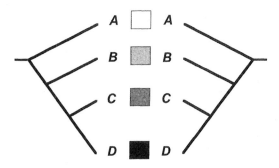

Fig. 3. Two side by side, hypothetical area cladograms with the geographic distributions of the terminal taxa represented by different shapes and given next to the species names. These clades show congruent biogeographic patterns of evolution across geographic space.

tives of two clades if they are pruned of their fossil representatives (Fig. 4). Such examples are said to be artificially incongruent (Lieberman, 2002b). The existence of some incongruence does not obviate the effectiveness of biogeographic analyses, but if enough incongruence accumulates, biogeographic patterns will be unresolved or only weakly supported. Artificial incongruence is one type of incongruence that should certainly be avoided whenever possible because it can lead to a loss of biogeographic resolution for reasons that have nothing to do with evolution or biogeography.

Lieberman (2002b) recognized that because of the artificial biogeographic incongruence in extant clades that can emerge due to extinction, there are times when the fossil record may be a more accurate repository of paleobiogeographic patterns than the extant biota. This is especially true for modern clades that have long durations or high extinction rates, which can lead to the accumulation of excessive extinction. In a similar vein, paleobiogeographic studies should avoid groups or time periods with a poor fossil record. The individual shortcomings of both paleobiogeographic and biogeographic studies suggest that it will often be advantageous to analyze fossil and living groups together to avoid the problem of artificial incongruence (Lieberman, 2002b). Again, the argument for increasing interdigitation between

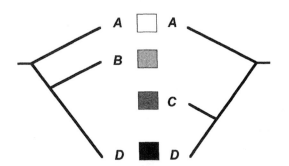

Fig. 4. A case of artificial incongruence involving the same two hypothetical clades shown in Fig. 3, but with the clade on the left having the species from the lightly shaded area unsampled and the clade on the right having the species from the darkly shaded unsampled. Such differences could plague a neontological study if one taxon from each clade was extinct and could not be sampled. These area cladograms are congruent if the entire history of each clade could be sampled, but appear incongruent due to artificial sampling biases, in a biogeographic framework based on BPA, because they imply different patterns of area relationship.

paleobiogeography and biogeography represents another compelling argument for a unified geobiological approach.

6. GIS and paleobiogeography

One promising new area of paleobiogeographic research is the application of Geographic Information Systems (GIS) (ESRI Press, 1999) to the study of the fossil record. Graham et al. (1996) have used these techniques as part of the FAUNMAP project to study shifts in Cenozoic terrestrial vertebrate communities through time and in relation to environmental changes. The way GIS is used in FAUNMAP allows the researchers to ask detailed questions about the dynamics of paleoecological change during a time of major climatic change.

GIS can also be combined with phylogenetic paleobiogeographic studies to focus on how species within clades change their geographic distributions through time and during cladogenesis. Rode and Lieberman (2000, 2001) have applied GIS to study paleobiogeographic patterns and processes during the Late Devonian biodiversity crisis. GIS can be used to create spatial distribution maps for many species and at many different time periods, thus enabling a large scale analysis of entire biotas. Furthermore, when integrated with PaleoGIS (Ross and Scotese, 2000), species ranges can be mapped onto Late Devonian paleogeography (Rode and Lieberman, 2000, 2001, 2004). GIS can also be used to calculate precisely and repeatedly the geographic ranges of fossils species; this is valuable because it makes it possible to quantitatively track changes in geographic range through time. When phylogenetic information is included in an analysis based on GIS, geographic ranges can be correlated with key parameters like speciation and extinction rates; it is then possible to picture the relationship between rates of evolution and extinction and changes in geographic range, allowing for a unique perspective on paleobiogeographic patterns and processes and macroevolution. Rode and Lieberman (2000, in press) found that when GIS and phylogenetics are combined together, they offer a comprehensive picture of the Late Devonian biodiversity crisis, building on the important work of McGhee (1996). Their results suggested that the causes for the biodiversity crisis involved, first, species expanding their geographic ranges due to tectonic collision and sea-level rise; and these range expansion were then correlated with declining speciation rates and rising extinction rates. The types of paleontological studies that integrate phylogenetics and GIS are only in the early stages, but this seems to be a promising new avenue for paleobiogeographic and macroevolutionary studies.

7. Conclusions

Geobiology is emerging as an important synthetic discipline embracing the geological and biological sciences. Paleobiogeography forms an important research area within geobiology that aims to address how Earth history change, especially tectonic and climatic changes, influences evolution. Recent advances in paleobiogeography point out the value of the synergistic geobiological approach, especially by emphasizing the way that studies of the fossil record and the extant biota can be mutually informative. Conceptual and methodological advances in paleobiogeography include the development of new analytical methods for the analysis of phylogenetic paleobiogeographic data, improved techniques for the extraction of ancient DNA, and revivified concepts such as geo-dispersal. These advances increase the opportunities for interaction between paleobiogeography and biogeography and also increase the analytical precision of this type of research. Furthermore, the growing interaction between these two subdisciplines and the important role that paleobiogeography plays as a bridge between the geological and biological sciences represents a compelling argument that geobiology is a research endeavor with great promise.

Acknowledgements

Thanks to A. Decho, R. Kaesler, N. Noffke, and A. Rode for comments on earlier versions of this manuscript, and thanks to N. Noffke for inviting me to participate in this volume. This research was supported by NSF OPP-9909302, EAR-0106885, NASA EXB03-0001-0001, and a Self Faculty Fellowship.

References

Austin, J.J, Arnold, E.N., 2001. Ancient mitochondrial DNA and morphology elucidate an extinct island radiation of Indian Ocean giant tortoises (cylindraspis). Proceedings of the Royal Society of London, Series B 268, 2515–2523.

Barrett, P.H., Gautrey, P.J., Herbert, S., Kohn, D., Smith, S., 1987. Charles Darwin's notebooks, 1836–1844. Cornell University Press, Ithaca, New York. 747 pp.

Beard, K.C., 1998. East of Eden: Asia as an important center of taxonomic origination in mammalian evolution. Bulletin of Carnegie Museum of Natural History 34, 5–39.

Bisconti, M., Landini, W., Bianucci, G., Cantalamessa, G., Carnevale, G., Ragaini, L., Valleri, G., 2001. Biogeographic relationships of the Galapagos terrestrial biota: parsimony analyses of endemicity based on reptiles, land bards and Scalesia land plants. Journal of Biogeography 28, 495–510.

Bowler, P.J., 1996. Life's Splendid Drama. University of Chicago Press, Chicago, IL. 525 pp.

Brooks, J.L., 1984. Just Before the Origin: Alfred Russel Wallace's Theory of Evolution. Columbia University Press, New York. 284 pp.

Brooks, D.R., 1985. Historical ecology: a new approach to studying the evolution of ecological associations. Annals of the Missouri Botanical Garden 72, 660–680.

Brooks, D.R., McLennan, D.A., 1991. Phylogeny, Ecology, and Behavior. University of Chicago Press, Chicago, IL. 434 pp.

Brooks, D.R., McLennan, D.A., 2002. The Nature of Diversity. University of Chicago Press, Chicago, IL. 668 pp.

Browne, J., 1983. The Secular Ark: Studies in the History of Biogeography. Yale University Press, New Haven, CT. 273 pp.

Buffon, G.L., 1749–1804. Histoire Naturelle, Générale et Particulière. Imprimerie Royale, Puis Plassan, Paris

Conti, E., Eriksson, T., Schonenberger, J., Systsma, K.J., Baum, D.A., 2002. Early tertiary out of-India dispersal of crypteroniaceaea: evidence from phylogeny and molecular dating. Evolution 56, 1931–1942.

Croizat, L., 1964. Space, Time, and Form, the biological synthesis. Author, Caracas. 881 pp

Croizat, L., Nelson, G., Rosen, D.E., 1974. Centers of origin and related concept. Systematic Zoology 23, 265–287.

Dalziel, I.W.D., 1997. Neoproterozoic–paleozoic geography and tectonics: review, hypothesis, and environmental speculations. Geological Society of America Bulletin 109, 16–42.

Darwin, C., 1839. The Voyage of the Beagle. Penguin Books, New York, NY. 432 pp.

Darwin, C., 1859. On the origin of species by means of natural selection; or the preservation of favored races in the struggle for life, Reprinted 1st edition. Harvard University Press, Cambridge, MA. 502 pp.

Darwin, C., 1872. On the origin of species by means of natural selection; or the preservation of favored races in the struggle for life, Reprinted 6th edition. New American Library, New York. 479 pp.

de Candolle, A.P., 1817. Mémoire sur la géographie des plantes de France, considéré dans ses rapports avec la hauteur absolue.

Mémoirs de Physique et de Chimie de la Société d'Arcueil 3, 262–322.

de Candolle, A.P., 1820. Géographie Botanique, in Dictionare des Sciences Naturelles 18, 359–422.

Eldredge, N., Gould, S.J., 1972. Punctuated equilibria: an alternative to phyletic gradualism. In: Schopf, T.J. (Ed.), Models in Paleobiology. Freeman, Cooper, San Francisco, pp. 82–115.

ESRI Press, 1999. Getting to Know Arcview GIS. ESRI Press, Redlands, CA.

Funk, V.A., Brooks, D.R., 1990. Phylogenetic systematics as the basis of comparative biology. Smithsonian Contributions to Botany 73, 1–45.

Graham Jr., R.W., Graham, M.A., Schroeder III, E.K., Anderson, E., Barnosky, A.D., Burns, J.A., Churcher, C.S., Grayson, D.K., Guthrie, R.D., Harington, C.R., Jefferson, G.T., Martin, L.D., McDonald, H.G., Morlan Jr., R.E., Webb, S.D., Werdelin, L., Wilson, M.C., 1996. Spatial response of mammals to late quaternary environmental fluctuations. Science 272, 1601–1606.

Grinnell, G., 1974. The rise and fall of Darwin's first theory of transmutation. Journal of the History of Biology 7, 259–273.

Haddrath, O., Baker, A.J., 2001. Complete mitochondrial DNA genome sequences of extinct birds: ratite phylogenetics and the vicariance biogeography hypothesis. Proceedings of the Royal Society of London, Series B 268, 939–945.

Hallam, A., 1977. Jurassic bivalve biogeography. Paleobiology 3, 58–73.

Hallam, A., 1981. Great Geological Controversies. Oxford University, New York. 182 pp.

Hallam, A., 1983. Early and mid-Jurassic molluscan biogeography and the establishment of the central Atlantic seaway. Palaeogeography, Palaeoclimatology, Palaeoecology 43, 181–193.

Hallam, A., 1992. Phanerozoic Sea-Level Changes. Columbia University Press, New York. 266 pp.

Hallam, A., 1994. An Outline of Phanerozoic Biogeography, vol. 10. Oxford University, Oxford. 246 pp.

Hooker, J.D., 1853. The Botany of the Antarctic Voyage of H.M. Discovery Ships "Erebus" And "Terror" in the Years 1839–1843: II. Flora Novae-Zelandiae: Part I. Flowering Plants. Lovell Reeve, London. 466 pp.

Humphries, C.J., Parenti, L., 1986. Cladistic biogeography. Oxford Monographs on Biogeography 2, 1–98.

Huxley, T.H., 1870. Anniversary address. In: Foster, M., Lankester, E.R. (Eds.), The Scientific Memoirs of Thomas Henry Huxley. Macmillan, London, pp. 510–550.

Knoll, A.H., 1991. End of the Proterozoic eon. Scientific American 265, 64–73.

Kottler, M.J., 1978. Charles Darwin's biological species concept and theory of geographic speciation: the transmutation notebooks. Annals of Science 35, 275–297.

Lieberman, B.S., 1997. Early Cambrian paleogeography and tectonic history: a biogeographic approach. Geology 25, 1039–1042.

Lieberman, B.S., 1999. Testing the Darwinian legacy of the Cambrian radiation using trilobite phylogeny and biogeography. Journal of Paleontology 73, 176–181.

Lieberman, B.S., 2000. Paleobiogeography. Plenum/Kluwer Academic Press, New York. 208 pp.

Lieberman, B.S., 2001. A test of whether rates of speciation were unusually high during the Cambrian radiation. Proceedings of the Royal Society of London, B Biological Science 268, 1707–1714.

Lieberman, B.S., 2002a. Phylogenetic analysis of some basal early Cambrian trilobites, the biogeographic origins of the Eutrilobita and the timing of the Cambrian radiation. Journal of Paleontology 76, 692–708.

Lieberman, B.S., 2002b. Phylogenetic biogeography with and without the fossil record: gauging the effects of extinction and paleontological incompleteness. Palaeogeography, Palaeoclimatology, Palaeoecology 178, 39–52.

Lieberman, B.S., 2003a. Unifying theory and methodology in biogeography. Evolutionary Biology 33, 1–25.

Lieberman, B.S., 2003b. Biogeography of the Cambrian radiation: deducing geological processes from trilobite evolution. Special Papers in Palaeontology 70, 59–72.

Lieberman, B.S., 2003c. Taking the pulse of the Cambrian radiation. Journal of International and Comparative Biology 43, 229–237.

Lieberman, B.S., Eldredge, N., 1996. Trilobite biogeography in the Middle Devonian: geological processes and analytical methods. Paleobiology 22, 66–79.

Lyell, C., 1832. Principles of Geology, vol. 2, 2nd ed. University of Chicago Press, Chicago, IL. 512 pp.

Mayr, E., 1942. Systematics and the Origin of Species. Dover Press, New York. 334 pp.

Mayr, E., 1976. Evolution and the Diversity of Life: Selected Essays. Harvard University Press, Cambridge, MA. 721 pp.

Mayr, E., 1982. The Growth of Biological Thought. Harvard University Press, Cambridge, MA. 974 pp.

McGhee Jr., G.R., 1996. The Late Devonian Mass Extinction. Columbia University Press, New York. 303 pp.

McKenna, M.C., 1975. Fossil mammals and Early Eocene North Atlantic Land Continuity. Annals of the Missouri Botanical Garden 62, 335–353.

McKenna, M.C., 1983. Holarctic landmass rearrangement, cosmic events, and Cenozoic terrestrial organisms. Annals of the Missouri Botanical Garden 70, 459–489.

Miller, A.I., Mao, S., 1995. Association of orogenic activity with the Ordovician radiation of marine life. Geology 23, 305–308.

Murray, A., 1866. The Geographical Distribution of Mammals. Day, London.

Nelson, G., 1978. From Candolle to Croizat: comments on the history of biogeography. Journal of the History of Biology 11, 269–305.

Nelson, G., Platnick, N.I., 1981. Systematics and Biogeography: Cladistics and Vicariance. Columbia University Press, New York. 567 pp.

Owen, A.W., Crame, J.A., 2002. Palaeobiogography and the ordovician and Mesozoic–Cenozoic biotic radiations. In: Crame, J.A., Owen, A.W. (Eds.), Palaeobiogeography and Biodiversity Change: The Ordovician and Mesozoic–Cenozoic Radiations, Special Publication- Geological Society of London, vol. 194, pp. 1–11.

Patterson, C., 1983. Aims and methods in biogeography. In: Sims, R.W., Price, J.H., Whalley, P.E.S. (Eds.), Evolution, Time and Space: The Emergence of the Biosphere. Academic Press, New York, pp. 1–28.

Richardson, R.A., 1981. Biogeography and the genesis of Darwin's ideas on transmutation. Journal of the History of Biology 14, 1–41.

Riddle, B.R., 1996. The molecular phylogeographic bridge between deep and shallow history in continental biotas. Trends in Ecology & Evolution 11, 207–211.

Rode, A., Lieberman, B.S., 2000. Using GIS and phylogenetics to study the role of invasive species during the late Devonian biodiversity crisis. Abstract with Programs- Geological Society of America Annual Meeting vol. A368.

Rode, A., Lieberman, B.S., 2001. Assessing the role of invasive species in mediating mass extinctions: a case study using Devonian phyllocarids. North American Paleontological Convention Abstracts with Programs, PaleoBios 21 (2), 109.

Rode, A., Lieberman, B.S., 2004. Using GIS to study the biogeography of the Late Devonian biodiversity crisis. Palaeogeography, Palaeoclimatology, Palaeoecology 211, 345–359.

Rode, A., Lieberman, B.S., 2004. Integrating biogeograpy and evolution using phylogenetics and PaleoGIS: a case study involving Devonian crustaceans. Journal of Paleontology 79 (2) (in press).

Ronquist, F., 1998. Phylogenetic approaches in coevolution and biogeography. Zoologica Scripta 26, 313–322.

Ross, M.I., Scotese, C.R., 2000. PaleoGIS/Arcview 3.5. PALEO-MAP Project. University of Texas, Arlington.

Sanmartin, I., Enghoff, H., Ronquist, F., 2001. Patterns of animal dispersal, vicariance and diversification in the Holarctic. Biological Journal of the Linnean Society 73, 345–390.

Scotese, C.R., 1997. Paleogeographic Atlas. Paleomap. Project. Arlington, Texas.

Sereno, P.C., 1997. The origin and evolution of dinosaurs. Annual Review of Earth and Planetary Sciences 25, 435–489.

Sereno, P.C., 1999. The evolution of dinosaurs. Science 284, 2137–2147.

Sereno, P.C., Dutheil, D.B., Iarochene, M., Larsson, H.C.E., Lyon, G.H., Magwene, P.M., Sidor, C.A., Varricchio, D.J., Wilson, J.A., 1996. Predatory dinosaurs from the Sahara and Late Cretaceous faunal differentiation. Science 272, 986–991.

Signor, P.W., Lipps, J.H., 1992. Origin and early radiation of the Metazoa. In: Lipps, J.H., Signor, P.W. (Eds.), Origin and Early Evolution of the Metazoa. Plenum Press, New York, pp. 3–23.

Sulloway, F.J., 1979. Geographic isolation in Darwin's thinking: the vicissitudes of a crucial idea. In: Coleman, W., Limoges, C. (Eds.), Studies in the History of Biology. Johns Hopkins University Press, Baltimore, MD, pp. 23–65.

Tiffney, B.H., Niklas, K.J., 1990. Continental area, dispersion, latitudinal distribution and topographic variety: a test of correlation with terrestrial plant diversity. In: Ross, R.M., Allmon, W.D. (Eds.), Causes of Evolution. University of Chicago Press, Chicago, IL, pp. 76–102.

Valentine, J.W., Moores, E.M., 1970. Plate tectonic regulation of faunal diversity and sea level: a model. Nature 228, 657–659.

Valentine, J.W., Moores, E.M., 1972. Global tectonics and the fossil record. Journal of Geology 80, 167–184.

Wallace, A.R., 1855. On the law which has regulated the introduction of new species. Annals & Magazine of Natural History, 2nd Ser. 16, 184–196.

Wallace, A.R., 1857. On the natural history of the Aru Islands. Annals & Magazine of Natural History, 2nd Ser. 20, 473–485.

Wells, M.L., Vallis, G.K., Silver, E.A., 1999. Tectonic processes in Papua New Guinea and past productivity in the eastern equatorial Pacific Ocean. Nature 398, 601–604.

Wiley, E.O., 1988. Vicariance biogeography. Annual Review of Ecology and Systematics 19, 513–542.

Wiley, E.O., Mayden, R.L., 1985. Species and speciation in phylogenetic systematics, with examples from the North American fish fauna. Annals of the Missouri Botanical Garden 72, 596–635.

Wiley, E.O., Siegel-Causey, D., Brooks, D.R., Funk, V.A., 1991. The Compleat Cladist. University of Kansas Press, Lawrence, Kan. 158 pp.

Available online at www.sciencedirect.com

Palaeogeography, Palaeoclimatology, Palaeoecology 219 (2005) 35–51

ELSEVIER

www.elsevier.com/locate/palaeo

The contributions of biogeomorphology to the emerging field of geobiology

Larissa A. Naylor*

School of Geography and the Environment, University of Oxford, Mansfield Road, Oxford. OX1 3TB and Honorary Fellow, Tyndall Centre for Climate Change Research, University of East Anglia, Norwich, NR4 7TJ, United Kingdom

Received 13 August 2003; accepted 29 October 2004

Abstract

Biogeomorphology has developed into a well-established research field over the past 15 years, with studies examining a range of two-way interrelations between organisms and geomorphology in a variety of terrestrial and marine environments. This paper starts by defining the core biogeomorphological processes—bioerosion, bioprotection and bioconstruction. Particular emphasis is placed on the study of bioconstructional forms; providing a clear definition, examples of bioconstructions, and, crucially, examining important interactions between bioconstruction and bioerosion. Three key areas where biogeomorphological research can directly contribute to the emerging field of geobiology are identified: (1) the use of biogeomorphological approaches, combined with palaeoecological investigations and predictive modelling, in the growing field of carbon sequestration and climate change mitigation; (2) providing geomorphological expertise to support the emerging field of astrobiology and (3) the potential contributions of biogeomorphology to the blossoming field of bio-geoengineering. Some key research directions and challenges for future biogeomorphological research include: exploring how biogeomorphological studies can contribute to and benefit from collaborations with other fields of geobiology, and assisting in the development of a useful, effective interdisciplinary toolbox of methods to improve quantification of geobiological processes. Importantly, suggestions are made for potentially fruitful collaborations with geomicrobiologists, geochemists and palaeoecologists.
© 2004 Elsevier B.V. All rights reserved.

Keywords: Biogeomorphology; Geobiology; Biocomplexity; Bioconstruction; Interdisciplinary approaches; Ecosystem engineers

1. Introduction

"There is no getting away from the fact that good ecological work cannot be done in an atmosphere of cloistered calm, of smooth concentrated focussing upon clean, rounded and elegant problems. Any ecological problem which is really worth working on at all, is constantly leading the worker on to

* Ecosystems Team, Science Group, Environment Agency, Evenlode House, Howberry Park, Wallingford, Oxon, OX10 8BD, UK. Tel.: +1 44 1491 828545; fax: +1 44 1491 828427.
E-mail address: larissa.naylor@environment-agency.gov.uk.

0031-0182/$ - see front matter © 2004 Elsevier B.V. All rights reserved.
doi:10.1016/j.palaeo.2004.10.013

neighbouring subjects, and is constantly enlarging his view of the extent and variety of the animal".

C.S. Elton in Animal Ecology, 1927, p. 188.

Nearly a century after Elton's writing, earth systems scientists from a range of disciplines are developing his insightful suggestion into an emerging field: geobiology. Although the term geobiology was first coined in the 1950s by Koch (1957), and was based on the premise of integrating biogeographical and ecological approaches, there has been little emphasis on geobiology as a distinct field of study until very recently. Geobiology is based on the premise that biological and geological activities are integrated, with complex interactions occurring between the biotic and abiotic systems at a range of spatial and temporal scales (Nealson and Ghiorse, 2001). Thus, geobiology is primarily concerned with exploring the interface and complex interactions between the biosphere and geosphere (Nealson and Ghiorse, 2001). It forms the scientific core of yet another emerging interdisciplinary field called biocomplexity.

Biocomplexity is primarily concerned with exploring between the margins of systems, to understand and begin to decouple the complex interdependencies between organisms and the environments which sustain their populations, affect them and/or are modified by the organisms themselves (www.hsf.gov/ od/lpa/news/media/99/fsbioenv.htm). Biocomplex systems are typically non-linear and chaotic which makes prediction difficult; the interactions within a given system can often span multiple spatial and temporal scales (esa.sdsc.edu/factsheetbiocomplexity.htm). This is an interdisciplinary area where biological, chemical and physical scientists work alongside social scientists, economists and computer modellers to grapple with the complex interactions between different components of natural systems and/or natural systems and the human environment (Mervis, 1999). Biocomplexity is viewed here as a form of integrated assessment that focuses on biotic processes and interactions. Interpreted this way, biocomplexity forms the intellectual envelope within which geobiology (and therefore biogeomorphology) sits (Fig. 1). As such, geobiology can form an integral part of such studies, as the interdisciplinary scientific basis for examining systems responses to environmental pressures and change.

One of the subdisciplines contributing to geobiology is biogeomorphology; an approach to geomorphology that focuses on the two-way interplay between ecological and geomorphological processes (Viles, 1988a). This field is based on the premise that the distribution of species is often related to underlying geomorphological forms, while surface morphology may in turn be altered by organisms. There is general agreement that biogeomorphology has much to con-

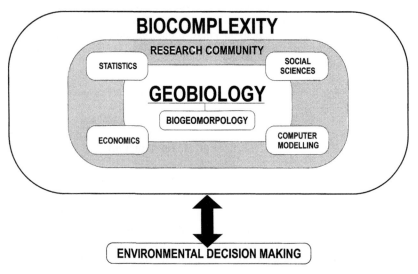

Fig. 1. Conceptual diagram illustrating some of the links between geobiology, integrated environmental assessments (e.g., biocomplexity) and environmental decision-making.

tribute to geobiology, with a large number of biogeomorphological papers having been included in the first few editions of Elsevier's new 'Virtual Journal of Geobiology' (http://earth.elsevier.com/geobiology).

1.1. Aims

Over the past 15 years, several papers and edited volumes have provided good reviews of the state of biogeomorphological research (e.g., Thornes, 1990; Viles, 1988b; Hupp et al., 1995; Viles and Naylor, 2002). Such synopses of biogeomorphology are not replicated here. This paper provides insight into the nature of biogeomorphological research, and importantly, it contextualises biogeomorphology in relation to the cognate subdisciplines of geobiology. Moreover, this paper illustrates how a biogeomorphological approach can provide a theoretical and methodological basis for a wider incorporation of geomorphological principles and techniques into other aspects of geobiology.

The paper is divided into three main sections: (1) defining and exemplifying the range of biogeomorphological processes, with emphasis on bioconstructions; (2) examining the linkages between different biogeomorphological processes; between biogeomorphological and inorganic processes and the links between, and potential contributions of, biogeomorphology to geobiology and; (3) three case studies which examine:

(a) Biogeomorphological inputs into climate change impacts research and carbon sequestration studies;
(b) Links between astrobiology and extraterrestrial geomorphology;
(c) Geomorphological contributions to bio-geoengineering.

2. Biogeomorphology defined

Biogeomorphology is a relatively new subdiscipline of geomorphology, which emerged in the late 1980s following a resurgence of multidisciplinary approaches to geomorphology. Viles (1988a) argued that ecological and geomorphological processes are on opposite ends of a continuum and that two different approaches to biogeomorphology can be considered:

(1) The influence of landforms/geomorphology on the distributions and development of plants, animals and microorganisms;
(2) The influence of plants, animals and microorganisms on earth surface processes and the development of landforms (Viles, 1988a, p. 6).

Biogeomorphology was intended to be a short-lived subdiscipline, its main aim being to illuminate the importance of biogenic agents in geomorphological systems at the geomorphological end of the spectrum, and the influence of geomorphological forms on the distribution of organisms at the ecological end of the spectrum (Viles, 1988a,b). Since its inception, there has been considerable work by geomorphologists (and others) on particular aspects of biogeomorphology, such as the influence of riparian vegetation on river channels; the role of biofilms in sediment stabilization and; biological contributions to rock weathering. Little general synthesis has occurred, however.

Biogeomorphology is a loosely defined term, where any study that examines the interplay between ecology and geomorphology constitutes 'biogeomorphological' research. Recent biogeomorphological research has focused on some aspects of the two-way linkages between ecological and geomorphological processes (see Naylor et al., 2002 for a recent review). Studies have explored a range of environments and organisms, examples of which include the role of: (a) lichens as weathering agents on limestone buildings (Carter and Viles, 2004), (b) microbialites in the development of micritic mud mounds (Saint Martin, 2001), (c) salt marshes in baffling wave energy (Möller et al., 1999), (d) bivalves, molluscs and echinoderms in eroding limestone coasts (Andrews and Williams, 2000; Kleeman, 1996; Vita-Finzi and Cornelius, 1973; Trudgill et al., 1987, respectively) and (e) beaver dams on sedimentation rates (Butler, 1995; Butler and Malanson, 1995). All of these studies fit within the broader field of geobiology. As such, biogeomorphology can usefully be viewed as a subdiscipline of geobiology which is fundamentally interdisciplinary in nature, and which focuses on exploring relationships at the margins between typically disparate groups such as geomicrobiology, geochemistry and ecology. There is much room for interactions with geobiologists. For example, theories could be shared between biogeomorphologists and other geobiologists, novel models could be developed, and new and mutually useful data collection techniques could be created.

There is a wide array of approaches and subjects that constitute biogeomorphological studies. They range from studies primarily concerned with evaluating links between biotic and geomorphological systems, such as through the processes of bioerosion, bioconstruction, biostabilization, bioweathering and bioprotection. Essentially, most of these are biologically mediated versions of characteristic earth surface processes such as physical and chemical weathering of rock surfaces, or fluvial processes, developing river channels. These species and processes have been referred to by others as bioengineers and/or physical ecosystem engineers (Reise, 2002; Jones et al., 1997, respectively). Physical ecosystem engineers are defined as, "organisms that directly or indirectly control the availability of resources to other organisms by causing physical state changes in biotic or abiotic materials" (Jones et al., 1997, p. 1947). Thus, biogenic agents active in biogeomorphological processes are essentially the same as the physical ecosystem engineers defined by Jones et al. (1997). The terms bioerosion, bioconstruction and bioprotection are used here as umbrella terms to provide general definitions which encapsulate and distinguish between the various types of biologically mediated earth system processes (that are defined in Sections 2.1.1–2.1.3).

Bioerosion has been the dominant mode of biogeomorphology research until very recently (Naylor and Viles, 2000; Naylor, 2001; Spencer and Viles, 2002). This is likely due to the fact that it is easier to conceptualise and measure the amount of material being removed from a surface, rather than measure the rates and amounts of material accumulating in a given system. The bioprotective role of species in fluvial environments has also received considerable attention in recent years (Thornes, 1990; Gurnell, 1998). This research has focused on the role of plant species in mediating fluvial processes and improving bank stabilization—processes that are also relatively easy to conceptualise and measure. Yet, many of the

interesting geomorphological forms and processes with strong linkages to other aspects of geobiology are of the lesser studied type broadly known as bioconstructions. This paper focuses on these.

2.1. Biogeomorphological processes

2.1.1. Bioerosion

The term bioerosion has been used to refer to many different types and scales of processes in the literature and generally can be thought of as the weathering and/or removal of material by organic agency (see for example, Spencer, 1992). Bioerosion is typically described as the active or passive, mechanical and/or chemical erosion of the land surface by organic means. It is the most well-studied of all aspects of biogeomorphology, with papers examining bioerosion by organisms ranging from molluscs to large ungulates.

2.1.2. Bioprotection

Bioprotection can be broadly defined as the active or passive, direct or indirect roles of organisms in preventing or reducing the action of other earth surface processes. Bioprotectors can be plant, animal and/or microorganism which encourage accretion and sedimentation, and/or reduce erosion. Bioprotectors range from near microscopic biofilms which can reduce the effects of physical processes such as wave action on sand flats, to small, inconspicuous organisms, such as epilithic coralline algae which cover a rock surface, thereby buffering it against the extremes of wetting and drying and/or sand abrasion in intertidal areas (Trudgill, 1985; Naylor, 2001), up to much larger species and more complex ecosystems which act to reduce erosion of sediment (such as the role of vegetation in controlling island bar development in fluvial systems). For example, in shallow soft-sediment marine environments, microbial mats aid in sediment stabilization and provide habitats for other species (see Underwood and Paterson, 1993 or; Noffke, and Visscher and Stolz, this volume).

2.1.3. Bioconstruction

Bioconstructors build films, crusts, mounds or reefs of material that they either produce internally (e.g., biogenic carbonate deposition), bind from other sources (using organic cement) or develop from a combination of the two. Some bioconstructions are relatively short-lived phenomenon, subject to non-linear cyclic growth and decay patterns, while other bioconstructions are subject to secondary diagenesis and lithification (Focke, 1978). Bioconstructions can persist for 1s to 1000s of years and may often play a bioprotective role.

There are three main ways in which bioconstructions form (adapted from Naylor, 2001; Naylor et al., 2002):

(1) where organisms produce minerogenic material themselves, such as the secretion of carbonate skeletons of *Dendropoma petraeum* (Novastoa) vermetid and *Negoniolithon notarisisii* (Dufour) coralline algal bioherms;

(2) where organisms actively accrete material by chemically fixing particulate matter as is the case with Sabellariidae reefs, microbial mats and biofilms;

(3) where inorganic cementation of organic debris results in the formation of bioconstructions (such as in many fluvial tufa barrages).

In biogeomorphology research and reviews there has been little attention placed on bioconstructions. This may be partly due to the fact that many bioconstructions are inconspicuous, with effects on environmental systems that are subtle and often difficult to detect. Yet, bioconstructor species often play a crucial role in the functioning of many environmental systems (Jones et al., 1997; Naylor et al., 2002). The limited interest in bioconstructions and particularly smaller, inconspicuous bioconstructional forms, is evidenced by the limited reviews on the subject. For example, in Viles's (1984) review paper on biokarst, attention is concentrated on bioerosional processes and only a small section is devoted to biodeposition of calcium carbonate. Meanwhile, Spencer's (1988) review of coastal biogeomorphology concentrates on intertidal and shallow subtidal soft-sediment environments, with attention focused primarily on bioerosion, and description limited to only a few bioconstructive species, such as biofilms and sea grass beds. Kelletat's (1989) paper on the zonality of rocky shores identifies many small and large bio-constructional forms and highlights the dearth of research on the smaller scale bioconstructions.

Most references to bioconstruction or bioprotection are marginal descriptions of morphological or distribution features. Consequently, only a limited amount of scientific research has focussed specifically on the geomorphological contributions of bioconstructions themselves; most researchers have used bioconstructions as a tool to date past events or predict large-scale environmental changes, rather than as a subject of direct study (see for example, Dermitzakis, 1973, Laborel et al., 1979, 1994; Thommeret et al., 1981 Laborel, 1986; Pirazzoli, 1986; Kelletat, 1991; Laborel and Laborel-Deguen, 1994; Antonioli et al., 1999). One exception is coral reefs—some of the largest, most persistent and most captivating biocon-structional forms.

Corals have been the subject of in-depth geo-morphological and ecological research evaluating the form, function, development and persistence of bioconstructions themselves (see for example, Spencer, 1988; Kelletat, 1989; Chazottes et al., 1995; Chazottes, 1996; Spencer and Viles, 2002). Meanwhile, the majority of other bioconstructional forms are the 'poor cousins' or 'unsung heroes' with little research or interest existing with regard to their geomorphological and ecological importance or the ways in which they contribute to the biocomplexity of environmental systems (Kelletat, 1989). As such, there is tremendous need and potential for research on bioconstructions both in terms of their intrinsic functions such as growth, development, resistance and resilience as well as their importance in larger scale ecosystem functioning. This research could be used to help improve our understanding of systems responses to large-scale environmental changes and, over time, to develop predictive models of biocom-plex responses to predicted sea-level rise and global climate change.

2.1.3.1. Locating the process. The majority of bioconstructions occur at or near the earth's surface. These include terrestrial bioconstructions such as beaver dams or other large woody debris accumulations which often stabilize or modify river banks, fluvial sediments and/or local hydrological processes (see the recent special issue on this topic, Montgomery and Piégay, 2003). In karst and fluvial environments, biofilms, tufa, travertine and directed speleotherms are typical bioconstructions. In coastal environments, bioconstructions are readily observed in the form of organic reefs, biofilms, microbial mats, sea grass beds, oyster beds and Sabellariidae reefs.

2.1.3.2. Defining the process. The main attributes of bioconstructions are (adapted from Naylor et al., 2002):

(1) In classifying bioconstructions, function takes precedence over form, size and persistence. As such, any bioaccumulation of material which stabilizes, enlarges or covers the surface material through biogenic means can be considered a bioconstruction; it is not necessary to develop a morphology unique from the underlying substrata.

(2) Persistence for periods ranging from days, as is the case with kelp rafts (Hobday, 2000), to thousands of years, as is the case with stromatolites (Thayer, 1979).

(3) Broad-ranging organic forms—from very thin (<10 μm), short-lived features such as stabilizing-sediment biofilms, to more intermediate, but short-lived forms such as beaver dams, to massive long-lived structures such as coral, serpulid and vermetid reefs.

(4) Where the dynamic and cyclical nature of the development of some bioconstructions (such as kelp rafts or Sabellaridae reefs) are viewed as part of the ecological functioning of the species, rather than an attribute which contests their classification.

(5) The potential to influence energy regimes, sediment dynamics or ecological functioning of systems (such as tufa barrages altering the hydrodynamics of streams).

(6) Bioconstruction is the only biogeomorphic term to have a landform connotation as well as a process meaning.

(7) Bioconstructions can perform multiple roles—they often either inhibit or enhance other ecosystem processes while developing a morphological structure. These functions may not always directly influence the substrate upon which they are affixed, they may also influence larger scale geomorphological or ecological processes (Wootton, 2002).

Relatively short-lived Sabellariidae reefs, mounds or crusts (i.e., worms existing up to 10 years, reefs existing for several decades) usefully illustrate some of the key attributes of bioconstructions. *Sabellaria alveolata* (Linné) (Polychaeta: Sabellariidae) is a filter-feeding sedentary polychaete that produces wave-resistant reefs from sand-sized (between 63 μm and 2 mm on the Wentworth Scale) mineral grains and shell debris. These typically form large reef colonies (up to 1.5 m in height) or encrusting reefs (up to 30 cm in height) which cover extensive areas of the mid-eulittoral, lower eulittoral and sublittoral zones and are particularly exposed during LWN–LWS (Wilson, 1971). Sabellariidae also preferentially sort sand grains and accumulate heavy minerals (Fager, 1964; Gram, 1968; Multer and Milliman, 1967). Thus, the reefs perform indirect and direct bioprotectional roles as they physically build structures that might influence local hydrodynamic or energy regimes, and preferentially store sediment that would otherwise be 'loose' in the system, available to physically abrade shore platforms (Naylor and Viles, 2000; Naylor, 2001). See Naylor and Viles (2000) and Naylor (2001) for a more detailed review and investigation of the geomorphological importance of *Sabellaria* sp.

2.2. Linking biogeomorphological processes

The processes of bioconstruction, bioprotection and bioerosion are not mutually exclusive, and their interrelationships are varied, complex and dynamic (Fig. 2).

Two examples illustrate the complex, interdependent nature of these processes. On many bare rock and soft sediment surfaces a community of microorganisms, commonly referred to as biofilms, play a series of key roles in altering the surface: (1) biofilms can themselves directly protect the surfaces of rock material and loose sediment deposits from weathering and erosive activity; (2) indirectly, biogeochemical transformations within biofilms can produce a range of secondary minerals which build up into a crust, often referred to as case-hardening or desert varnish and; (3) biofilms can also become an active biological weathering agent, thereby facilitating physical and chemical weathering through bio-mediated processes.

Although the biofilm example does provide some insight into the interdependence of bioprocesses and

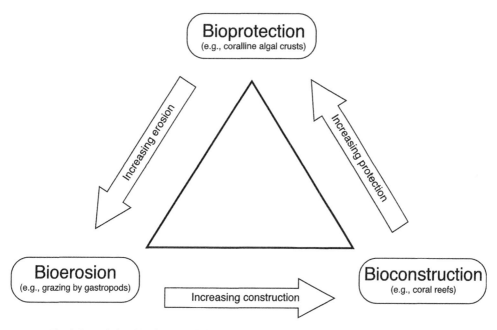

Fig. 2. Interrelationships between different biogeomorphological processes (after Naylor, 2001).

other inorganic agents, it fails to provide a quantitative assessment of how different bioprocesses interact. The recent research on the relationships between grazers, macroborers and microborers on coral reefs in French Polynesia and the Indian Ocean (Réunion) provides a clear example of such interrelationships and how they might be quantified (Chazottes et al., 1995; Chazottes, 1996, respectively). On experimental substrata, these researchers found that differences in macroalgal colonisation strongly influenced the activity of grazing as well as micro- and macroboring activity. On *Porites* blocks with low densities of coralline algal colonisation, there were high rates of grazing and macroboring coupled with low rates of microboring. Conversely, on blocks with thick, high-density crusts of coralline algae, grazing and macroboring rates were significantly reduced while microboring rates increased substantially. Moreover, the mean grain size of sediment produced was also affected by the colonisation patterns where the mean grain size diameter decreased in areas with reduced grazing intensity. This example illustrates the interdependent nature of biogenic processes themselves and their strong relationship with biological community dynamics.

2.3. Interrelationships between biogeomorphological processes and other inorganic earth system processes

The bioprocesses described in the previous section also interact with, and are affected by, a suite of exogenic processes (inorganic modes of erosion, transport and deposition) as well as a range of endogenic earth surface processes such as tectonic activity, geological controls and climatic factors.

It has been suggested that climate influences the rates, occurrence and intensity of particular biogeomorphological processes (Kelletat, 1989). For example, Fischer (1983) has suggested that bioerosion processes in the lower latitudes occur on a range of hard and soft rocks where biological community dynamics are more important than lithological controls over bioerosion rates, while in more northerly climates lithological controls become increasingly dominant so that as latitude increases only limestone coasts are subject to active bioerosion. At the microclimate scale, researchers have recently determined that air climate is often very different than the adjacent rock surface climate, and as such microorganisms can persist on rock surfaces where the air temperature has historically been considered too cold (see for example, Hall and André, 2001). Moreover, Dornieden et al.

(2000) have argued that microorganisms experience very different life history characteristics depending on the climate in which they are found, a phenomenon termed the 'poikilotrophic concept'. Meanwhile, many bioconstructional forms can alter the microscale climate of the local environment to provide habitats for a range of more sensitive species, such as the way in which microbial mats provide a habitat for burrowing species (see Noffke, this volume, for more details).

Geological processes and landforms can exert a strong control on the types of biotic processes operating in a given environment, including the form and functioning of ecological systems. Some rock types, and structural conditions, are more physically resistant to weathering and erosion than others. For example, the chemical composition of limestone increases the potential for chemical and biological weathering and erosion as compared to other rock types (Trudgill, 1985). Meanwhile, the degree of jointing, thickness of beds, surface elevation (relative to current mean sea level) and dip of platforms strongly influence the types of processes occurring on rocky shore platforms (Trenhaile, 1987). For example, platforms with a higher surface elevation will be subject to longer emersion times, which will increase the potential for chemical, salt or frost weathering, due to prolonged wetting and drying cycles, while shore platforms with a lower surface elevation and more residual water, provide more opportunities for bioprocesses to help shape landforms.

3. Biogeomorphology as part of geobiology

Biogeomorphology should be viewed as an evolving, dynamic science that is not bound by disciplinary lines—it should be viewed as part of a growing body of earth systems science where multidisciplinary approaches and partnerships are the norm rather than the exception. As such, biogeomorphology is only one subdiscipline of the wider field of geobiology and it joins a host of other research foci in present day science which are tackling other aspects of geobiology—such as astrobiology, palaeoecology, paleontology, geomicrobiology and biosedimentology. Biogeomorphological processes also strongly affect and are mediated by biological community dynamics,

and as such there is great potential for fruitful collaborations and exciting research advances through collaboration with other geobiologists. Until recently, most biogeomorphological studies (and also most ecological studies) have failed to investigate the crucial links between bioprocesses, biological community dynamics and 'inorganic' earth surface processes. Exceptions to this are recent doctoral studies, including: an examination of the spatial and temporal variations of bioeroding species on different coral reefs in Australia (Tribollet, 2001); an investigation of the role of fungi in cold climate rock weathering (Etienne, 2001); and an examination of the interrelationships between biological community dynamics, rock pool geomorphology and bioerosion rates on limestone coasts (Naylor, 2001). Exploring these interlinkages between bioprocesses and other subdisciplines of geobiology and inorganic earth systems processes remains one of the biggest challenges and largest opportunities for biogeomorphological research in the coming years.

As numerous subdisciplines contribute to geobiology, and the range of biogeomorphological processes contributing to this emerging field vary in space and time, it is perhaps useful to conceptualise some of these differences. Fig. 3 illustrates how several of the subdisciplines relate to each other along axis of spatial and temporal scales. This diagram can be applied to any planetary system—for example, geomicrobiological research would probably operate at a similar spatial and temporal scale on both the Earth and Mars.

4. Case study 1: climate change impacts research and carbon sequestration

4.1. Monarch Project

The Monarch project is funded by a consortium of governmental and non-governmental organisations (led by English Nature) and covers all of the UK and the Republic of Ireland. This research project forms an integral part of the United Kingdom's Climate Impacts Program (UKCIP). The Monarch study includes models of how British habitats and species could respond to predicted climate changes in the 21st century. Many species response models are

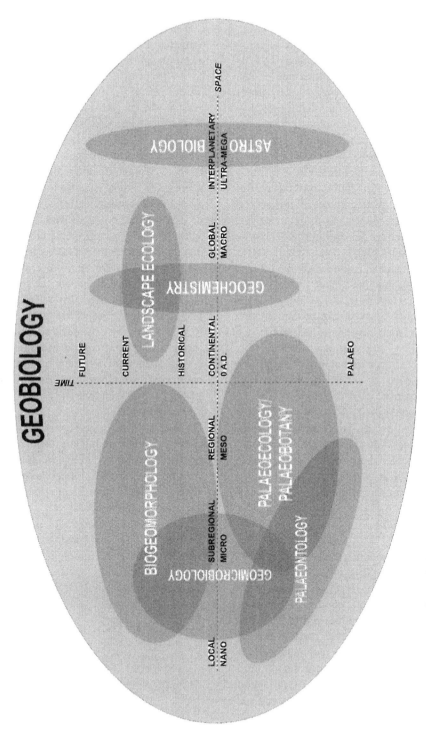

Fig. 3. Placing biogeomorphology in the context of other elements of geobiology, on a space–time continuum.

innovative as they combine state-of-the-art climate models with palaeoecological and current day ecological distributions (see for example, Pearson et al., 2002). These models often include only sparing information on geomorphological parameters such as soil properties, geological conditions and/or the distribution of landforms. For example, the authors of the first Monarch report note that geological conditions may restrict the distribution of particular species and ecosystems, such as chalk grasslands (Pearson et al., 2002). However, geomorphological and geological parameters were not explicitly included as variables in the model. Fortunately, this limitation has been recognized and in the second phase of the Monarch project there is increasing emphasis on land cover in the refined, higher resolution species model (Berry, pers. comm., 2003). For example, in Monarch 2.1 researchers have found 'land cover' to be an important input variable in the model; importantly, where 'land cover' is included as an input variable, an improved relationship between the actual and simulated vegetation distribution has been achieved (Berry, pers. comm., 2003). Geomorphologists and biogeomorphologists can provide specialist expertise based on a thorough understanding of landform characteristics and their rates of change, to help improve the type(s) of geomorphological information included in species envelope models. This knowledge can help further refine the geomorphological variables used in integrated model frameworks, thereby improving the geobiological baseline in vegetation change models.

4.2. Carbon sequestration

Climate change mitigation is the process by which humans seek to reduce the amount of carbon they release into the atmosphere through measures which actively reduce our carbon emissions by reducing production, developing alternatives and/or by storing carbon that would otherwise enter the atmosphere. The latter of these options is commonly called 'carbon sequestration' and biological sequestration particularly plays an important role in overall carbon mitigation. At present, terrestrial vegetation and soils absorb (i.e., sequester) approximately 40% of global CO_2 emissions from human activities. It is in understanding this aspect of climate change mitiga-

tion that biogeomorphology can contribute most effectively.

The importance (and potential mitigating role) of terrestrial carbon land sinks is recognized by and incorporated into the Kyoto protocol[1]. Under the Kyoto protocol, individual countries are eligible to offset some of their carbon consumption by developing land-based carbon sinks (Royal Society, 2001). Thus, developing successful terrestrial carbon sequestration methods will be a new growth area as individual countries grapple with meeting their Kyoto targets. The application of biogeomorphological techniques, such as providing a biogeomorphic assessment of suitable species for sequestration and/ or identifying the geomorphological conditions characteristic of long-term Quaternary carbon sinks, can help governments meet these targets, while providing an exciting area for new biogeomorphological and geobiological research. For example, Ferris et al. (1994) have suggested that microbial crusts in high latitudes might represent part of the 'missing sink' for CO_2 in these areas. If that is the case, biogeomorphological assessment can be used to identify potential carbon sinks and importantly, to devise methods for colonising such sinks elsewhere, to help increase the global terrestrial carbon sink.

Biogeomorphologists can assist in understanding the ecological and geomorphological basis for successful clean development mechanisms (i.e., 'clean', non-polluting carbon mitigation tools), such as 'Kyoto forests'. For example, biogeomorphologists can work with palaeoecologists to provide an interdisciplinary perspective of the ecological, geomorphological and climate conditions that enabled land carbon sinks to persist in the geological past. This information, coupled with geomorphological assessments of current conditions and climate change predictions, could be used to help identify the most appropriate sites and/ or species for land management projects designed to

---

[1] The Kyoto Protocol was adopted at the Third Session of the Conference of the Parties (COP) to the UN Framework Convention on Climate Change (UNFCCC) in 1997 in Kyoto, Japan. ... Countries included in Annex B of the Protocol (most OECD countries and EITs) agreed to reduce their anthropogenic emissions of greenhouse gases (CO_2, CH_4, N_2O, HFCs, PFCs, and SF6) by at least 5% below 1990 levels in the commitment period 2008 to 2012 (definition from IPCC, 2001).

increase terrestrial carbon sinks. Meanwhile, biogeo-morphologists can also assist in assessing the potential for and quantifying the effects of erosion processes in releasing stored terrestrial carbon. For example, assessments could be made of the varying amounts of carbon stored in peat mires and afforesting mineral soils and then used to quantify potential variations in carbon release based on different scenarios of erosion. Such studies could provide insight into the potential stability of different types of carbon sinks and be used to identify landscapes which are more geomorpho-logically stable—therefore providing longer, more secure CDMs. Importantly, researchers have advo-cated that climate prediction models need to better account for the interactions between vegetation, soils and climate that underpin them, so that more realistic climate prediction models are generated (Royal Society, 2001). Such recommendations will hopefully pave the way for more biogeomorphological and geobiological inputs into climate change impacts modelling research.

4.3. Wider benefits of geoengineering to improve carbon sequestration

Wild et al. (2001) illustrated that the carbon storage function in peatlands can be restored by encouraging the growth of tall helophytes, such as *Typha* spp. Planting techniques such as these can be further developed as a means of increasing the amount of carbon sequestration in upland peatlands and mires, thereby helping reach Kyoto protocol targets. The colonisation of peat mires with carbon sequestering species also had some knock on biocomplex responses as mire water quality was substantially improved with the addition of these plants (Wild et al., 2001). In this scenario, the colonising plants act as an ecosystem engineer that facilitates improved ecolog-ical conditions for other species. The use of bio-geomorphological techniques may increase the rate of success of colonisation trials as a quantitative assess-ment of the processes responsible for the function of natural peat and mire systems could be used to develop suitable techniques for restoring and creating peat and mire bogs. For example, on a moderately sloping hillside, a stepped structure with water flowing through the system will be much more successful in creating a suitable habitat for colonisa-

tion than one such as a closed pond system, which ignores the natural landform characteristics. Recently, the eminent wetlands ecologist, Prof. Joy Zelder, has determined that wetland recreation projects are much more successful when the geomorphological and hydrological properties of a given system are taken into account during the creation of an artificial wetland.

5. Case study 2: extraterrestrial geomorphology–astrobiology links

Large-scale geomorphological assessments of Mar-tian landscapes could help focus and increase the economic viability of research efforts for geobiolog-ical studies, through the use of satellite imagery. As many landforms are visible on remotely sensed images, one geomorphological contribution to astro-biological studies could be to provide broad-brush surveys of the landforms of Mars and identify a selected subset for detailed astrobiological study based on an assessment of landforms likely to have been biologically mediated (see, for example Gulick, 2001 and Zimbelman, 2001 in the edited volume by Baker, 2001). For example, remote sensing has been used by researchers at the University of Colorado at Boulder to examine the potential for life on Mars (Bortman, 2002), and other geomorphologists and earth scientists have recently questioned the assump-tion that Mars is waterless, suggesting that perhaps the planet is merely hydrologically dormant (Cabrol and Grin, 2001). Even more recently, scientists at NASA's Jet Propulsion Laboratory have determined that there is liquid water on the surface of Mars during summer months and that Mars has a liquid core—these findings leading astrobiologists to a greater certainty regarding the possibility of life on Mars (JPL, 2003). As such, there is an increasing desire to quantify these speculations and the knowledge and toolkits of biogeomorphologists and geomorphologists may con-tribute substantially toward a definite answer to the question of whether there is life on Mars.

There are several reasons why geomorphologists are well-suited to help identify landforms on Mars where there may be life. First, the primary focus and historical grounding of geomorphological studies has been on examining the types of landforms found on the earth's surface and determining how such landforms develop and evolve (Small, 1978). Second, geomorphologists have a proven track record of deciphering the landscape, using a combination of descriptive studies, process-form investigations, satellite image analyses, palaeo-environmental reconstruction techniques and measurements of rates of change and laboratory experiments to consistently unravel the complex process-form responses of many landscapes (see for example, Masson et al., 2001). Thirdly, a biogeomorphological approach to geomorphology studies provides the necessary combination of geomorphological expertise and theoretical grounding, coupled with an understanding of ecological and biological requirements of species, which in combination can provide a unique lens for evaluating likely locations of life on Mars.

Research specialisations within geomorphology can also contribute toward understanding the processes (i.e., biological, chemical and physical) which are responsible for shaping Martian landscapes, and developing techniques to quantify rates of Martian landscape change. Glacial geomorphology studies have an extensive track-record of determining the nature and rates of processes which have shaped landforms—both through past glacial episodes, such as the Late Wisconsin period in North America, and as a result of current activity on mountain or ice sheet glaciers. These studies can decouple process-form responses to help elucidate which types of processes are likely responsible for the creation of particular landforms. For example, several cold climate, rock weathering geomorphologists have spent over 25 years decoupling the nature of frost weathering and its relationship to freeze–thaw cycles in areas currently glaciated on earth (see Hall and André, 2001). Such studies, particularly when coupled with satellite analyses of current Martian landscapes, can be used to identify landforms which are partly formed by rock weathering processes where the resultant landscapes require alternating periods of freezing and thawing. The use of current process geomorphological techniques and knowledge can thus help identify regions on Martian landscapes that may be particularly amenable to biota. Moreover, the techniques used to monitor rock temperatures or solute discharge rates in glacial streams or subglacial lake processes on earth, could probably be adapted

and included as part of proposed investigations on Mars (for recent applications of these techniques see Hall and André, 2001, and Siegert et al., 2001, respectively). For example, as many quantitative instruments are now completely automated, it may be possible to deploy them on Mars at the beginning of a Robotic and/or probe investigation of the land surface and collect them at the end of the survey. Such measures could provide a relatively inexpensive, high-resolution set of data on processes such as rock surface temperature fluctuations.

6. Case study 3: bio-geoengineering applications

Organisms are increasingly being valued and developed as integral components of environmental engineering and habitat restoration projects. In many fields, ranging from coastal flood defenses to river management and sustainable urban drainage systems to urban building stone decay, the environmental and economic benefits of using organisms in engineering projects are being recognized. Coupled with this industrial enthusiasm for biotic processes in environmental engineering applications has been the realization that the basis for successful habitat restoration and recreation projects lies with a sound understanding of the hydrological and geomorphological processes of a given system and their interplay with biota (Nienhuis et al., 2002). Hence, a biogeomorphological approach. Yet, not all spheres of geobiology research are adopting an interdisciplinary, biogeomorphological approach at present. For example, although there has been a spurt of recent research on the importance of bio- and/or geoengineering for ecological restoration projects (see, for example, the edited volume by Verhoeven, 2001 or Shroder and Bishop, 1995), much of this research has tended along disciplinary lines rather than examining integrated, biocomplex responses (i.e., biogeomorphological approaches) of these techniques.

There is much potential to be gained from drawing on biogeomorphological and geomorphological techniques in the design and application of environmental engineering projects. Many 'soft engineering' schemes for environmental management involve the recreation or encouragement of biogeomorphological systems. For example, salt marsh restoration schemes and the creation of salt marshes from previously reclaimed land, commonly referred to as 'managed realignment', require a thorough understanding of the two-way interplay between sedimentation and biological growth (see for example, Navas et al., 2002), as well as the interaction of both of these with water flows (Burd, 1995; Möller et al., 1999). Moreover, recent hydrodynamic modelling research on the Blyth Estuary has highlighted the importance of using a catchment-scale, geomorphological assessment of the effects of proposed managed realignment schemes on the tidal prism of estuaries (French, 2001). Recently, the Department of Food and Rural Affairs (DEFRA) in the UK has commissioned two projects where geomorphological knowledge was an integral and pivotal part of the investigation. The first project, called Emphasys, involved an assessment made of our current state of knowledge regarding geomorphological processes and models which can aid our understanding of estuary systems. Meanwhile, the second project, 'Future Coast', used a combination of historical geomorphological assessments, current trends, and future predictions, to develop coastal tendencies for the entire UK coastline. This baseline geomorphological assessment will be used to underpin and guide the second phase of coastal engineering planning in the UK, called Shoreline Management Plans, and serve as a template upon which biodiversity responses to climate change in coastal habitats can be related. The need for biogeomorphological knowledge in this area is likely to escalate in future years as the UK government is now actively encouraging coastal management that 'works with the sea', with managed realignments being an integral part of this scheme (DEFRA, 2002).

Similarly, attempts to rehabilitate eroding desert landscape and arable land require a biogeomorphological understanding, such as the propagation of microbial communities to create bioprotective crusts in desert environments (see Kidron et al., 2000). Furthermore, the techniques commonly applied in river bank stabilization and riparian habitat recreation (see Goodson et al., 2002) are increasingly being applied in other areas, such as testing the use of vegetated spillways rather than concrete structures in highway developments, or the development of soakways as part of sustainable urban drainage systems,

or the restoration of disturbed wetland systems (see Escarameria et al., 2002; Environment Agency, 2001; Nienhuis et al., 2002, respectively). All of these techniques have been found to improve habitat availability and water quality, while reducing peak flow runoff or river flow rates. When underpinned by a geomorphological understanding of systems dynamics, many of these 'softer' engineering techniques result in more resilient recreated systems that better emulate natural conditions. Thus, the use of biogeomorphological techniques in applied settings can have much wider impacts and help governments meet sustainable development and biodiversity targets.

Although this research area appears very applied in scope, such biological or geoengineering projects can also serve as important field trials and laboratory experiments which could yield new ideas on biogeomorphological inputs to geobiology, thereby advancing the development of the discipline. Recent examples of this include the surfacing of research questions during managed realignment programs. In most cases, managed realignment schemes are designed to provide a longer term, more sustainable and less expensive means of coastal defense—yet the effects of a managed realignment on biogeochemical cycling or geomorphological properties of estuaries can be immense. As such, many new research projects have developed as an offshoot of applied projects. For example, researchers at the Tyndall Centre for Climate Change Research are examining the effects of managed realignment on biogeochemical cycling and sedimentation processes in estuaries.

7. Developing an interdisciplinary toolbox

A new emphasis on developing an interdisciplinary, quantitative toolbox should help answer many of the questions emerging in geobiology. As geobiology itself is a multidisciplinary field of study, which brings together ideas from an array of fields, there is a need for new tools which can help answer research questions along disciplinary margins. A series of new applications for existing techniques coupled with new protocol has led to the development of several innovative biogeomorphological methods. A few of these methods are illustrated here to highlight the potential for applying techniques in different contexts and/or developing new techniques as geobiological studies explore new research topics.

The novel techniques in biogeomorphological studies tend to be those where: (1) laboratory techniques from another field are used to improve quantification of biogeomorphological studies; (2) laboratory methods are used in conjunction with field experiments, or; (3) existing geomorphological methods are modified and/or enhanced by those of another discipline.

Recently, French researchers have used microbiological techniques to aid in the identification of microboring fungi in Iceland. Samuel Etienne explored the role of organisms in producing weathering rinds in periglacial environments and greatly enhanced his analytical results through collaboration with a botanist, J. Dupont, from the Museum of Natural History, Laboratory of Cryptogamy, Paris. Together the combined field and laboratory analyses enabled the microboring fungi to be isolated, inoculated and identified so that the specific species responsible for creating Icelandic weathering rinds were found (Etienne, 2002; Etienne and Dupont, 2002). Similarly, researchers at the University of Oxford have successfully colonised cyanobacteria collected in the field and subsequently reestablished cyanobacterial communities on freshly exposed limestone quarry faces (Courtney, 2001). A third recent project examined whether microcatchments, which are typically used to monitor chemical weathering of limestone in urban areas, could be used to determine the effects of lichens on the weathering of limestone in urban environments (Carter, 2002).

In field biogeomorphology studies, techniques have typically been adapted and/or expanded upon to assess the landscape from a wider field of view. This has ranged from analysing the functional importance of a species using an interdisciplinary geomorphological, hydrodynamical and ecological assessment (Naylor, 2001; Reise, 2002) to combined bioerosion and bioconstruction studies which have sought to decouple the interrelationships between these two processes (Schneider and Le-Champion Alsumard, 1999; Naylor and Viles, 2002). The latter type of study has highlighted the benefits of assessing landscape change from both erosion and construction, as more realistic rates of change have been measured through such studies. Similar to this has been the application of ecological theory to geomorphological investigations (Naylor, 2001; Jackson et al., 2002). Here, the pivotal role of community dynamics in influencing biogeomorphological change rates has been recognized, along with a seed change in some ecological studies where more emphasis is being placed on geomorphological control of community dynamics (Hawkins pers. comm., 2002).

8. Conclusions

Biogeomorphology is an active component of geobiological research, which can contribute different theoretical and methodological perspectives to help develop the emerging field of geobiology and collaborate with cognate disciplines to answer complex research questions in an innovative, interdisciplinary fashion. Biogeomorphology should thus been seen as an integral part of an evolving, holistic earth systems science which deals with the biological, chemical and physical characteristics of earth surface systems over a range of temporal and spatial scales.

Acknowledgements

The author would like to express sincere thanks for the financial and intellectual support of NSERC and the University of Oxford. A special I am grateful to Heather Viles, Samuel Etienne and David Forest for their valuable comments on early drafts, which greatly improved this manuscript. Christine Russell of Komex is also appreciated for improving the quality of the figures presented in this report. The comments of two anonymous referees and the journal editor are also much appreciated. Please note that the ideas presented in this paper are the views of the author, rather than the institutions to which she is affiliated.

References

Andrews, C., Williams, R.B.G., 2000. Limpet erosion of chalk shore platforms in southeast England. ESPL 25 (12), 1371–1381.

Antonioli, F., Chemello, R., Improta, S., Riggio, S., 1999. *Dendropoma* lower intertidal reef formations and their palaeoclimatological significance, NW Sicily. Mar. Geol. 161, 155–170.

Baker, V.R., 2001. Extraterrestial geomorphology: an introduction. Geomorphology 37, 175–178.

Bortman, Henry, "Mars: tilting towards life", Astrobiology Magazine [online], www.astrobio.net/news/article383.html, February 26, 2002, NASA Office of Space Sciences.

Burd, F., 1995. Managed retreat: a practical guide. Engl. Nat. (28 pp.).

Butler, D.A., 1995. Zoogeomorphology: Animals as Geomorphic Agents. Cambridge University Press, Cambridge. 231 pp.

Butler, D.R., Malanson, G.P., 1995. Sedimentation rates and patterns in beaver ponds in a mountain environment. Geomorphology 13, 255–269.

Cabrol, N.A., Grin, E.A., 2001. The evolution of lacustrine environments on Mars: is Mars only hydrologically dormant? Icarus 149, 291–328.

Carter, N.E.A., 2002. Bioprotection explored: lichens on limestone. Thesis, D. Phil., University of Oxford.

Carter, N.E.A., Viles, H.A., 2004. Lichen hotspots: raised rock temperatures beneath Verrucaria nigrescens on limestone. Geomorphology 62 (1–2), 1–16.

Chazottes, V., 1996. Étude expérimentale de la bioérosion et de la sédimentogénèse en milieu récifal: effets de l'eutrophisation (île de la Réunion, Océan Indien). C. R. Acad. Sci., Paris 323, 787–794.

Chazottes, V., Le Campion-Alsumard, T., Peyrot-Clausade, M., 1995. Bioerosion rates on coral reefs: interactions between macroborers, microborers and grazers (Moorea, French Polynesia). Palaeogeogr. Palaeoclimatol. Palaeoecol. 113, 189–198.

Courtney, A., 2001. Accelerated recovery: bioremediation of the visual impact of limestone quarry scarps. Quarry Manag., 47–48 (December).

DEFRA, 2002. Working with the grain of nature: a biodiversity strategy for England. DEFRA Publications. 92 pp.

Dermitzakis, M.D., 1973. Recent tectonic movements and old strandlines along the coasts of Crete. Bull. Geol. Soc. Greece 10 (1), 48–64.

Dornieden, T., Gorbushina, A.A., Krumbein, W.E., 2000. Biodecay of cultural heritage as a space/time-related ecological situation—an evaluation of a series of studies. Int. Biodeterior. Biodegrad. 46, 261–270.

Elton, C.S., 1927. Animal ecology. Sidgwick and Jackson, London.

Environment Agency (EA), 2001. Enhancing the Environment: Twenty Case Studies in London. 34 pp.

Escarameria, M., Gasowski, Y., May, R., 2002. Grassed drainage channels—hydraulic resistance characteristics. Wat. Mar. Eng. 154 (4), 333–341.

Etienne, S., 2001. Les processus de météorisation des surfaces volcaniques en Islande; approche épistémologique de la géomorphologie des milieux froids. Thesis, Sorbonne, Paris. 477 pp.

Etienne, S., 2002. The role of biological weathering in periglacial areas: a study of weathering rinds in south Iceland. Geomorphology 47 (1), 75–86.

Etienne, S., Dupont, J., 2002. Fungal weathering of basaltic rocks in a cold oceanic environment (Iceland): comparison between experimental and field observations. Earth Surf. Processes Landf. 27, 737–748.

Fager, E.W., 1964. Marine sediments: effects of a tube-building polychaete. Science 143, 356–359.

Ferris, F.G., Wiese, R.G., Fyfe, W.S., 1994. Precipitation of carbonate minerals by microorganisms: implications for silicate weathering and the global carbon dioxide budget. Geomicrobiol. J. 12, 1–13.

Fischer, R., 1983. Bioerosion, ein Gesteinsunabhängiger Küstenmorphologischer Prozess. Essen. Geogr. Arb. Bd. 6, 251–263.

Focke, J.W., 1978. Limestone cliff morphology on Curaço (Netherlands Antilles), with special reference to the origin of notches and vermetid/coralline algal surf benches ("cornices", "trottoirs"). Z. Geomorphol., N. F. 22 (3), 329–349.

French, J.R., 2001. Hydrodynamic modelling of the Blyth estuary: impacts of sea-level rise. Coastal and Estuarine Research Unit, University College London. Report for Environment Agency, Anglian Region. 113 pp.

Goodson, J.M., Gurnell, A.M., Angold, P.G., Morrissey, I.P., 2002. Riparian seed banks along the Lower River Dove, UK: their structure and ecological implications. Geomorphology 47 (1), 45–61.

Gram, R., 1968. A Florida Sabellaridae reef and its effect on sediment distribution. J. Sediment. Petrol. 38 (1), 863–868.

Gulick, V.C., 2001. Origin of the valley networks on Mars: a hydrological perspective. Geomorphology 37, 241–268.

Gurnell, A.M., 1998. The hydrogeomorphological effects of beaver dam-building activity. Prog. Phys. Geogr. 22, 167–189.

Hall, K., André, M.F., 2001. New insights into rock weathering from high-frequency rock temperature data: an Antarctic study of weathering by thermal stress. Geomorphology 41 (1), 23–35.

Hobday, A.J., 2000. Persistence and transport of fauna on drifting kelp (*Macrocystis pyrifera* (L.) C Agardh) rafts in the Southern California Bight. J. Exp. Mar. Biol. Ecol. 253, 75–96.

Hupp, C.R., Osterkamp, W.R., Howard, A.D. (Eds.), 1995. Biogeomorphology, terrestrial and freshwater systems, Geomorphology, vol. 13 (1–4), pp. 1–347.

IPCC, 2001. Climate change 2001 impacts, adaptation and vulnerability. http://www.ipcc.ch/pub/tar/wg2/689.htm.

Jackson, N.L., Nordstrom, K.F., Smith, D.R., 2002. Geomorphic–biotic interactions on beach foreshores in estuaries. J. Coast. Res. 36, 414–424.

Jet Propulsion Laboratory (JPL), 2003. Scientists Say Mars Has a Liquid Iron Core, Press Release, March 6, 2003. http://www.jpl.nasa.gov/releases/2003/32.cfm.

Jones, C.G., Lawton, J.H., Shachak, M., 1997. Positive and negative effects of organisms as physical ecosystem engineers. Ecology 78 (7), 1946–1957.

Kelletat, D., 1989. Zonality of rocky shores. Essen. Geogr. Arb. Bd. 18, 1–29.

Kelletat, D., 1991. The 1550 BP tectonic event in the Eastern Mediterranean as a basis for assessing the intensity of shore processes. Z. Geomorphol. N. F. Suppl. Bd. 81, 181–194.

Kidron, G.J., Barzilay, E., Sachs, E., 2000. Microclimate control upon sand micribiotic crusts, western Negev Desert, Israel. Geomorphology 36, 1–18.

Kleeman, K., 1996. Biocorrosion by bivalves. P.S.Z.N.I. Mar. Ecol. 17 (1–3), 145–158.

Koch, L.F., 1957. Index of biotal dispersity. Ecology 38 (1), 145–148.

Laborel, J., 1986. Vermetid gastropods as sea-level indicators. In: van de Plassche, O. (Ed.), Sea-level research: a manual for the collection and evaluation of data. Geo Books, Norwich, pp. 281–310.

Laborel, J., Laborel-Deguen, F., 1994. Biological indicators of relative sea-level variations and of co-seismic displacements in the Mediterranean region. J. Coast. Res. 10 (2), 395–415.

Laborel, J., Pirazzoli, P.A., Thommeret, J., Thommeret, Y., 1979. Holocene raised shorelines in Western Crete (Greece). 1978 International symposium on coastal evolution in the quaternary, São Paulo, Brazil.

Laborel, J., Morhange, C., Lafont, R., Le Campion-Alsumard, T., Laborel-Deguen, F., Sartoretto, S., 1994. Biological evidence of sea-level rise during the last 4500 years on the rocky coasts of continental southwestern France and Corsica. Mar. Geol. 120, 203–223.

Mervis, J., 1999. Biocomplexity blooms in NSF's research garden. Science 286, 2068–2069.

Masson, P., Carr, M.H., Costard, F., Greeley, R., Hauber, E., Jarmann, R., 2001. Geomorphic evidence of for liquid water. Space Sci. Rev. 96, 333–364.

Möller, I., Spencer, T., French, J.R., 1999. Wave transformation over salt marshes: a field and numerical modelling study from north Norfolk, England. Estuar. Coast. Shelf Sci. 49 (3), 411–426.

Montgomery, D.R., Piégay, H., 2003. Wood in rivers: interactions with channel morphology and processes. Geomorphology 51, 1–5.

Multer, H.G., Milliman, J.G., 1967. Geologic aspects of sabellarian reefs, Southeastern Florida. Bull. Mar. Sci. 17 (2), 257–267.

Naylor, L.A., 2001. An assessment of the links between biogenic processes and shore platform geomorphology. Glamorgan Heritage Coast, South Wales, UK. Thesis, D. Phil., University of Oxford.

Naylor, L.A., Viles, H.A., 2000. A temperate reef builder: an evaluation of the growth, morphology and composition of *Sabellaria alveolata* (L.) colonies on carbonate platforms in South Wales. In: Insalaco, E.R., Skelton, P.W., Palmer, T.J. (Eds.), Carbonate platform systems: components and interactions, Spec. Publ.-Geol. Soc. Lond., vol. 178, pp. 9–19.

Naylor, L.A., Viles, H.A., 2002. A new technique for evaluating short-term rates of coastal bioerosion and bioprotection. Geomorphology 47 (1), 31–45.

Naylor, L.A., Viles, H.A., Carter, N.E.A., 2002. Biogeomorphology revisited: looking towards the future. Geomorphology 47 (1), 3–14.

Navas, F., Malvarez, G.C., Jackson, D.W.T., Cooper, J.A.G., Portig, A.A., 2002. Geomorphological and biological monitoring of sensitive intertidal flat environments. J. Coast. Res. 36, 531–543.

Nealson, K., Ghiorse, W.A., 2001. Geobiology: a report from the American Academy of Microbiology. American Academy of Microbiology, Washington. 14 pp.

Nienhuis, P.H., Bakker, J.P., Grootjans, A.P., Gulati, R.D., de Jonge, V.N., 2002. The stare of the art of aquatic and semi-aquatic ecological restoration projects in the Netherlands. Hydrobiologica 478 (1–3), 219–233.

Pearson, R.G., Dawson, T.P., Berry, P.M., Harrison, P.A., 2002. SPECIES: a spatial evaluation of climate impact on the envelope of species. Ecol. Model. 154/3, 289–300.

Pirazzoli, P.A., 1986. The early Byzantine tectonic paroxysm. Z. Geomorphol. N. F. Suppl. Bd. 62, 31–49.

Reise, K., 2002. Sediment mediated species interactions in coastal waters. J. Sea Res. 48, 127–141.

Royal Society, 2001. The Role of Land Carbon Sinks in Mitigating Global Climate Change. The Royal Society, London.

Saint Martin, J.-P., 2001. Implications de la présence de *mud-mounds* microbiens au Messinien (Sicile, Italie). C. R. Acad. Sci. Paris, Sci. Terre Planètes 332, 527–534.

Schneider, J., Le-Champion Alsumard, T., 1999. Construction and destruction of carbonates by marine and freshwater cyanobacteria. Eur. J. Phycol. 34, 417–426.

Shroder, J.F., Bishop, M.P., 1995. Geobotanical assessment in the Great Plains, Rocky Mountains and Himalayas. Geomorphology 13, 101–119.

Siegert, M.J., Cynan Ellis-Evans, J., Tranter, M., Mayer, C., Petit, J.-R., Salamatin, A., Priscu, J.C., 2001. Physical and chemical processes in Lake Vostok and implications for life in Antarctic subglacial lakes. Nature 414, 603–609.

Small, R.J., 1978. The study of landforms: a textbook of geomorphology. Cambridge University Press, Cambridge. 501 pp.

Spencer, T., 1988. Coastal biogeomorphology. In: Viles, H.A. (Ed.), Biogeomorphology. Basil Blackwell, Oxford, pp. 255–318.

Spencer, T., 1992. Bioerosion and biogeomorphology. In: John, D.M., Hawkins, S.J., Price, J.H. (Eds.), Plant–animal interactions in the marine benthos, Systematics Association Special Volume, vol. 46. Clarendon Press, Oxford, pp. 493–509.

Spencer, T., Viles, H., 2002. Bioconstruction, bioerosion and disturbance on tropical coasts: coral reefs and rocky limestone shores. Geomorphology 48, 23–50.

Thayer, C.W., 1979. Biological bulldozers and the evolution of marine benthic communities. Science 203, 458–461.

Thommeret, Y., Thommeret, J., Laborel, J., Montaggioni, L.F., Pirazzoli, P.A., 1981. Late Holocene shoreline changes and seismo-tectonic displacements in western Crete (Greece). Z. Geomorphol. N. F. Suppl. Bd. 40, 127–149.

Thornes, J.B. (Ed.), 1990. Vegetation and erosion. John Wiley and Sons, Chichester.

Trenhaile, A.S., 1987. Geomorphology of rock coasts. Clarendon Press, Oxford.

Tribollet, A., 2001. Bioerosion processes in coral reefs (Great Barrier Reef, Australia)—importance of the boring microflora. thesis. PhD. Méditerranée University (France). 225 pp.

Trudgill, S., 1985. Limestone geomorphology. Longman, London.

Trudgill, S., Smart, P.L., Friedrich, H., Crabtree, R.W., 1987. Bioerosion of intertidal limestone, Co. Clare, Eire: 1. *Paracentrotus lividus*. Mar. Geol. 74, 85–98.

Underwood, G.J.C., Paterson, D.M., 1993. Seasonal changes in diatom biomass, sediment stability and biogenic stabilisation in the Severn Estuary, UK. J. Mar. Biol. Assoc. U.K. 73, 871–8879.

Verhoeven, J.T.A., 2001. Guest editorial: ecosystem restoration for plant diversity conservation. Ecol. Eng. 17, 1–2.

Viles, H.A., 1984. Biokarst: review and prospect. Prog. Phys. Geogr. 8, 523–542.

Viles, H.A., 1988a. Introduction. In: Viles, H.A. (Ed.), Biogeomorphology. Basil Blackwell, Oxford, pp. 1–8.

Viles, H.A., 1988b. Biogeomorphology. Basil Blackwell, Oxford. 364 pp.

Viles, H.A., Naylor, L.A., 2002. Introduction. Geomorphology 47, 1–2.

Vita-Finzi, C., Cornelius, P.F.S., 1973. Cliff sapping by molluscs in Oman. J. Sediment. Petrol. 43 (1), 31–32.

Wild, U., Kamp, T., Lenz, A., Heinz, S., Pfadenhauer, J., 2001. Cultivation of *Typha* spp. in constructed wetlands for peatland restoration. Ecol. Eng. 17, 49–54.

Wilson, D.P., 1971. *Sabellaria alveolata* (L.) at Duckpool, North Cornwall, 1961–1970. J. Mar. Biol. Assoc. U.K. 51, 509–580.

Wootton, J.T., 2002. Indirect effects in complex ecosystems: recent progress and future challenges. J. Sea Res. 48, 157–172.

Zimbelman, J.R., 2001. Image resolution and evaluation of genetic hypotheses for planetary landscapes. Geomorphology 37, 179–199.

Available online at www.sciencedirect.com

Palaeogeography, Palaeoclimatology, Palaeoecology 219 (2005) 53–69

www.elsevier.com/locate/palaeo

Temperature and salinity history of the Precambrian ocean: implications for the course of microbial evolution

L. Paul Knauth[*]

Department of Geological Sciences, Arizona State University, Tempe, AZ 85287-1404, United States

Received 18 May 2003; accepted 29 October 2004

Abstract

The temperature and salinity histories of the oceans are major environmental variables relevant to the course of microbial evolution in the Precambrian, the "age of microbes". Oxygen isotope data for early diagenetic cherts indicate surface temperatures on the order of 55–85 °C throughout the Archean, so early thermophilic microbes (as deduced from the rRNA tree) could have been global and not just huddled around hydrothermal vents as often assumed. Initial salinity of the oceans was 1.5–2× the modern value and remained high throughout the Archean in the absence of long-lived continental cratons required to sequester giant halite beds and brine derived from evaporating seawater. Marine life was limited to microbes (including cyanobacteria) that could tolerate the hot, saline early ocean. Because O_2 solubility decreases strongly with increasing temperature and salinity, the Archean ocean was anoxic and dominated by anaerobic microbes even if atmospheric O_2 were somehow as high as 70% of the modern level.

Temperatures declined dramatically in the Paleoproterozoic as long-lived continental cratons developed. Values similar to those of the Phanerozoic were reached by 1.2 Ga. The first great lowering of oceanic salinity probably occurred in latest Precambrian when enormous amounts of salt and brine were sequestered in giant Neoproterozoic evaporite basins. The lowering of salinity at this time, together with major cooling associated with the Neoproterozoic glaciations, allowed dissolved O_2 in the ocean for the first time. This terminated a vast habitat for anaerobes and produced threshold levels of O_2 required for metazoan respiration. Non-marine environments could have been oxygenated earlier, so the possibility arises that metazoans developed in such environments and moved into a calcite and silica saturated sea to produce the Cambrian explosion of shelled organisms that ended exclusive microbial occupation of the ocean.

Inasmuch as chlorine is a common element throughout the galaxy and follows the water during atmospheric outgassing, it is likely that early oceans on other worlds are also probably so saline that evolution beyond the microbial stage is inhibited unless long-lived continental cratons develop.

Keywords: Precambrian; Oxygen isotopes; Salinity; Paleotemperature; Neoproterozoic

* Tel.: +1 480 965 2867; fax: +1 480 965 8102.
  E-mail address: Knauth@asu.edu.

0031-0182/$ - see front matter © 2004 Elsevier B.V. All rights reserved.
doi:10.1016/j.palaeo.2004.10.014

1. Introduction

The Precambrian represents the first 3.5 billion years of Earth history and is the age of the microbes. Isotopic evidence (Mojzsis et al., 1996; Rosing, 1999; Rosing and Frei, 2004) and possible microfossil evidence (Schopf, 1993) hint that prokaryotic microbial life was already established at the time of the earliest preserved rock record at 3.5–3.8 Ga (billion years before present). Eukaryotes did not appear until about 2.7 Ga (Brocks et al., 1999) and metazoans until about 0.55 Ga. Considering the rapidity with which life can evolve, these delays in the appearance of more complex varieties pose outstanding challenges for understanding the course of evolution on Earth and for exploring the probable nature of life or past life elsewhere in the solar system and cosmos. As often noted, the Precambrian is an enormous interval of time in which evolutionary advances could have been made via internal advances in biological mechanisms, so it seems likely that environmental conditions rather than internal factors were responsible for the delay in the appearance of more complex life. The most popular current explanation for the end of microbial dominance is that atmospheric oxygen levels rose throughout the Precambrian until some critical threshold necessary for metazoan respiration was achieved (Runnegar, 1982 and references cited therein). Clearly, a minimum amount of atmospheric oxygen is needed, but the level of oxygen *dissolved in water* is the real issue. Temperature and salinity strongly govern oxygen solubility, and these are largely unconstrained variables which have probably varied significantly over geologic time.

In this paper, I explore aspects of the Precambrian ocean environment in which earliest microbial life evolved and explore the possibility that high temperature and salinity were major factors affecting microbial evolution, that the "Cambrian explosion" may represent movement of already evolved metazoans from non-marine environments into the sea after a major salinity decline resulting from deposition of enormous salt deposits in the Neoproterozoic, and that early hydrospheres on planets, rather than being "cradles of life", are probably so saline that early evolution is actually inhibited until, or unless, long-lived continental cratons with non-marine aqueous environments develop.

2. Analytical methods

Oxygen isotope data for cherts were measured using the conventional fluorine extraction method on 20-mg samples. CO_2 derived from this procedure was isotopically analyzed with a 15-cm isotope ratio mass spectrometer. δ-values were first determined relative to CO_2 derived from a carbonate working standard and then referred to the SMOW standard using α_{CO_2}–H_2O=1.0412 (O'Neil et al., 1975). Data for Precambrian cherts from the PPRG collection (Schopf and Klein, 1992) are given in Table 1. Other new data are given in Table 2.

3. Climatic temperature history

Inasmuch as different types of microbes have greatly different temperature ranges in which they thrive, major changes in climatic temperature over geologic time would certainly have affected the course of their evolution. The common occurrence of pillow lavas, cross-beds and other sedimentary structures indicative of liquid water throughout the Precambrian clearly indicate past temperatures in the approximate range 0–100 °C. The silicate weathering "thermostat" of Walker et al. (1981) is often invoked as a constraint on temperature variations. In this model, increases in atmospheric CO_2 from volcanic emissions are consumed during silicate weathering which then causes greenhouse heating to decline. This reduces the rate of weathering and allows CO_2 emitted from volcanoes to increase. A steady state is reached and the CO_2 level is thus regulated to keep climatic temperatures in the 0–100 °C range. The time interval in which this feedback cycle works and the actual temperature range of the "thermostat" cannot be specified. While useful for understanding why the Earth has been hospitable to life for so long, this feedback model cannot be used to constrain major temperature changes that may have occurred in the Precambrian within the broader temperature range of 0–100 °C.

$^{18}O/^{16}O$ ratios provide an approach for measuring past temperatures, but the method is not straightforward and varies depending upon the age of the materials. Measurements of $\delta^{18}O$ for unrecrystallized, original carbonate precipitates such as sea shells or fibrous aragonitic marine cements can yield the actual

Table 1
PPRG chert samples

PPRG #	Age	$\delta^{18}O$ (‰)	PPRG #	Age	$\delta^{18}O$ (‰)	PPRG #	Age	$\delta^{18}O$ (‰)	PPRG #	Age	$\delta^{18}O$ (‰)	PPRG #	Age	$\delta^{18}O$ (‰)
1475	550	26.8	1282	770	19.1	1309	1700	17.4	2027	2735	16.3	1409	3000	12.8
1475	550	26.4	1283	770	20.6	1261	1885	20.0	2027	2735	16.0	1633	3000	10.6
1476	550	27.1	1340	770	17.2	1261	1885	19.7	2028	2735	16.8	1436	3300	18.9
1477	550	26.4	1342	770	20.7	1261	1885	19.3	2029	2735	15.6	1444	3300	16.4
1477	550	26.2	1229	775	20.5	1587	1925	17.7	2031	2735	15.6	1465	3330	13.6
1318	610	24.3	1280	780	21.3	1845	2250	26.4	2031	2735	16.0	1466	3330	14.6
2245	640	23.5	1222	785	18.6	1415	2325	21.0	2031	2735	15.5	2718	3330	14.2
2245	640	22.9	1424	800	27.1	1415	2325	21.7	2032	2735	16.5	2718	3330	15.3
1473	650	20.3	1424	800	28.4	2526	2325	23.4	2033	2735	16.2	1458	3400	15.2
1474	650	22.7	1635	800	26.2	1417	2400	23.3	2033	2735	16.1	2002	3400	17.1
2333	680	24.4	1676	800	23.3	1412	2450	20.4	2035	2735	16.1	2002	3400	16.9
2332	685	29.3	2364	800	23.6	2515	2450	24.3	2038	2735	15.1	2002	3400	17.4
2332	685	30.3	1245	850	28.0	2519	2450	17.7	2039	2735	14.9	1440	3400	12.5
1478	740	27.5	1246	850	29.6	2520	2450	19.8	2040	2735	14.2	1440	3400	12.8
1479	740	27.4	1253	850	19.6	2520	2450	22.7	2041	2735	14.6	1441	3400	12.8
1480	740	27.2	1243	850	22.6	1293	2500	17.1	2041	2735	14.8	1998	3400	15.3
1481	740	27.2	1249	850	26.1	1293	2500	15.8	2044	2735	14.9	1999	3400	15.2
1481	740	23.6	1251	850	20.1	1294	2500	20.1	2044	2735	15.0	2001	3400	17.5
1481	740	24.7	1251	850	21.9	1372	2650	16.6	2044	2735	15.4	2644	3400	15.1
1482	740	27.3	1251	850	23.2	1372	2650	17.1	2045	2735	14.3	1447	3435	14.1
1483	740	26.8	1252	850	27.4	1416	2650	12.1	2046	2735	15.2	2010	3435	13.2
1248	760	29.2	1256	850	20.1	1907	2650	18.6	2048	2735	16.2	2010	3435	11.9
1255	760	24.3	1313	850	24.7	1907	2650	18.4	1354	2750	8.7	2010	3435	13.9
1257	760	23.5	1313	850	24.6	1921	2650	18.2	1354	2750	9.1	1448	3435	18.5
1258	760	25.6	1314	850	27.0	2435	2650	16.3	1354	2750	11.6	1438	3435	13.7
1259	760	25.9	1315	850	24.9	1467	2690	17.4	1354	2750	11.8	1439	3435	14.5
1322	760	22.6	1317	850	27.8	1467	2690	18.0	1355	2750	10.9	1439	3435	15.0
1322	760	21.8	1317	850	27.9	1467	2690	14.8	1355	2750	11.1	1439	3435	18.7
1325	760	31.3	1319	850	28.3	1467	2690	14.6	1355	2750	13.0	1439	3435	14.7
1230	770	27.8	1321	850	26.7	2690	2690	16.4	1356	2750	7.8	1442	3435	13.1
1230	770	26.9	1328	850	26.7	2690	2690	14.8	1356	2750	8.2	1451	3435	6.3
1230	770	24.2	1332	850	20.8	2690	2690	14.5	1356	2750	11.1	1451	3435	6.0
1231	770	26.4	1333	850	22.7	2019	2735	16.5	1356	2750	10.9	1453	3435	14.4
1232	770	22.2	1333	850	23.3	2020	2735	15.8	1455	2750	15.5	1453	3435	14.3
1232	770	22.3	1333	850	22.1	2021	2735	14.9	1457	2750	16.3	2637	3435	6.6
1233	770	25.4	1359	900	22.2	2021	2735	16.4	1461	2750	13.6	2638	3435	14.8
1233	770	24.9	1360	1050	24.7	2021	2735	14.6	1469	2750	11.2	2645	3435	15.4
1281	770	19.9	2242	1364	27.3	2022	2735	15.6	1469	2750	13.2	2646	3435	14.7
1281	770	20.3	1308	1700	16.6	2024	2735	16.6	1464	2768	12.8	1449	3450	10.0
1281	770	21.2	1309	1700	17.4	2025	2735	15.9	1464	2768	12.9	1449	3450	9.3
1281	770	22.2	1309	1700	17.5	2025	2735	16.0	1408	3000	11.0	1450	3525	12.7
1281	770	21.9	1309	1700	17.2	2026	2735	15.7	1408	3000	11.2			

temperature of the sea water, and measurements across a single shell can even record seasonal variations during the life of the organism (Urey et al., 1951). Thousands of paleotemperatures have now been measured for Cenozoic sediments with this method, but it is rare to find actual original precipitates in strata much older. All marine precip- itates eventually go through a period of mineralogical "stabilization" where the originally deposited microcrystalline high-Mg calcite, aragonite and calcite dissolve and reprecipitate into interlocking crystals of calcite and/or dolomite. In platform deposits, stabilization usually occurs within a few million years because meteoric waters enter the system along

Table 2
Miscellaneous chert data

Unit/age	Lab #	$\delta^{18}O$ (‰)	Unit/age	Lab #	$\delta^{18}O$ (‰)	Unit/age	Lab #	$\delta^{18}O$ (‰)	Unit/age	Lab #	$\delta^{18}O$ (‰)
Edwards Fm.	415	32.4	Edwards Fm.	447b	33.2	Belt Supergroup	700	20.7	Gunflint Fm.	547	19.9
Texas	413	31.8	Texas	446	31.3	Montana	701	20.4	Ontario, Canada	563	18.2
100 Ma	412	32.2	100 Ma.	431c	27.9	1350 Ma.	703	20.1	2090 Ma.	513	19.1
	428	33.0		431a	29.5		706	19.5		550	21.6
	406	33.2		432	30.5		707	18.9		549	22.2
	418	31.9		424	32.2		710	19.2		553	18.9
	409	32.9		425	31.2		717	19.9		507	25.9
	401	32.7		470	31.9		736	20.5		555	23.7
	466a	33.3		457	29.1	Gunflint Fm.	527	22.6		542	15.1
	466b	32.2		469	32.0	Ontario, Canada	525	22.3		541	20.7
	463	32.0		452	33.5	2090 Ma.	524	23.4		536	23.1
	404	34.7		465	30.5		519	23.6		545	22.9
	461	30.4		462	31.3		523	24.1		538	17.4
	421	33.1		458	31.5		522	21.8		526	22.7
	408	33.4		468	32.3		518	20.6		566	21.7
	422	33.7		449	33.0		516	23.7		544	23.1
	423	32.7		450	33.3		507	26.1		548	23.1
	424	30.8		467	32.2		515	21.7		565	25.2
	410	33.2		472	32.8		509	24.0		564	18.7
	407	34.0	Chuar Group	907	30.2		507	26.0			
	400	33.0	Arizona	901	26.8		501	23.1			
	434	33.6	750 Ma.	902	26.3		500	23.3			
	439	34.2		903	26.6		514	23.0			
	431b	34.9		906	25.2		517	22.2			
	433	32.5		906	26.1		517	22.9			
	431c	27.8		904	27.1		514	21.6			
	430	32.3		905	26.8		514	23.4			
	427	32.1		900	31.2		509	24.4			
	421	33.1		910	26.2		507	26.2			
	401	32.4		911	29.8		502	23.9			
	425	30.6		915	31.0		508	23.6			
	441	33.0	Bass Limestone	912	28.5		506	21.8			
	442	32.4	Arizona	913	26.4		504	23.2			
	443	33.4	1200 Ma.	916	30.3		505	22.7			
	445	33.5		909	20.4		541	21.1			
	435	32.6		917	26.5		543	24.3			
	451	33.3		914	23.0		507	25.6			
	447a	30.6					546	23.1			
	448	32.6					532	21.5			

coastal margins and vigorously promote solution/recrystallization. Original marine precipitates of Precambrian carbonates have never been documented; all have been stabilized into limestone or dolostones. Inasmuch as most Precambrian carbonates are platform carbonates (deep sea varieties having been long since scraped into subduction zones), the parent water for present crystals probably had a low ^{18}O meteoric water component, and O isotope variations thus reflect both temperature and also variable $\delta^{18}O$ of the

stabilizing fluids (see James and Choquette, 1990 for an excellent review). An additional problem is that carbonates undergo recrystallization rather easily in response to heating during deep burial and low grade metamorphism. For all these reasons, oxygen isotope paleothermometry of carbonates as practiced on Cenozoic material usually cannot be used in the Precambrian.

Cherts composed of microcrystalline quartz are very common in sedimentary rocks of all ages and typically form during stabilization of carbonates (see review by Knauth, 1994). The primary silica precipitate is usually opal and this metastable amorphous material is dissolved and reprecipitated into the intergrown microcrystalline quartz masses sampled as chert. In Phanerozoic examples, the initial opal is usually precipitated by diatoms, radiolarians, and sponges. In Precambrian examples, the silica may have been biologically produced by microorganisms, introduced as an inorganic precipitate from a silica-saturated ocean, released during stabilization of clay minerals, or derived from volcanic glasses in the flow path of the stabilizing pore fluids. Archean strata contain cherts representing wholesale silicification of sediments on a scale not present in younger rocks (Knauth and Lowe, 2003). Opal was apparently precipitated on a colossal scale in the Archean ocean and all materials deposited, whether volcanic or sedimentary, were subjected to silicification and conversion to chert during early diagenesis. Quartz itself is one of the most resistant minerals to later isotopic alteration (Taylor, 1974). Masses of chert have essentially no measurable permeability and strongly resist the later dissolution/reprecipitation episodes so common in carbonates. Recrystallization can occur during metamorphic heating but water/rock ratios at that point are usually so low that $\delta^{18}O$ is largely unchanged (Knauth and Lowe, 2003). The isotopic composition of cherts thus provides a possible approach for determining past stabilization temperatures. If stabilization occurs during shallow burial (as is generally the case for shallow water deposits), such temperatures approximate the climatic temperature during stabilization. An explanation of the method and early results were presented by Knauth and Epstein (1976) and Knauth and Lowe (1978).

Knauth and Lowe (2003) showed that Archean cherts of the Onverwacht Group, S. Africa have $\delta^{18}O$ values about 10‰ lower than those of Phanerozoic cherts. They argued from this that climatic temperatures at 3.5 Ga were on the order of 40 °C warmer than those of the Phanerozoic. Fig. 1 shows their data along with previously published data from the Precambrian and new data. The approximate 10‰ variation of $\delta^{18}O$ for cherts of a given age is not data scatter, but is caused by variable amounts of low ^{18}O meteoric water in the initial stabilization to form chert

or by later stabilization during deep burial at elevated temperatures (often millions of years later, but not resolvable on the time axis of Fig. 1).

Data plotted in Fig. 1 are nearly all for platform cherts where field evidence indicates stabilization to quartz during shallow burial (e.g. synsedimentary brecciation of chert nodules, host-rock laminae compacted around nodules, slumped nodules, preservation of organic-walled microfossils, etc.). The samples were further screened petrographically to eliminate cherts metamorphically recrystallized to mosaic quartz. The 3.8 Ga data (Perry et al., 1978) are for amphibolite-grade metamorphic rocks but are included here because they are the only known data for the oldest cherts and because the authors argued that the maximum values represent early diagenetic values. Because of the large amount of field work at vastly different localities required to sample and screen appropriate samples from the whole of geologic time, the data are necessarily still very limited in number. There is clearly an overall trend of cherts getting more ^{18}O-rich with time as initially proposed by Perry (1967), and this has been attributed to an overall climatic cooling of the Earth with time (Knauth and Epstein, 1976; Knauth and Lowe, 1978). Alternative explanations include increasing $\delta^{18}O$ of the ocean with time and later alteration due to metamorphism and/or long-term isotopic exchange with low ^{18}O ground waters. Knauth and Lowe (2003) evaluated these alternative explanations and argued that they were incompatible with the data for the 3.5 Ga Onverwacht cherts and thus not likely explanations for the younger data. Unless the arguments of Knauth and Lowe (2003) can be refuted, the overall trend of the data in Fig. 1 must be due primarily to an Earth that has cooled down over the past 3.5 Ga.

Deducing an actual temperature for a given data point in Fig. 1 is generally not possible from $\delta^{18}O$ values alone. $\delta^{18}O$ of a chert depends only upon $\delta^{18}O$ of the parent water and the temperature, but it is rarely possible to specify $\delta^{18}O$ of ancient early diagenetic waters. An additional problem is that there are several quartz-water isotopic fractionation with temperature expressions proposed for temperatures <200 °C (Matheney and Knauth, 1989). What can be said is that there has been a general increase of about 10‰ in $\delta^{18}O$ of cherts between the Archean and the Phanerozoic with possible rises and falls in the maximum

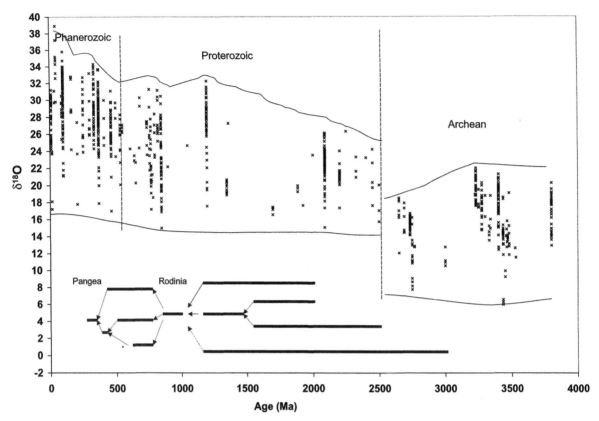

Fig. 1. Compilation of new and previously published oxygen isotope data for screened chert samples over geologic time. All samples except the 3800 Ma examples are for unmetamorphosed cherts or cherts metamorphosed no higher than Greenschist Facies. The history of long-lived continental cratons is taken from Rogers (1996) and the horizontal lines represent periods of stability of specific cratonic masses identified by him. The 10‰ variation in $\delta^{18}O$ for cherts at any given time is caused by the presence of low ^{18}O meteoric waters present during diagenetic chert formation and/or formation of cherts during deeper burial at elevated temperatures. The overall increase in $\delta^{18}O$ with time is interpreted as global cooling over the past 3500 Ma. New data are given in Tables 1 and 2. Published data are from Knauth and Lowe (1978, 2003), Force et al. (1986), Gao and Land (1991), Suchecki and Hubert (1984), Kenny and Knauth (1992), Beeunas and Knauth (1985), Winter and Knauth (1992) and references cited therein.

values along the way, as suggested from the overall pattern in Fig. 1. Regardless of which quartz-water fractionation expression is used, this 10‰ increase corresponds to an overall decrease in temperature of about 40 °C (Knauth and Lowe, 2003). If global temperatures in the Tertiary (extreme left side of Fig. 1) were on the order of 10–20 °C, then the lower $\delta^{18}O$ values at approximately 3.5 Ga corresponds to temperatures on the order of 55–85 °C (see Knauth and Lowe, 2003 for a detailed discussion of this). The cause of such large climatic temperature variations is unknown, but variations in the amount of greenhouse gases such as water vapor, CO_2 and CH_4 are commonly invoked (e.g. Kasting and Siefert, 2002).

Although possibly premature, there is strong suggestion of a major jump of about 3–5‰ to higher $\delta^{18}O$ somewhere around 2.5 Ga (Fig. 1). Upper Archean cherts from several localities are strongly depleted in ^{18}O relative to those of the Paleoproterozoic. Most of these are from the Fortesque Group, Western Australia, and are likely non-marine deposits (Walter, 1983) that formed in lower ^{18}O waters. Existing values at 2.65 Ga are below +18.4 and are more ^{18}O depleted than almost all younger cherts. Going forward in time, values jump to as high as +24.5 by 2.45 Ga and never return to values similar to those of the Archean. This change in $\delta^{18}O$ is one of the striking aspects of the data in Fig. 1 and corresponds to an approximate general temperature decrease on the order of 20 °C, suggesting Paleoproterozoic temperatures on the order of 40 °C. The change toward lower climatic temperatures suggested by these data could be related to drawdown of CO_2 during weathering of the first truly extensive, long-lived continental cratons thought to have developed at this time (Goodwin, 1996; Rogers, 1996) and/or by the decay of a methane greenhouse (Kasting and Siefert, 2002).

$\delta^{18}O$ values appear to generally rise throughout the Proterozoic and become more similar to Phanerozoic values by 1.2 Ga. The interval between 1.2 and 0.685 Ga yields values >+30, similar to post-Ordovician cherts. Data are frustratingly scarce during the important Neoproterozoic–Lower Paleozoic transition period. For time intervals where there are too few data to establish the approximately 10‰ spread in $\delta^{18}O$ inherent in cherts as they form, there is always a possibility that significantly more positive or negative values will be found as new data are obtained. However, the combined δD–$\delta^{18}O$ approach (e.g. Knauth and Epstein, 1976; Kenny and Knauth, 1992; Fallick et al., 1985; Sharp et al., 2002) allows a rough assessment of temperature or temperature changes without knowledge of $\delta^{18}O$ of the parent water, something not possible with only one isotope. The combined δD, $\delta^{18}O$ data for Cambrian strata indicate that the low $\delta^{18}O$ values of the lower Paleozoic are not due simply to the inadequacy of sampling (Knauth and Epstein, 1976). There is thus a tantalizing suggestion in Fig. 1 that climatic temperatures were not much different from most of the lower Paleozoic in the Mesoproterozoic and Neoproterozoic, but "jumped" on the order of 10–15 °C (the value necessary to increase $\delta^{18}O$ about 2‰) sometime between about 685 and 550 Ma, possibly associated with termination of late Proterozoic glaciation events. This is roughly coincident with the advent of metazoans in the ocean, but the isotope data are presently far too few to clarify climatic temperature changes during this important interval. Platform cherts probably do not form during glaciations when carbonate production drops, and the time between deposition and diagenetic fixing of the isotopic signal in stable microcrystalline quartz may be too great to detect short intervals of glaciation and the possible sudden

post-glacial warmings proposed by Hoffman et al. (1998). The same considerations apply to the likely early Paleoproterozoic glaciations. So far, no chert of any kind that is known to have actually formed as microquartz during a glacial time has been identified or isotopically analyzed.

If the isotopic data for cherts are related to overall climatic temperatures, then there were very hot conditions throughout most of the Archean, an overall decline to cooler temperatures throughout the Proterozoic, and a possible major warming somewhere near the Precambrian/Cambrian boundary. Although still sparse, the data are perhaps sufficient enough to consider implications for the history of microbial life.

Thermophilic and hyperthermophilic microbes are deeply rooted in the rRNA phylogenetic tree suggesting that the earliest organisms on Earth were high temperature forms. This has supported speculation that earliest life originated at sea floor hydrothermal vents (e.g. Baross and Hoffman, 1985). However, recent results suggest that the most deeply rooted bacteria are thermophilic but not hyperthermophilic (Gaucher et al., 2003). The optimal temperature is about 65 °C, essentially that deduced from the O isotope record for the time of the earliest sedimentary rock record (Knauth and Lowe, 2003). Earliest thermophilic life could have thus been global in extent rather than a phenomenon localized only around hydrothermal vents.

As first noted by Hoyle (1972), the fossil record suggests that organisms appeared on Earth sequentially in order of tolerance to high temperatures. Schwartzman (1999) has independently explored this concept in detail and argues that the overall declining temperatures over geologic time have largely determined the course of evolution. In this view, the Precambrian was the age of microbes because of overall higher temperatures and the ability of microbes to thrive under such conditions. It follows that metazoans and more complex forms had to await the cooling of the Earth. However, additional factors are likely because the emerging data (Fig. 1) suggest that only the Archean was at temperatures typically greater than about 40 °C. $\delta^{18}O$ of the limited data from the Proterozoic are similar to those of the Phanerozoic, hinting that Phanerozoic-like temperatures were already present in much of the upper Proterozoic. There is no evidence for sudden increase in $\delta^{18}O$ (indicating cooling) at the Precambrian/

Cambrian boundary; the available data indicate a lowering of $\delta^{18}O$ values suggesting a major *warming* around this time. While temperature was likely a major factor in affecting the course of evolution in the Precambrian as argued by Schwartzman (1999), it appears unlikely as the sole explanation for the overall late arrival of the metazoans.

4. Ocean salinity

Although life has now adapted to a wide range of salinities, the salt concentration is clearly an important environmental variable that controls and limits the nature of biologic activity. For example, restricted lagoons of the Persian Gulf turn into "faunal deserts" when salinities approach $2\times$ the open ocean values (Hughes Clarke and Keij, 1973). Although certain bacteria can thrive at higher salinities, populations of marine bacteria generally decrease with increasing salinity and diversity is exceptionally low in hypersaline water masses (Larsen, 1980). Salinity can also have a strong effect on the solubility of atmospheric gases, especially oxygen. The salinity history of sea water is thus a crucial parameter that has received very little attention.

Although significant progress has been made on determining cation ratios in Phanerozoic sea water from fluid inclusions in halite (e.g. Lowenstein et al., 2001), Cl^- variations over time cannot presently be determined from the rock record. DeRonde et al. (1997) and Channer et al. (1997) attempted to do so by analyzing fluid inclusions 3.5 Ga quartz/goethite deposits interpreted as Archean sea-floor hydrothermal chimneys. However, these goethite deposits have now been reinterpreted as Cenozoic spring deposits (Lowe and Byerly, 2003). At present, there are no compelling examples of inclusion-bearing crystals that grew in Precambrian sea water and remained unaltered with time.

Enormous salt beds, the so-called "saline giants" are observed in the geologic record, and these were deposited at specific times rather than continuously (Fig. 2). The episodic nature of the deposition and subsequent episodic erosion of these units during tectonic uplift has surely led to major salinity variations in the oceans over time. Salinity of the ocean is thus an unconstrained variable that depends

upon the frequency of deposition and erosion of these giant salt beds.

Cl^- is an incompatible element and was probably initially outgassed as HCl along with water during the earliest history of the Earth (Holland, 1984). The entire Cl inventory would have been in the ocean and could not have been reduced until evaporitic salt deposits began to form. This could not have happened until long-lived continental cratons developed in the Paleoproterozoic. Salt deposited in coastal environments is eventually recycled to the sea via dissolution and runoff. At any given time, some amount of salt is stored in supratidal deposits causing ocean salinity to be lower than its initially high, pre-continent value. However, the issue of whether a steady state for the removal–return cycle of evaporites has ever been achieved is overshadowed by the larger issue of the saline giants, the *enormous* deposits of salt that occur at specific times in Earth history.

There appear to be two large time intervals in which most of the known salt deposits were deposited (Fig. 2). About 40% of the Phanerozoic inventory was sequestered in the interval 180–250 Ma. Much of the rest was deposited in the Neoproterozoic and Cambrian. The Neoproterozoic inventory has never been estimated, but the enormous accumulations in Australia, Oman, Saudi Arabia, Iran and Pakistan clearly rival the Luann Salt (USA, 180 Ma) in geographic extent and thickness and are therefore shown in Fig. 2 as approximately equal to, or greater than, the Luann inventory. Holser et al. (1980) and Hay et al. (2001) have argued that major salinity changes occurred in the Phanerozoic in response to the imbalance between deposition and erosion that allowed the net deposition of salt at certain times as clearly illustrated by the figure.

Assuming that the Earth's hydrosphere was outgassed early (rather than continuously), Holland (1978) estimated the initial salinity by adding back the halite that has been removed by deposition and is currently stored in sedimentary basins on the continents. Using an early estimate by Zharkov (1981) for the amount of halite, the salinity would have been only about $1.2\times$ the modern value. An ocean this saline would have been a problem for most metazoans today, but its effect on bacteria cannot be considered significant. Cyanobacteria, for example, have little problem with this enhanced salinity. However, this estimate did not take into account many huge new

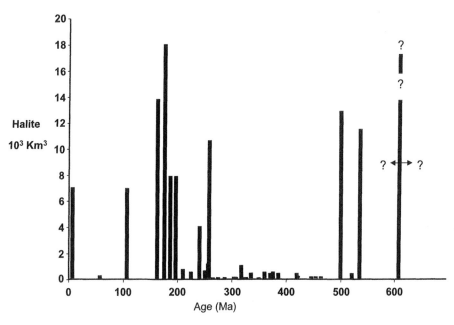

Fig. 2. Halite accumulations in the Phanerozoic. Data were originally listed by geologic time period (Zharkov, 1981; Land et al., 1988) and are centered here on the absolute age of these designated periods. Precambrian halite has never been inventoried. The Neoproterozoic value estimated here is based on the assumption that the enormous accumulations in Australia, Oman, Iran and Pakistan are spread over basins larger in extent than that hosting the Luann salt, USA (180 Ma). Most of the known salt was clearly deposited in two large pulses, one in the interval 180–250 Ma and one in the interval 500–700 (?) Ma. Sequestration of these "saline giants" requires a rare combination of geographic, geologic, climatic, oceanographic and depositional conditions.

subsurface salt deposits that have been subsequently discovered or were not inventoried by Zharkov (1981). These include all Precambrian examples, all African and South American occurrences, and the Gulf Coast Luann salt. More importantly, there is possibly an equal or even much greater amount of NaCl that was previously in sea water but is now present in the ubiquitous deep basin brines (Land, 1995). The salinity of ground water almost everywhere increases with depth (with notable exception when shales dewater to form overpressured reservoirs) and many large sedimentary basins have vast amounts of hypersaline brine with Cl^- concentrations in excess of 100,000 mg/l. The brines are remnants of evaporite fluids that deposited the salts (Carpenter, 1978) and also develop when subsurface salt is dissolved by groundwater. Once lodged in sedimentary basins, brines are exceedingly difficult to flush out with meteoric waters (Domenico and Robbins, 1985) and persist for hundreds of millions of years (Knauth, 1988; Rose and Dresel, 1990). The amount of brine currently sequestered within continents has never been estimated, but it likely contains at least as much

NaCl as the halite deposits, and this must be added to the solid salt inventory when estimating how much NaCl has been removed from sea water. Land (1995) recognized this problem and suggested an amount of NaCl in subsurface brine possibly two times greater than the halite value. It is thus possible that ocean salinity could rise to values of 1.6–2× the modern value, or even higher, if all the salt and brine currently sequestered on the continents were returned to the ocean from which it was extracted by sedimentation on continents (Knauth, 1998). The value is certainly much higher than the previously proposed value of 1.2× (Holland, 1978), which was considered inconsequential. It seems likely that the early ocean was significantly more saline and significantly warmer than previously thought.

5. Consequences of a hot, super-saline early ocean

A hot, super-saline early ocean puts significant environmental constraints on the nature and diversity of early marine life. Cyanobacteria can tolerate both the high temperatures and high salinities suggested here, so there is no environmental objection to interpretations of early Archean microfossils as possible cyanobacteria (Schopf, 1993). The salinity tolerance of thermophilic and hyperthermophilic organisms deep in the phylogenetic tree is not known, so it is presently not clear if their inferred antiquity contradicts the salinity scenario suggested here. Halophilic bacteria can thrive in salinities >10× sea water, but most of these aerobic microbes are clearly recent adaptations that could only have appeared after the rise of atmospheric oxygen. One variety, *Halobacterium* species NRC-1, surprisingly, appears to be one of the most ancient archeans (Ng et al., 2000). Halophiles therefore may have existed in marine or non-marine evaporitic settings on the early Earth, but they could not have thrived in the lower salinities of the open ocean. The high temperature and salinity proposed here therefore do not preclude microbial life in marine environments, but these environmental variables could have limited its diversity.

Estuaries and coastal environments diluted by meteoric water runoff as well as non-marine environments would all have the high temperatures, but salinity there could range from fresh water in lakes and rivers to hypersaline in evaporitic environments. Diversity and evolutionary adaptation should have been enhanced in these ever changing, diverse environments although the small amounts of emergent continental crust and lack of long-lived cratons would have limited the environments available. Targeted searches for a microfossil record in these types of environments are highly warranted, but such environments are difficult to identify in the oldest sedimentary rocks. One possible example is where the uppermost strata of the 3.5 Ga Onverwacht Formation, S. Africa, culminate in a shallow water sequence complete with conglomerates and O isotope evidence suggesting progressive input of meteoric water (Knauth and Lowe, 1975). Possible coastal or lacustrine strata in the even older Warawoona cherts in Australia have been suggested (Lowe, 1983). Unfortunately, low grade, extensive metamorphism lowered $\delta^{18}O$ of most Australian cherts (Richards et al., 2001), and areas of best preservation of microquartz have not yet been identified as has been done for the African examples (Knauth and Lowe, 2003). Considering the rarity of preserved cherts that formed in coastal or non-marine environments at this time together with the low probability of finding microfossils in any rock, the lack of a fossil record from these environments is therefore not surprising. However, the extraordinary rarity of claimed microfossils in any of the marine 3.5 Ga rocks despite relentless searches is consistent with the idea that life in the hot, salty Archean ocean was a very limited affair.

The effect of high salt on the origin of life itself is a crucial issue that has received little attention. Most origin of life experiments and considerations have been done assuming normal sea water as the medium. However, experiments by Apel et al. (2002) suggest that fresh water was more conducive to the origin of life and Ricardo et al. (2004) have argued that evaporitic borate-rich environments are favorable for formation and preservation of ribose in an "RNA world". It is clear that consideration of early molecular and biologic evolution in the "primordial soup" should at least consider the possibility that marine fluids were hot and saline and ask whether this matters in the various origin of life scenarios.

6. Oxygen solubility

A hot ocean at 1.5–2× modern salinity is a significantly different fluid from the modern ocean with respect to many physical and chemical properties. Geochemical models of the early Earth usually use the dilute solution chemistry of the textbooks at temperatures of 25 °C, but dilute solution chemistry is only an approximation even for solutions as saline as modern sea water. It is even more inappropriate for the much higher salinities that must have existed on the early Earth. There have been no attempts to geochemically model such an Archean ocean and the task is not within the scope of this paper. However, one chemical aspect of the decline in salinity with time merits consideration and that is the effect of salinity on O_2 solubility in sea water.

Fig. 3 shows the solubility of oxygen in sea water as a function of salinity and temperature under the present atmospheric level (PAL) of oxygen (data from Weiss, 1970). At the present salinity value of 35‰ and temperatures less than 25 °C, dissolved O_2 is >5 ml/l. At Archean temperatures of >55 °C and 2× modern salinity (70‰) dissolved O_2 would be less than 3 ml/l

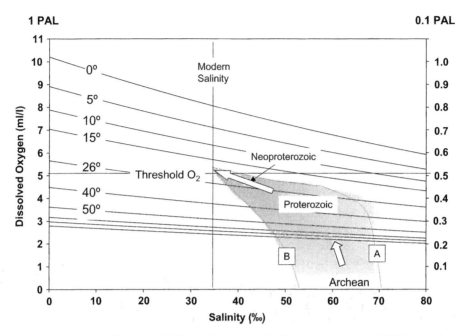

Fig. 3. Dissolved oxygen in seawater as a function of salinity and temperature for the present atmosphere (1 PAL O_2) and an ancient one with 10% of the present atmospheric level of O_2 (0.1 PAL) using experimentally determined algorithms by Weiss (1970). An example threshold level of 0.5 ml/l for dissolved O_2 required for Neoproterozoic metazoans is shown together with example trajectories for the evolution of dissolved O_2 in seawater with time. Curve B invokes a minimum salinity in the Archean and a relatively steady decrease throughout the Precambrian. Curve A is for high initial salinity with little decline until deposition of the great Neoproterozoic saline giants. Under an atmosphere of 0.1 PAL O_2 and temperatures of 15–25 °C, reduction of ocean salinity by deposition of these salt deposits could shift dissolved O_2 to values above that required for metazoan respiration, as shown by the arrow labeled "Neoproterozoic".

even if atmospheric O_2 were at the modern value. If the Archean atmosphere had just 30% less oxygen than the modern value, the oceans would have been anaerobic due to the high temperature and salinity alone, even if these values were somewhat less than the 55 °C, 70‰ example values used here. In today's ocean, sinking organic matter gets respired and an oxygen minimum zone with dissolved O_2 often <1 ml/l forms at several hundred meters depth before deep, oxygenated polar waters mix back into the system at the deeper depths. A hot, saline early ocean would not have the cold, oxygenated deep waters even if atmospheric O_2 levels were high. In other words, the high salinity and high temperatures alone guarantee a basically anaerobic ocean environment as long as atmospheric O_2 levels were below about 0.7 PAL.

It has long been argued that atmospheric O_2 was very low in the Archean although alternative interpretations involving significant O_2 levels have persisted up to Ohmoto's (2003) recent advocacy. The debate has been temporarily settled with the discovery

of mass independent fractionation of S isotopes in Archean sedimentary rocks (Farquhar et al., 2000) that can currently only be explained in terms of UV fluxes sufficient to make ozone from top to bottom of the atmosphere that would keep O_2 levels $<10^{-5}$ PAL (Pavlov and Kasting, 2002). The already low atmospheric O_2 level inferred in this way for the Archean atmosphere would be over three times lower in the ocean because of the reduced solubility from high temperature and salinity (Fig. 3). The cause of precipitation of oxidized iron or partially oxidized iron in iron formations and jasper (chert with fine red hematite) in such anoxic fluids remains problematic. An intriguing possibility is that estuaries and coastal waters were diluted enough with meteoric waters to reduce salinity enough to allow longer retention of oxygen generated by photosynthesizers than in the open ocean. Oxidation in such environments would help certain cases of iron clearly oxidized in the Archean, such as the hematitic clasts in the Onverwacht conglomerates (Lowe and Knauth, 1977). $\delta^{18}O$

of the associated cherts in this example decreases going up-section suggesting progressively increasing amounts of low ^{18}O non-marine waters in this depositional and diagenetic environment.

7. The Proterozoic salinity decline

The only natural mechanism for getting salt out of sea water is evaporation. The initially high salinity of the ocean must have declined in parallel with the development of sedimentary basins and climatic conditions capable of forming and preserving the great saline giants. The conditions necessary to do this actually occur only rarely and require vast inland shallow seas on continents mainly centered on the equator as well as possible special coastline geo-metries, ocean currents and other variables that converge to produce these improbable deposits (Sonnenfeld, 1984). The history of stable cratons is a different question from the history of when and how continental crust formed in the geochemical sense. Continental crust, per se, may have segregated early in Earth history (Armstrong, 1991) or may have devel-oped largely after 2.5 Ga (e.g. Goodwin, 1996). The history of stable cratons, however, is something that cannot be investigated geochemically but must be reconstructed from the geology and age-dates of ancient terrane remnants still preserved, as done by Rogers (1996). When compared with this history of continental platforms (Fig. 1), there is strong sugges-tion that the two great episodes of salt deposition in the late Precambrian and in the Phanerozoic occurred during and after the breakup of the two great supercontinents in Earth history, Rodinia and Pangea. Numerous enclosed arid equatorial basins associated with these supercontinents apparently developed during their breakup to account for the deposition of these salt (and associated brine) deposits.

Large salt deposits older than Neoproterozoic are not known in the geologic record. One interpretation of this is that such deposits existed but have been dissolved away by younger ground waters and/or squeezed out of the older sequences by deformation, the so-called "vanished evaporites" (Garrels and MacKenzie, 1981). Salt is certainly highly soluble and physically mobile; deep burial usually induces diapirs such as the thousands of examples in the Gulf

Coast region of the United States. Salt is so soluble that it is incredible that any remains at all in pre-Tertiary sequences and much of the deep brine in sedimentary basins is undoubtedly derived from the "vanished evaporites". However, the inventory shown in Fig. 2 does not show a progressive decrease in the amount of subsurface salt with increasing age of the sediments. Certainly, the older Precambrian strata have had ample time and exposure to deformation to insure that vast amounts of salt were removed and recycled into the sea. However, the continental history of Rogers (Fig. 1) suggests that the amount of long-lived craton was less prior to 1.0 Ga and essentially absent prior to 2.0 Ga. This, together with the absence of any inferred supercontinents suggests that the amount of salt sequestered on continents in the Meso/Paleoproterozoic was vastly less than in later times and exceptionally low in the Archean. The inference is that significant lowering of the ocean's initially high salinity did not begin until about 2.5 Ga and probably did not undergo its first really big decline until the Neoproterozoic, the time of the oldest strata bearing any evidence of saline giants.

A remarkable aspect of the microbial record in the Precambrian is the limited diversity of the eukaryotes throughout the Proterozoic (Schopf, 1992). High salinity could be an additional factor to others proposed for this limited diversity such as the scavenging of nutrients in a stratified, sulfidic ocean (Anbar and Knoll, 2002). Basically, only salt tolerant organisms could have been present in the open ocean.

Metazoan life is impeded not only because of the direct effects of salt on cell functions, but also because of the decreased solubility of oxygen with increased temperature and salinity. Threshold values have been proposed for a number of primitive animals based on biochemical considerations, diffusion rates and empir-ical size/weight requirements. Runnegar (1982) reviewed these efforts and derived a probable mini-mum oxygen requirement for *Dickinsonia*, a probable Neoproterozoic metazoan, on the order of 0.5 ml/l. This threshold value is shown as an example in Fig. 3 together with calculated curves for dissolved O_2 in sea water as a function of temperature and salinity in an atmosphere with O_2 levels of 0.1 PAL, a lower estimate for the Neoproterozoic (Canfield, 1998). Several scenarios are shown in this figure for the evolution of dissolved O_2 in the Precambrian. All start with an essentially anoxic Archean ocean with high salinity in the range 50–70‰. Curve A is for an extreme case where the first major reduction in ocean salinity occurs in the Neoproterozoic. Temperatures are allowed to decline in the Mesoproterozoic to about 25 °C, possibly somewhat lower than suggested by the existing (but still inadequate) O isotope data. The salinity decline then follows the oxygen solubility curve for 25 °C to higher dissolved O_2 levels until the threshold for *Dickinsonia* is reached. In this case, a major salinity decline alone associated with Neoproterozoic salt deposition is a major factor in allowing metazoans to occupy the ocean.

Curve B is similar to A, but allows for a less salty Archean ocean and a less episodic sequestration of salt over time. Salinity decline could still be a major factor in oxygenating the ocean, but it would have to be coupled with a dramatic late Proterozoic temperature decrease not apparent in the existing O isotope data (Fig. 1). There are many variables that have to be estimated to construct these example curves and the threshold level for Dickinsonia is only one rather uncertain example. The important point is that, if the great salt deposits of the Neoproterozoic produced the first great salinity decline in the ocean, then dissolved O_2 could have moved up one of the temperature curves as shown to cross whatever threshold level was required for oceanic metazoans. Lower temperatures associated with the late Precambrian glaciations would deflect example curves A and B strongly upward in addition to the slide to the left (Fig. 3). The combined effects of Neoproterozoic salt deposition and glaciation cannot be ignored when considering a rise in oxygen as the cause of the rise of metazoans.

8. End of the age of microbes—the Cambrian explosion of metazoan life

The cause of the Cambrian explosion of organisms with mineralized hard parts is a long-standing puzzle with many proposed explanations. Molecular clocks suggest metazoan evolution was well under way in the upper Precambrian (Wray et al., 1996; Lynch, 1999; Benton and Ayala, 2003), but the marine strata are essentially barren of such organisms. One unexplored resolution to this contradiction is the possibility that animals evolved in the upper Precambrian and have

not been found yet as fossils because they thrived in environments that have a low preservation potential. The deep sea and the non-marine are the two depositional environments with the least preservation potential relative to the marine shallow water platform environments where fossils are most favorably preserved. Deep sea sediments are usually carried to subduction zones in less than 200 Ma and become severely metamorphosed. Non-marine environments are high on continents and most are eroded away in short order. However, some, such as the Neoproterozoic Bitter Springs Formation and the 2.7 Ga Fortesque Group in Australia, have survived and do contain microfossils (Schopf and Klein, 1992). Geochemical evidence suggests microbial occupation of soil environments as far back as 2.6 Ga (Watanabe et al., 2000). Microfossils have been found in paleokarst as far back as 1.2 Ga (Horodyski and Knauth, 1994) and Kenny and Knauth (2001) presented isotopic evidence indicating extensive photosynthesizing communities in humid karst terranes in the Neoproterozoic. Lakes, soils and karst pits appear to have been the site of microbial activity throughout much of the Precambrian. Hedges (2003) has even argued that the Neoproterozoic ice ages were brought on by a greening of the land surface so extensive that biotically enhanced weathering brought the CO_2 level of the atmosphere down so low that glaciation set in. Very few deposits from these non-marine environments have been identified and explored for a fossil record in the uppermost Precambrian. The possibility remains that these are the stealth environments where metazoan evolution could have progressed before an explosion into the sea.

As argued above, salinity and temperature of the ocean may not have achieved Phanerozoic-like values until deposition of the great Neoproterozoic salt deposits. There is an interval of 100 million years, or more, prior to the Cambrian explosion during which metazoans could have evolved in lakes, rivers, streams and soils. Non-marine environments have an enormous range of salinities, from very low values in lakes and rivers to hypersaline values in playas, which would have allowed ample opportunity for experimentation at optimal salinity levels. With oxygenation of the atmosphere, the more dilute waters could oxygenate well ahead of the ocean. With deposition of the Neoproterozoic salt deposits, ocean salinity declined enough to allow adaptation of non-marine organisms to this vast new niche. The appearance of shells may thus be simply the result of soft bodies moving into fluids saturated with respect to calcite, silica and phosphate (?). This is the reverse direction of the conventionally assumed sea to land migration of life, but it is equally plausible and perhaps more so if the salinity issue (as well as its effect on dissolved O_2) is considered.

The test of this heterodox idea is a thorough exploration of Neoproterozoic non-marine environments wherever they can be identified. Floodplain deposits and lake deposits need to be identified and explored for possible bioturbation. Paleokarst on the 1.2 Ga Mescal Limestone has been exhaustively searched for disruption to karst pit and cave deposit laminae, and none has been found. If metazoans were in that non-marine setting that deep in the Proterozoic, they did not leave their calling cards. Similar searches in younger paleokarst or any non-marine deposits in the upper Precambrian may prove more fruitful, especially considering molecular clock indications that metazoans were *somewhere* in the upper Precambrian.

9. Implications for astrobiology

"Follow the water" is the current slogan for the new science of astrobiology. Inasmuch as the only form of life we know requires water to exist, the logical first step in astrobiological prospecting in the solar system and elsewhere in the galaxy is to identify targets that have, or had, oceans of water. There is even hope of someday imaging a planet around a distant star that appears as a "pale blue dot" indicating a water-covered planet similar to Earth. Such an image would be an encouraging sign that life is possible throughout the cosmos and would provide justification for research efforts in astrobiology. However, Cl also "follows the water" because it does not fit into silicate mineral structures and would outgas as HCl along with water to produce very saline early hydrospheres.

One initial astrobiology target is Mars because there is considerable geomorphic evidence (e.g. tributary networks, outflow channels) that a significant hydrosphere existed early in the planet's history (Baker, 2001 and references cited therein). As much as 90% of the early hydrosphere was lost, probably due in large part to photodissociation of water vapor followed by hydrogen loss (Jakosky and Philips, 2001). This water loss would have effectively evapoconcentrated any early martian hydrosphere into a brine which would have evolved into a concentrated Ca–Na–Mg–Cl brine after reaction with the basaltic megaregolith and become enormously more concentrated during eutectic freezing when the climate turned cold (Knauth and Burt, 2002). Any early large bodies of water on Mars would have started salty and become vastly more salty with time. Although rare forms of life eventually adapted to rather salty water masses on Earth (as in the Dead Sea), it is not known whether concentrated brines are conducive to the formation and early evolution of living systems. Considering that such brines are generally lethal to most known forms of life, the prospect of evolution of life in early martian oceans seems unlikely. Abundant geomorphic evidence (Baker, 2001) suggests rains in the higher elevations with probable formation of short-lived lakes, rivers and other less saline water masses. Caliche would have formed on the ubiquitous martian basalts and could have trapped any early microbes that somehow developed in these upland, less salty water masses (Knauth et al., 2003).

Europa, a satellite of Jupiter, is another major astrobiology target because there is evidence of an ocean below its icy crust. If Europa assembled out of materials with a Cl/H_2O ratio similar to that of the Earth, then this putative ocean would have had a similarly salty early ocean. Eutectic freezing of salty water bodies produces H_2O ice (see Knauth and Burt, 2002 for a discussion of eutectic freezing), which moves outward in the gravity field and leaves a more concentrated solution behind. The thick icy crust of Europa thus suggests extensive removal of water from the initially salty solution and further suggests that the subsurface ocean is probably a very concentrated brine. Some of the subsurface fluid is interpreted as having debouched along "spreading centers" where the icy crust is thought to have split, and remote sensing suggests the presence of large amounts of sulfates in these areas (McCord et al., 1998). Cl-bearing phases to be expected there at these low temperatures are hydrohalite ($NaCl \cdot 2H_2O$) and antarctictite ($CaCl_2 \cdot 6H_2O$). There is no indication that such phases have been considered in the remote sensing techniques and there are no know laboratory calibration curves to test the obvious prediction of Cl phases along with the sulfates. As in the case of early Mars, the issue is whether an early hydrosphere that starts salty and becomes even more salty with time is good or bad for the origin of life.

Cl is produced during oxygen burning in stars just prior to supernova explosions (Wallerstein et al., 1997) and is strewn into space during the subsequent explosion along with other "metals" (anything heavier than He in astronomical parlance) created in stars. There is therefore every reason to expect that Cl follows the water throughout the galaxy and that there are no regions basically free of this element. Early hydrospheres throughout the cosmos are thus likely to have been initially salty. On Earth, the evolution of life beyond the microbial stage was retarded for 3.5 Ga, possibly in large part due to high temperatures and high salinity. The development of continents was crucial to regulating temperatures to life-tolerant values and to the sequestration of salt and brine, the only known natural mechanism that can significantly reduce salinity in oceans. The discovery of a "pale blue dot" elsewhere in our galaxy would be encouraging for astrobiology, but finding a "pale blue dot" with brown (or green) spots would significantly enhance the prospects of metazoan evolution somewhere else than on Earth.

Ward and Brownlee (2000) recently argued that microbial life could be widespread elsewhere, but that more complex life forms are probably exceptionally rare. Certainly on Earth, something retarded evolution of metazoans for over 3.5 billion years. The thesis presented here is that high oceanic Cl and high temperature were major factors and that the development of large stable continental platforms as well as chance geographic configurations during their drift histories to allow huge evaporite deposits were necessary factors for oxygenation of the ocean and the breakthrough to metazoan life. Inasmuch as the Earth is the only planet in the solar system with continents and plate tectonics, it is likely that similar planets are equally scarce in other solar systems, even if they are within the "habitable zone" with regard to stability of water. Microbiology was the paleontology of the Precambrian, and the simple considerations presented here support the idea that microbial life dominates the history of any water-based life elsewhere in the cosmos.

Acknowledgments

This work was funded by NASA Exobiology Grants NAG513441 and NNG04GJ47G. I thank J.W. Schopf for generous access to Precambrian chert samples in the PPRG collection at the University of California, Los Angeles. Very helpful reviews of the initial manuscript were provided by Sherry Cady and Tomas Hode. Stan Klonowski assisted in the laboratory.

References

Anbar, A.D., Knoll, A.H., 2002. Proterozoic ocean chemistry and evolution: a bioinorganic bridge? Science 297, 1137–1142.

Apel, C., Monnard, P.-A., Deamer, D.W., 2002. Fresh water or marine environments for the earliest forms of life? Part I: Stability of primitive membranes. Astrobiology Science Conference. NASA Ames Research Center, Moffett Field, CA, p. 255.

Armstrong, R.L., 1991. The persistent myth of crustal growth. Aust. J. Earth Sci. 38, 613–630.

Baker, V.R., 2001. Water and the martian landscape. Nature 412, 228–236.

Baross, J.A., Hoffman, S.E., 1985. Submarine hydrothermal vents and associated gradient environments as sites for the origin and evolution of life. Orig. Life Evol. Biosph. 15, 327–345.

Beeunas, M.A., Knauth, L.P., 1985. Preserved stable isotopic signature of subaerial diagenesis in the 1.2-b.y. Mescal Limestone, Central Arizona—implications for the timing and development of a terrestrial plant cover. Geol. Soc. Amer. Bull. 96, 737–745.

Benton, M.J., Ayala, F.J., 2003. Dating the tree of life. Science 300, 1698–1700.

Brocks, J.J., Logan, G.A., Buick, R., Summons, R.E., 1999. Archean molecular fossils and the early rise of eukaryotes. Science 285, 1033–1036.

Canfield, D.E., 1998. A new model for Proterozoic ocean chemistry. Nature 396, 450–453.

Carpenter, A.B., 1978. Origin and chemical evolution of brines in sedimentary basins. Circ.-Okla. Geol. Surv. 79, 60–77.

Channer, D.M.D., deRonde, C.E.J., Spooner, E.T.C., 1997. The Cl–Br–I-composition of 3.23 Ga modified seawater: implications for the geological evolution of ocean halide chemistry. Earth Planet. Sci. Lett. 150, 325–335.

DeRonde, C.E.J., Channer, D.M.D., Faure, K., Bray, C.J., Spooner, E.T.C., 1997. Fluid chemistry of Archean seafloor hydrothermal vents: implications for the composition of circa 3.2 Ga seawater. Geochim. Cosmochim. Acta 61, 4025–4042.

Domenico, P.A., Robbins, G.A., 1985. The displacement of connate water from aquifers. Geol. Soc. Amer. Bull. 96, 328–335.

Fallick, A.E., Jocelyn, J., Donnelly, T., Guy, M., Behan, C., 1985. Origin of agates in volcanic-rocks from Scotland. Nature 313, 672–674.

Farquhar, J., Bao, H.M., Thiemens, M., 2000. Atmospheric influence of Earth's earliest sulfur cycle. Science 289, 756–758.

Force, E.R., Back, W., Spiker, E.C., Knauth, L.P., 1986. A groundwater mixing model for the origin of the Imini manganese deposit (Cretaceous) of Morocco. Econ. Geol. 81, 65–79.

Gao, G.Q., Land, L.S., 1991. Nodular chert from the Arbuckle Group, Slick Hills, SW Oklahoma—a combined field, petrographic and isotopic study. Sedimentology 38, 857–870.

Garrels, R., MacKenzie, F.T., 1981. Evolution of sedimentary rocks. W.W. Norton and Co, New York.

Gaucher, E.A., Thomson, J.M., Burgan, M.F., Benner, S.A., 2003. Inferring the palaeoenvironment of ancient bacteria on the basis of resurrected proteins. Nature 425, 285–288.

Goodwin, A.M., 1996. Principles of Precambrian geology. Academic Press, New York.

Hay, W.W., Wold, C.N., Soeding, E., Floegel, S., 2001. Evolution of sediment fluxes and ocean salinity. In: Merriam, D.F., Davis, J.C. (Eds.), Geologic modeling and simulation; sedimentary systems. Kluwer Academic/Plenum Publishers, New York, pp. 163–167.

Hedges, S.B., 2003. Molecular clocks and a biological trigger for the Neoproterozoic snowball Earth and Cambrian explosion. In: Donoghue, P., Smith, P. (Eds.), Telling evolutionary time: molecular clocks and the fossil record. CRC Press, London, pp. 27–40.

Hoffman, P.F., Kaufman, A.J., Halverson, G.P., Schrag, D.P., 1998. A Neoproterozoic snowball earth. Science 281, 1342–1346.

Holland, H.D., 1978. The chemistry of the atmosphere and oceans. Wiley, New York.

Holland, H.D., 1984. The chemical evolution of the atmosphere and oceans. Princeton University Press, Princeton.

Holser, W.T., Hay, W.W., Jory, D.E., O'Connel, W.J., 1980. A census of evaporites and its implications for oceanic geochemistry. The geological society of america, 93rd annual meeting, Abstracts with Programs-Geological Society of America vol. 12, no. 7. The Geological Society of America, Boulder, CO, United States, p. 449.

Horodyski, R.J., Knauth, L.P., 1994. Life on land in the Precambrian. Science 263, 494–498.

Hoyle, F., 1972. History of Earth. Q. J. R. Astron. Soc. 13, 328–345.

Hughes Clarke, M.W., Keij, A.J., 1973. Organisms as producers of carbonate sediment and indicators of environment in the southern Persian Gulf. In: Purser, B.H. (Ed.), The Persian Gulf. Springer-Verlag, Berlin, pp. 33–56.

Jakosky, B.M., Philips, R.J., 2001. Mars volatile and climate history. Nature 412, 237–244.

James, N.P., Choquette, P.W., 1990. Limestones—the meteoric diagenetic environment. In: McIlreath, I.A., Morrow, D.W. (Eds.), Diagenesis, Reprint Series, vol. 4. Geological Association of Canada, pp. 35–73.

Kasting, J.F., Siefert, J.L., 2002. Life and the evolution of Earth's atmosphere. Science 296, 1066–1068.

Kenny, R., Knauth, L.P., 1992. Continental paleoclimates from delta-D and delta O-18 of secondary silica in paleokarst chert lags. Geology 20, 219–222.

Kenny, R., Knauth, L.P., 2001. Stable isotope variations in the Neoproterozoic Beck Spring Dolomite and Mesoproterozoic Mescal Limestone paleokarst: implications for life on land in the Precambrian. Geol. Soc. Amer. Bull. 113, 650–658.

Knauth, L.P., 1988. Origin and mixing history of brines, Palo Duro Basin, Texas, USA. Appl. Geochem. 3, 455–474.

Knauth, L.P., 1994. Petrogenesis of Chert. Rev. Miner. 29, 233–258.

Knauth, L.P., 1998. Salinity history of the Earth's early ocean. Nature 395, 554–555.

Knauth, L.P., Burt, D.M., 2002. Eutectic brines on Mars: origin and possible relation to young seepage features. Icarus 158, 267–271.

Knauth, L.P., Epstein, S., 1976. Hydrogen and oxygen isotope ratios in nodular and bedded cherts. Geochim. Cosmochim. Acta 40, 1095–1108.

Knauth, L.P., Lowe, D.R., 1978. Oxygen isotope geochemistry of cherts from onverwacht group (3.4 billion years), Transvaal, South-Africa, with implications for secular variations in isotopic composition of cherts. Earth Planet. Sci. Lett. 41, 209–222.

Knauth, L.P., Lowe, D.R., 2003. High Archean climatic temperature inferred from oxygen isotope geochemistry of cherts in the 3.5 Ga Swaziland Supergroup, South Africa. Geol. Soc. Amer. Bull. 115, 566–580.

Knauth, L.P., Brilli, M., Klonowski, S., 2003. Isotope geochemistry of caliche developed on basalt. Geochim. Cosmochim. Acta 67, 185–195.

Land, L.S., 1995. The role of saline formation water in crustal cycling. Aquat. Chem. 1, 137–145.

Land, L.S., Kupecz, J.A., Mack, L.E., 1988. Louann salt geochemistry (Gulf of Mexico Sedimentary Basin, USA)—a preliminary synthesis. Chem. Geol. 74, 25–35.

Larsen, H., 1980. Ecology of hypersaline environments. In: Nissenbaum, A. (Ed.), Hypersaline brines and evaporitic environments. Elsevier, Amsterdam, pp. 23–40.

Lowe, D.R., 1983. Restricted shallow-water sedimentation of early Archean stromatolitic and evaporitic strata of the Strelley Pool Chert, Pilbara Block, Western Australia. Precambrian Res. 19, 239–283.

Lowe, D.R., Byerly, G.R., 2003. Ironstone pods in the Archean Barberton greenstone belt, South Africa: Earth's oldest seafloor hydrothermal vents reinterpreted as Quaternary subaerial hot-springs. Geology 31, 909–912.

Lowe, D.R., Knauth, L.P., 1977. Sedimentology of the Onverwacht Group (3.4 billion years), Transvaal, South-Africa, and its bearing on characteristics and evolution of early earth. J. Geol. 85, 699–723.

Lowenstein, T.K., Timofeeff, M.N., Brennan, S.T., Hardie, L.A., Demicco, R.V., 2001. Oscillations in Phanerozoic seawater chemistry: evidence from fluid inclusions. Science 294, 1086–1088.

Lynch, M., 1999. The age and relationships of the major animal phyla. Evolution 53, 319–325.

Matheney, R.K., Knauth, L.P., 1989. Oxygen-isotope fractionation between marine biogenic silica and seawater. Geochim. Cosmochim. Acta 53, 3207–3214.

McCord, T.B., Hansen, G.B., Fanale, F.P., Carlson, R.W., Matson, D.L., Johnson, T.V., Smythe, W.D., Crowley, J.K., Martin, P.D., Ocampo, A., Hibbitts, C.A., Granahan, J.C., 1998. Salts an

Europa's surface detected by Galileo's near infrared mapping spectrometer. Science 280, 1242–1245.

Mojzsis, S.J., Arrhenius, G., McKeegan, K.D., Harrison, T.M., Nutman, A.P., Friend, C.R.L., 1996. Evidence for life on Earth before 3800 million years ago. Nature 384, 55–59.

Ng, W.V., Kennedy, S.P., Mahairas, G.G., Berquist, B., Pan, M., Shukla, H.D., Lasky, S.R., Baliga, N.S., Thorsson, V., Sbrogna, J., Swartzell, S., Weir, D., Hall, J., Dahl, T.A., Welti, R., Goo, Y.A., Leithauser, B., Keller, K., Cruz, R., Danson, M.J., Hough, D.W., Maddocks, D.G., Jablonski, P.E., Krebs, M.P., Angevine, C.M., Dale, H., Isenbarger, T.A., Peck, R.F., Pohlschroder, M., Spudich, J.L., Jung, K.H., Alam, M., Freitas, T., Hou, S.B., Daniels, C.J., Dennis, P.P., Omer, A.D., Ebhardt, H., Lowe, T.M., Liang, R., Riley, M., Hood, L., DasSarma, S., 2000. Genome sequence of *Halobacterium* species NRC-1. Proc. Natl. Acad. Sci. U. S. A. 97, 12176–12181.

Ohmoto, H., 2003. Gast lecture—chemical and biological evolution of the early Earth: a minority report. Geochim. Cosmochim. Acta 67 (Suppl. 1), A2.

O'Neil, J.R., Adami, L.H., Epstein, S., 1975. Revised value for the ^{18}O fractionation between CO_2 and H_2O at 25 °C. J. Res. U.S. Geol. Surv. 3, 623–624.

Pavlov, A.A., Kasting, J.F., 2002. Mass-independent fractionation of sulfur isotopes in Archean sediments: strong evidence for an anoxic Archean atmosphere. Astrobiology 2, 27–41.

Perry, E.C., 1967. Oxygen isotope chemistry of ancient cherts. Earth Planet. Sci. Lett. 3, 62–66.

Perry Jr., E.C., Ahmad, S.N., Swulius, T.M., 1978. The oxygen isotope composition of 3800 M.Y. old metamorphosed chert and iron formation from Isukasia, West Greenland. J. Geol. 86, 223–239.

Ricardo, A., Carrigan, M.A., Olcott, A.N., Benner, S.A., 2004. Borate minerals stabilize ribose. Science 303, 196.

Richards, I.J., Gregory, R.T., Ferguson, K.M., Douthitt, C.B., 2001. Archean hydrothermal alteration and metamorphism of the Pilbara Block, Western Australia. Abstr. Programs-Geol. Soc. Am. 33, 400.

Rogers, J.J.W., 1996. A history of continents in the past three billion years. J. Geol. 104, 91–107.

Rose, A.W., Dresel, P.E., 1990. Deep brines in Pennsylvania. In: Majumdar, S.K., Miller, E.W., Parizek, R.R. (Eds.), Water resources in Pennsylvania: availability, quality and management. The Pennsylvania Academy of Science, Easton, PA, pp. 420–431.

Rosing, M.T., 1999. C-13-depleted carbon microparticles in >3700-Ma sea-floor sedimentary rocks from west Greenland. Science 283, 674–676.

Rosing, M.T., Frei, R., 2004. U-rich Archaean sea-floor sediments from Greenland—indications of >3700 Ma oxygenic photosynthesis. Earth Planet. Sci. Lett. 217, 237–244.

Runnegar, B., 1982. The Cambrian explosion—animals or fossils. J. Geol. Soc. Aust. 29, 395–411.

Schopf, J.W., 1992. Patterns of Proterozoic microfossil diversity: an initial, tentative, analysis. In: Schopf, J.W., Klein, C. (Eds.), The Proterozoic biosphere. Cambridge University Press, New York, pp. 529–552.

Schopf, J.W., 1993. Microfossils of the Early Archean Apex Chert—new evidence of the antiquity of life. Science 260, 640–646.

Schopf, J.W., Klein, C. (Eds.), 1992. The Proterozoic biosphere. Cambridge University Press, New York.

Schwartzman, D., 1999. Life, temperature, and the earth. Columbia University Press, New York.

Sharp, Z.D., Durakiewicz, T., Migaszewski, Z.M., Atudorei, V.N., 2002. Antiphase hydrogen and oxygen isotope periodicity in chert nodules. Geochim. Cosmochim. Acta 66, 2865–2873.

Sonnenfeld, P., 1984. Brines and evaporites. Academic Press, Orlando, FL.

Suchecki, R.K., Hubert, J.F., 1984. Stable isotopic and elemental relationships of ancient shallow-marine and slope carbonates, Cambro–Ordovician Cow Head Group, Newfoundland—implications for fluid flux. J. Sediment. Petrol. 54, 1062–1080.

Taylor, H.P., 1974. Application of oxygen and hydrogen isotope studies to problems of hydrothermal alteration and ore deposition. Econ. Geol. 69, 843–883.

Urey, H.C., Epstein, S., McKinney, C.R., 1951. Measurement of paleotemperatures and temperatures of the Upper Cretaceous of England, Denmark, and the southeastern United States. Geol. Soc. Amer. Bull. 62, 399–416.

Walker, J.C.G., Hays, P.B., Kasting, J.F., 1981. A negative feedback mechanism for the long-term stabilization of Earths surface-temperature. J. Geophys. Res.—Oceans Atmos. 86, 9776–9782.

Wallerstein, G., Iben, I., Parker, P., Boesgaard, A.M., Hale, G.M., Champagne, A.E., Barnes, C.A., Kappeler, F., Smith, V.V., Hoffman, R.D., Timmes, F.X., Sneden, C., Boyd, R.N., Meyer, B.S., Lambert, D.L., 1997. Synthesis of the elements in stars: forty years of progress. Rev. Modern Phys. 69, 995–1084.

Walter, M., 1983. Archean stromatolites: evidence of the Earth's earliest benthos. In: Schopf, J.W. (Ed.), Earth's earliest biosphere. Princeton U. Press, Princeton, pp. 187–213.

Ward, P.D., Brownlee, D., 2000. Rare earth. Copernicus, New York.

Watanabe, Y., Martini, J.E.J., Ohmoto, H., 2000. Geochemical evidence for terrestrial ecosystems 2.6 billion years ago. Nature 408, 574–578.

Weiss, R.F., 1970. Solubility of nitrogen, oxygen and argon in water and seawater. Deep-Sea Res. 17, 721–735.

Winter, B.L., Knauth, L.P., 1992. Stable isotope geochemistry of cherts and carbonates from the 2.0-ga Gunflint iron formation—implications for the depositional setting, and the effects of diagenesis and metamorphism. Precambrian Res. 59, 283–313.

Wray, G.A., Levinton, J.S., Shapiro, L.H., 1996. Molecular evidence for deep Precambrian divergences among metazoan phyla. Science 274, 568–573.

Zharkov, M.A., 1981. History of Paleozoic salt accumulation. Springer-Verlag, Berlin.

Available online at www.sciencedirect.com

ELSEVIER

Palaeogeography, Palaeoclimatology, Palaeoecology 219 (2005) 71–86

PALAEO

www.elsevier.com/locate/palaeo

Production and cycling of natural microbial exopolymers (EPS) within a marine stromatolite

Alan W. Decho[a,*], Pieter T. Visscher[b], R. Pamela Reid[c]

[a]Department of Environmental Health Sciences, Arnold School of Public Health, University of South Carolina,
Columbia, SC 29208, United States
[b]Department of Marine Sciences, University of Connecticut, 1084 Shennecossett Rd., Groton, CT 06340, United States
[c]Rosenstiel School of Marine and Atmospheric Science, Division of Marine Geology and Geophysics, University of Miami,
4600 Rickenbacker Causeway, Miami, FL 33149-1098, United States

Received 3 June 2004; accepted 29 October 2004

Abstract

Extracellular polymeric secretions (EPS) that are produced by cyanobacteria represent potential structuring agents in the formation of marine stromatolites. The abundance, production, and degradation of EPS in the upper layers of a microbial mat forming shallow subtidal stromatolites at Highborne Cay, Bahamas, were determined using ^{14}C tracer experiments and were integrated with measurements of other microbial community parameters. The upper regions of a Type 2 [Reid, R.P., Visscher, P.T., Decho, A.W., Stolz, J., Bebout, B., MacIntyre, I.G., Dupraz, C., Pinckney, J., Paerl, H., Prufert-Bebout, L., Steppe, T., Des Marais, D., 2000. The role of microbes in accretion, lamination and early lithification of modern marine stromatolites. Nature (London) 406, 989–992] stromatolite mat exhibited a distinct layering of alternating "green" cyanobacteria-rich layers (Layers 1 and 3) and "white" layers (Layers 2 and 4), and the natural abundance of EPS varied significantly depending on the mat layer. The highest EPS abundance occurred in Layer 2. The production of new EPS, as estimated by the incorporation of ^{14}C-bicarbonate into EPS, occurred in all layers examined, with the highest production in Layer 1 and during periods of photosynthesis (i.e., daylight hours). A large pool (i.e., up to 49%) of the total ^{14}C-bicarbonate uptake was released as low molecular-weight (MW) dissolved organic carbon (DOC). This DOC was rapidly mineralized to CO_2 by heterotrophic bacteria. EPS degradation, as determined by the conversion of ^{14}C-EPS to [14]CO_2, was slowest in Layer 2. Results of slurry experiments, examining O_2 uptake following additions of organic substrates, including EPS, supported this degradation trend and further demonstrated selective utilization by heterotrophs of specific monomers, such as acetate, ethanol, and uronic acids. Results indicated that natural EPS may be rapidly transformed post-secretion by heterotrophic degradation, specifically by sulfate-reducing bacteria, to a more-refractory remnant polymer that is relatively slow to accumulate. A mass balance analysis suggested that a layer-specific pattern in EPS and low-MW DOC turnover may contribute to major carbonate precipitation events within stromatolites. Our findings represent the first estimate of EPS

---

* Corresponding author. Fax: +1 803 777 6584.
  E-mail addresses: awdecho@gwm.sc.edu (A.W. Decho), visscher@uconnvm.uconn.edu (P.T. Visscher), preid@rsmas.miami.edu
(R.P. Reid).

turnover in stromatolites and support an emerging idea that stromatolite formation is limited by a delicate balance between evolving microbial activities and environmental factors.

Keywords: EPS; Bacteria; Cyanobacteria; Stromatolite; Production; Degradation; Biofilm

1. Introduction

Marine stromatolites represent a biogenic system that is of significant interest to geologists, microbiologists, and ocean chemists. Stromatolites are laminated sedimentary structures produced by microbial organisms (Awramik, 1992). Fossil stromatolites represent the earliest macroscopic evidence of life in the fossil record (Schopf, 1996). The microbial communities that produced these structures may have been instrumental in the generation of atmospheric oxygen (Des Marais, 1991; Knoll, 1992). The microorganisms forming stromatolites, thought to be mainly cyanobacteria and associated heterotrophic bacteria, were the dominant life forms for over 80% of the history of life on earth, forming extensive microbial reefs throughout the shallow waters of Precambrian oceans (Awramik, 1992). Stromatolites abruptly declined in the fossil record with the concurrent emergence of multicellular life approximately 600 my b.p. (Grotzinger, 1990; Knoll, 1992; Grotzinger and Knoll, 1999). Modern stromatolites, with living cyanobacterial surface mats, were discovered in a hypersaline environment in Shark Bay in Western Australia (Logan, 1961; Davis, 1970; Hoffman, 1976) and more recently in open marine conditions along the margins of Exuma Sound in the Bahamas (Dravis, 1983; Dill et al., 1986; Reid and Browne, 1991; Reid et al., 1995). These microbial mats provide a valuable system for examining the microbiogeochemical interactions involved in stromatolite formation and the precipitation of $CaCO_3$ (Stolz, 2000).

Recent studies of a wide range of the well-laminated, shallow, subtidal stromatolites at Highborne Cay, Bahamas (76° 49′ W; 24° 43′ N), have revealed that three major microbial mat types, representing a continuum of growth stages, can be defined (Reid et al., 2000). Type 1 mats are characterized by sparse populations of the cyanobacterium *Schizothrix* and resemble pioneer communities (Stal, 1991, 1995). These dominate during periods of rapid sediment accretion. Type 2 mats represent a more mature surface community characterized by the development of a continuous surface film of extracellular polymeric secretions (EPS) and the development of a spatially-organized biofilm community. Type 3 mats are more fully-developed and in a more advanced developmental stage and include an abundant population of the boring coccoid cyanobacterium *Solentia* sp. This represents a climax community of the stromatolite system, and the microboring and calcification activities of the *Solentia* sp. result in laterally-cohesive carbonate crusts, which supports the longer-term preservation of the stromatolite (Reid et al., 2000).

High molecular-weight (MW) extracellular polymeric secretions (EPS) that are produced by cyanobacteria (Decho, 1990) may represent potentially important structuring agents in marine stromatolites (Decho, 2000). EPS serve to physically stabilize microbial cells against the high-energy environments (e.g., waves, tidal currents) that these structures commonly experience (De Winder et al., 1999; De Brouwer et al., 2002). Further, they may provide a chemically protective microenvironment for cells. They serve to bind and concentrate Ca^{2+} and Mg^{2+} ions from the surrounding seawater. The high abundance of EPS may effectively chelate large amounts of dissolved Ca^{2+} ions, perhaps preventing the precipitation of $CaCO_3$ (Kawaguchi and Decho, 2002a). Similarly, the degradation of EPS by heterotrophic bacteria may release Ca^{2+} and influence $CaCO_3$ precipitation.

The activities of different microbial functional groups within stromatolites in the Exuma Cays, Bahamas, appear to influence the precipitation of microcrystalline calcium carbonate (micrite) in distinct horizons (Pinckney et al., 1994; Visscher et al., 1998; MacIntyre et al., 1996, 2000). This precipitation results in the formation of lithified layers.

Confocal scanning laser microscopy (CSLM) and vacuum and environmental scanning electron microscopy (ESEM) studies indicate that lithified micritic horizons within Exuma stromatolites are in a proximate spatial association with dense layers of the filamentous cyanobacterium, *Schizothrix* sp. (Decho and Kawaguchi, 1999; Reid et al., 1999). However, recent studies using microautoradiography (Paerl et al., 2001) indicated that the aragonite needles (in the laminae) are closely associated with sulfate-reducing bacterial activities (Visscher et al., 1999, 2000). Some micritic horizons are also characterized by an abundance of the coccoid cyanobacterium, *Solentia* sp. (Visscher et al., 1998; MacIntyre et al., 2000). Micrite precipitation within these cyanobacteria-rich layers may be closely coupled with the activities of heterotrophic bacteria, either aerobic or anaerobic, and the production and degradation of EPS.

The purpose of this study was to examine the abundance, production, and degradation of microbial EPS within living marine stromatolites at Highborne Cay, Bahamas. The abundance of EPS material in different layers of a stromatolite mat was determined to establish levels of EPS with respect to resident photosynthetic organisms. The production of "newly-secreted EPS" was examined over a diel cycle using ^{14}C tracer experiments and slurry experiments to determine when highest EPS production occurred. The concomitant degradation of cyanobacterial EPS by heterotrophic bacteria was also examined. These data were used to establish a mass balance of EPS material within the stromatolite mat.

2. Methods

2.1. Sample descriptions

Samples of microbial mat were collected from the surface of a shallow subtidal stromatolite (Sample # 6/97 NS.8n) on the east beach of Highborne Cay, Bahamas (24 42.596°N; 76 49.372°W). The Highborne Cay study site was described in detail by Reid et al. (1999, 2000). The stromatolite mats were subsampled into 1 cm × 1 cm (diameter × depth) cores, which were used for all experiments. Under low magnification microscopy, four major layers were easily distin-

guished (Fig. 1b). See Reid et al. (2000) for a more detailed description of the mat layers. These were: (1) *Layer 1*, an unlithified caramel layer approximately 1 mm thick with an upper EPS-rich film (referred to as Layer 1A) underlain by a thin but dense layer of cyanobacteria (referred to as Layer 1B); (2) *Layer 2*, a white layer, 3–5 mm thick, which was unlithified and very cohesive and spongy in texture; (3) *Layer 3*, a dark, crusty, grey–green layer, which was relatively thin (1–2 mm) and was dominated by *Schizothrix* sp., but also contained abundant coccoid cyanobacteria *Solentia* sp., *Microcoleus* sp., and *Oscillatoria* sp.; and (4) *Layer 4*, a white layer which was relatively thick (3–5 mm) and was cohesive and spongy in texture. The microbial composition of this mat is described by Stolz et al., (in prep.).

2.1.1. Direct counts of bacteria

After the separation of the individual Layers 1–5 (see above), samples were fixed in buffered formaldehyde (2% final concentration, pH 7.5). The fixed samples of known volume were then subjected to dilute HCl (1 N) treatment, until the $CaCO_3$ was dissolved. The solution was then filtered onto destained Nuclepore membrane filters (0.45 μm) under mild vacuum. During filtration, acridine orange (163 μM) was added. Counts of fluorescent cells were made using an Olympus Vanox microscope. At least 1400 cells and two replicate samples per layer were counted.

2.1.2. Quantification of natural EPS material

The seawater (38 ppt salinity) and stromatolite mats used in experiments were collected from Highborne Cay, Bahamas. All glassware used in the below-described procedures was acid cleaned (10% HCl), then rinsed five times in deionized water and air-dried. Three replicate samples were used for the determinations of EPS abundance from each mat layer. Mat layer samples were placed into separate scintillation vials, then lightly homogenized into separated sand grains using a blunt plastic probe. EPS material was initially extracted by mixing from the mat material with ethylenediamine tetra-acetic acid (EDTA in seawater; 4 mM final concentration) and heating to 40 ± 1 °C for 15 min, with occasional stirring. This solubilization of EPS did not lyse the cyanobacterial cells. The suspended material, containing cells, detritus, and

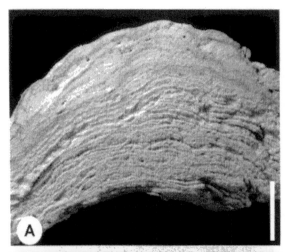

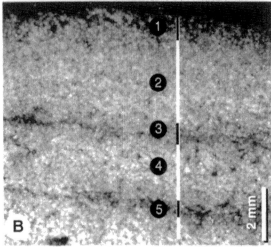

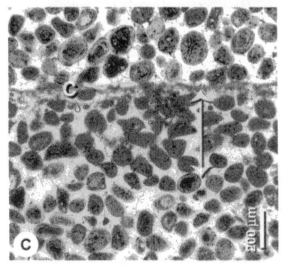

EPS, was placed in a microcentrifuge tube and centrifuged (8000×g, 6 min) to pellet cells and particulate detritus. The resulting supernatant, containing the EPS, was removed, was mixed with 70% (final concentration) cold (4 °C) ethanol for 8 h. The pellet and remaining sediment were extracted using EDTA and heating as described above. This procedure was repeated three times. Preliminary experiments showed that no extractable EPS remained after three extractions. Precipitated material from the supernatant was collected by centrifugation, then redissolved in deionized H_2O. The ethanol precipitation was repeated twice, then the EPS was dialyzed (12,000 MWCO) against several changes of deionized H_2O using "Slid-A-Lyzer" microdialysis cassettes (Pierce Biochem.) for 24 h. EPS was lyophilized to dryness and stored at −70 °C until use. The sediments from the sample were washed with deionized H_2O to remove salts and loose detritus and were dried for the determination of sediment dry weights. All EPS samples were standardized according to sediment dry weight.

The quantification of EPS material was conducted by two different methods: the phenol–sulfuric acids method (Dubois et al., 1956; Liu et al., 1973; Underwood et al., 1995) and by total EPS dry weight. D-Glucose and alginic acid were used to obtain the standard curves for spectrophotometric determinations. C:N ratios of EPS material were also determined using Lehman Labs Model CE440 automated organic elemental analyzers (Dr. R. Petty, Univ. Calif., Santa Barbara). Results were reported as weight percent of the element.

2.1.3. Diel production of new EPS material

The production of "New EPS" material in the stromatolite mat was examined over a 24-h (diel)

Fig. 1. Stromatolites, Highborne Cay. (A) Sawed section showing lamination, which is defined by lithified horizons, approx. 1 mm thick, that stand out in relief on the cut surface. Sample 6/97.NS.8p. (B) Microbial mat from the surface of stromatolite 6/97.NS.8n used in the experiments in this work. Four major layers are indicated: Layer 1 is subdivided into an upper caramel colored section (1A) and a lower section that is green–yellow (1B). Layer 3 is grey–green and crusty. Layers 2 and 4 are grey–white and soft and cohesive. (C) Thin-section photomicrograph showing characteristic microstructure of the lithified layers: A micritic crust (c) overlies a zone of micritized grains that are cemented together at point contacts (arrow). Sample 6/97.NS.8f.

cycle by tracing the uptake of ^{14}C-bicarbonate and its subsequent incorporation into the EPS. For these experiments, twenty-one individual 1 cm diameter cores of intact stromatolite mat were placed in precleaned 22 ml borosilicate scintillation vials, with polypropylene caps. Clean sand (cleaned with sodium hypochlorite, $3\times$ rinsed in distilled water) was added to the vials to completely surround, but not cover, the stromatolite cores. The vials were filled with seawater (i.e., no head space). At 6:00 am (i.e., sunrise), 50 μl of $NaH^{14}CO_3$ (1.20 μCi; 30–60 mCi/mmol) was added to each vial. The sealed vials were randomly arranged and placed in a circulating water bath (to maintain ambient seawater temperature conditions) under ambient sunlight conditions. It was assumed that the added $NaH^{14}CO_3$ tracer reached all layers of the stromatolite equally during the incubations. A factor of 1.06 was used to correct for isotope discrimination against heavier ^{14}C during bicarbonate uptake (Paerl, 1997). A set of triplicate dark vials (wrapped in foil) was used as controls. Sediment dark-controls, containing clean sand and a section of stromatolite mat, were used to control for the non-specific uptake of ^{14}C (i.e., sorption of radiolabeled bicarbonate) to the sediment. Dark-controls were wrapped in aluminum foil to prevent light exposure and were collected at 12 and 24 h. All samples were analyzed for ^{14}C on a Packard TR2400 liquid scintillation system, using 10 ml Ecolite Scintillation cocktail (ICN BioChem.). Sample counting times varied and were continued until 10,000 total counts were obtained for each sample. Counting efficiency quenching was corrected for using the external standards ratio method. For each time treatment, triplicate samples were removed from the water bath after incubation times (T) of 4, 8, 12,16, 20, and 24 h and were injected with buffered Formalin (2% final concentration v/v). Several parameters were measured for each sample.

2.1.3.1. Samples of overlying water.

Aliquots of the overlying sample water, collected from stromatolite incubation vials, were removed, and the pH was determined to ensure that it was above 8.2. The water samples were then centrifuged ($10,000\times g$; 10 min) to pellet the particulates. Activities, expressed as disintegrations per minute (DPM), were determined for both pellet and supernatant. The following parameters were measured on the water (supernatant) and particulates (centrifuged pellet):

(1) Total ^{14}C in water=^{14}C including bicarbonate, measured after the addition of 0.01 M NaOH.
(2) ^{14}C-DOC in water=^{14}C present in water after acidification (HCl for 24 h).
(3) Centrifuged pellet=uptake by water-column bacteria and detritus.

2.1.3.2. Stromatolite Layers.

In the laboratory, each mat sample was carefully separated into individual layers using microscopy (as previously described). The EPS were extracted from each stromatolite layer using the methods described above. The following parameters were measured for each stromatolite layer:

(1) ^{14}C-EPS Activity=ethanol-precipitated fraction from stromatolites extracted as previously described above.
(2) Low-MW ^{14}C-DOC Activity=collected as ethanol-soluble fraction derived from stromatolite EPS extract (after centrifugation).
(3) Sediment dry weight:=represented the dry weight of sediment collected from each layer. This was used to standardize EPS concentrations.

The free sediment remaining in the sample vials (which was used to surround the stomatolite core during incubations) was also assayed for ^{14}C activity. Dark-control DPM were subtracted from the treatment DPM.

2.1.4. Heterotrophic degradation of EPS

2.1.4.1. Isolation of ^{14}C-EPS.

Radiolabeled EPS, to be used in the degradation experiments, was extracted from laboratory cultures of a *Schizothrix* sp. previously isolated from stromatolite mats collected at Highborne Cay. Cultures were grown in 100 ml of CHU-10 medium (Rippka et al., 1979) in "Sigma seawater" (Sigma, St. Louis; collected outer Gulf stream; 32 ppt salinity; pH 8.2; 33–35 °C) using a 12 h/12 h light/dark cycle. Cultures were grown under fluorescent lamps (approx. 300 lx) at 23–24 °C. During mid log-phase of growth,

$NaH^{14}CO_3$ (50 μCi; activity=30–60 mCi/mM; ICN Radiochemicals) was added to the cultures and incubated for 14 days. EPS were extracted and purified using the methods described above. The composition of this EPS is described elsewhere (Kawaguchi and Decho, 2000, 2002b). The activities of labeled EPS were described as DPM of ^{14}C per μg EPS.

2.1.4.2. ^{14}C-EPS degradation experiments.

EPS degradation experiments were performed to examine the aerobic heterotrophic decomposition of EPS material within specific layers of the stromatolite. The degradation of ^{14}C-labeled EPS substrates was assayed by the generation of $^{14}CO_2$. Intact stromatolite samples were prepared as described above, then ^{14}C-labeled *Schizothrix* EPS (activity=2.6×10^9 DPM per μg EPS), or [U-^{14}C] D-glucose (250–360 mCi per mM; ICN Radiochemicals) as a control, was carefully injected into either Layer 2 or Layer 3 of the stromatolite using a 26-gauge needle. For each sample, two injections of 5 μl EPS solution were made, yielding a total activity of approximately 2.60×10^6 dpm (1.18 μCi) per stromatolite. The cores were placed in precleaned scintillation vials and surrounded with loose sand. The vials were filled with 10 ml seawater, which provided a 9–10 ml headspace for air exchange. Scintillation vial lids were fitted with input and output ports using teflon septa and were sealed with silicon. A fresh supply of air was slowly bubbled into the seawater using an electric aquarium pumps fitted with Tygon tubing and silicon sealant to a multiport adjustable airvalve, to regulate gas exchange to the vessels. This provided a slow, uniform bubbling of air within all incubation vials.

$^{14}CO_2$ was collected sequentially from each sample by bubbling the outflow air through CO_2 traps. The $^{14}CO_2$ trap for each stromatolite sample consisted of two sequentially arranged vials, each containing 5 ml of 1 M KOH. The outflow from the incubation vials was bubbled into the bottom of the first trap. Air collected in the first trap was bubbled into the bottom of the second trap. The KOH trap vials were removed at different time intervals (2, 4, 6, 8, 10, 12, 18, 24, 30, 36, 40, and 48 h) and were quickly replaced with new vials containing KOH solution. This allowed the contin-

uous and efficient capture by the two sequential trapping vials of radiolabeled CO_2 gas produced during respiration. The vials then were individually assayed for $^{14}CO_2$ activity by mixing with Ecolite scintillation cocktail, chilled (4 °C) in darkness for 24 h, then assayed by LSC as described above. The CO_2 trapping efficiency of the two sequential vial set-up was determined by injecting ^{14}C-bicarbonate directing into triplicate incubation vials containing sterile seawater. Because of the substantial quenching activity of the KOH solution, external standards were used for conversion to DPM. The DPM derived from the two sequential vials collected at a given time interval were then combined to yield the $^{14}CO_2$ activity for that time interval. Results were plotted cumulatively as the percentage of added DPM (carbon) that was recovered as $^{14}CO_2$ over time.

2.1.4.3. Degradation of organic carbon in slurries.

Mat Layers 1–3 were separated and slurries were prepared by homogenizing the individual layers and mixing the homogenate with an equal volume of filter-sterilized seawater collected from the site. The slurries (10–25 ml) were incubated at ambient temperature (26–28 °C) while stirring in a dark PVC vessel without a headspace, and O_2 uptake upon organic substrate additions was measured for up to 1 h. Oxygen was measured with a glass microelectrode with a guard cathode (Unisense, Arhus, Denmark), and stirring was briefly stopped during O_2 readings. The O_2 consumption was calculated from the initial slope of oxygen uptake and corrected for endogenous respiration. All organic substrate additions were repeated twice. Substrates (glucose, xylose, mannose, acetate, and ethanol) were added to a final concentration ranging from 2 to 35 μM, except for *Schizothrix* EPS, of which 1.67 μg ml^{-1} was added.

2.1.5. Statistical analyses

EPS abundance data were analyzed using a one-way analysis of variance (ANOVA) (SAS, 1985) to examine the effects of mat layers on EPS abundance. Total EPS weights and hexose-equivalent weights for each mat layer were compared using a two-way ANOVA. Multiple comparisons of significant treatment effects among means were determined *aposteriori* using Bonferroni tests. A two-way factorial design ANOVA was used to examine the effects of mat layer and incubation time on EPS production (Piegorsch and Bailer, 1997). If a two-way ANOVA result indicated that as a significant factor, then the factor was compared using simple linear contrasts, followed by Bonferroni's multiple contrast for modifying alpha values (SAS, 1985). Appropriate transformations of data values were conducted to ensure the validity of the parametric assumptions. All statistical analyses were conducted using the GLM module of the SAS statistical software package (SAS, 1985).

3. Results

3.1. Bacterial counts

Epifluorescence microscopy of total cell counts (Fig. 2) revealed that the highest bacterial population was associated with Layer 1 and that Layers 1 and 3 (4×10^6 and 7×10^5 cells cm^{-3} sediment, respectively) contained higher biomass than Layers

2 and 4 did (2×10^4 and 8×10^3 cells cm^{-3} sediment, respectively).

3.2. Natural EPS abundance

The mean concentrations of EPS varied depend on the stromatolite layer. A high abundance of EPS occurred in Layers 1a, 2, and 4, while low abundance was found in Layers 1b and 3. The highest mean abundance occurred in Layer 2 (0.707 ± 0.232 mg EPS/g sed dry wt; average $\pm$ S.D.; $n=3$) (Table 1; Fig. 3). The lowest polymeric abundance occurred in Layer 3 (0.137 ± 0.055 mg EPS/g sed; $n=3$). This pattern of EPS abundance within each stromatolite layer was consistent across all replicate samples. Analysis of variance showed that the total weight of extractable EPS was significantly ($P=0.0024$) greater in all layers than hexose-equivalent weights, determined using the phenol–sulfuric acid method. The C:N ratio of EPS from different layers also varied with the highest C:N ratios (6.3, 8.1, 7.1) being observed in the top caramel layers (1a) and white layers (2 and 4), respectively (Table 1). The C:N ratios of *Schizothrix* sp. EPS extracted from the laboratory cultures, which were grown under similar conditions to the ^{14}C-EPS used in the degradation experiments, were 9.2 ± 0.1 ($0 \pm S.E._{.0}$; $n=3$).

3.3. Diel production of "New" EPS material

Total bicarbonate uptake and production of labeled EPS were closely coupled to the light period. Maximum bicarbonate uptake occurred at the 6:00 pm sampling time period, with 53% of the added bicarbonate label having been incorporated in cells, DOC, or EPS (Fig. 4). The percent recovery of added ^{14}C-label ranged between 84% and 96% for all incubations. The uptake of ^{14}C by dark controls accounted for 0.003% of added ^{14}C-label.

The production of newly-secreted EPS, as evidenced by the accumulation of ^{14}C-label in the EPS, occurred within 4 h after the initial addition of ^{14}C-bicarbonate label. The net EPS production peaked at 2:00 pm and decreased thereafter (Fig. 5). The decrease coincided with an increase in ^{14}C detected as CO_2 (Fig. 6). Maximum EPS production represented 7.56% of the total ^{14}C-bicarbonate taken up by

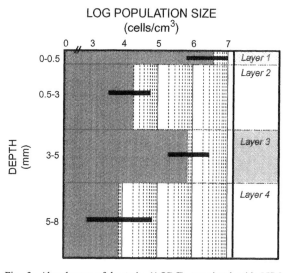

LOG POPULATION SIZE
(cells/cm³)

Fig. 2. Abundances of bacteria (AODC) associated with NS.8 stromatolite mat layers. The average of two replicate samples and error bar ($\pm$ standard deviation) is provided for the top 4 layers. Cell densities (logarithmic scale) peak at the surface and are two to three orders of magnitude higher in lithified Layers 1 and 3 than in non-lithified Layers 3 and 4.

Table 1
Abundance and C:N ratios of natural extracellular polymeric material (EPS) collected from stromatolite mat layers

Stromatolite layer/color	EPS abundance (mg EPS/g dry sed)			Ratio Hx.Eq/ EPSweight[a]	EPS C:N ratio
	Depth (cm)	Dry weight	Hexose-equivalent		
1 A (caramel)	0–0.1	0.620±0.030	0.453±0.06	0.73	6.3±0.02
1 B (green–yellow)	0.1–0.3	0.340±0.090	0.101±0.01	0.29	3.6±0.01
2 (white)	0.3–1.0	0.707±0.013	0.434±0.08	0.61	8.1±0.02
3 (dark–green)	1.0–1.5	0.137±0.031	0.032±0.02	0.23	2.4±0.01
4 (white)	1.5–2.0	0.551±0.020	0.346±0.10	0.63	7.1±0.01

EPS abundances and C:N ratios are expressed as mean±S.E. (n=3). EPS abundances are expressed as "Total EPS weight (mg)/g dry sediment", and as "hexose-equivalent wt./g dry sediment" as estimated by the phenol–sulfuric acid method.

[a] Ratio of hexose-equivalent EPS/EPS dry wt.

cells. The ^{14}C in the extractable EPS decreased progressively between 6:00 pm and 6:00 am (i.e., during darkness) (Fig. 5).

3.4. Heterotrophic decomposition of EPS

The results of the degradation experiments to examine the decomposition of ^{14}C-EPS added to Layers 2 and 3 in the stromatolite mat are shown in Fig. 7. The recovery efficiencies of added ^{14}C-bicarbonate by dual KOH traps were 92% to 95%. During the first 6 h of incubation, only a low fraction (<7%) of the added ^{14}C-EPS was degraded to ^{14}CO$_2$.

The degradation proceeded rapidly after 6-h incubation, with a cumulative increase in ^{14}CO$_2$ until 40 h. After 40 h, no major increase in cumulative ^{14}CO$_2$ generation was observed. Approximately 50% of EPS was degraded within Layer 3 (as indicated by ^{14}CO$_2$ production), and 37% within Layer 2 (Fig. 7). Controls, using ^{14}C D-glucose, were more rapid and efficient (c.a., 70–80% respired) in conversion to ^{14}CO$_2$ than EPS treatments were.

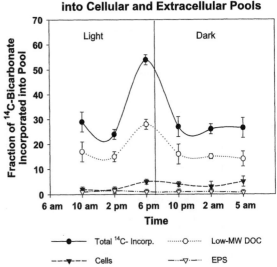

Fig. 4. Partitioning of ^{14}C-bicarbonate into major cellular and extracellular pools during the 24-h diel production experiment. Note that a relatively large fraction of ^{14}C incorporation is released as low-MW DOC.

EPS Abundance in Stromatolite Layers

Fig. 3. Abundance of EPS material in stromatolite mat layers as determined by total dry weight EPS, and as glucose equivalent mass using the phenol–sulfuric acid method. Error bars represent standard error of mean (n=3).

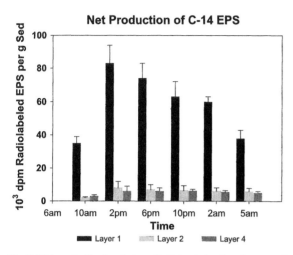

Fig. 5. Diel production by photosynthetic bacteria of newly-secreted ^{14}C-EPS in three upper layers of a stromatolite mat. ^{14}C-bicarbonate was added to stromatolite mats at 6 am (i.e., sunrise; 0-h incubation time), and its incorporation into extractable EPS was followed over a 24-h period; dpm=disintegrations per minute.

The addition of organic substrates resulted in a 5–10 fold increase of aerobic respiration in slurried mat samples (Table 2). Oxygen uptake rates were the highest in the slurries of Layer 1, and the lowest in the slurries of Layer 2. The stimulation of oxygen

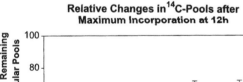

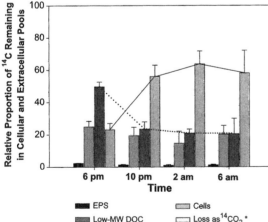

Fig. 6. Relative changes in cellular and extracellular ^{14}C pools after maximum incorporation at 12-h incubation time. Note the rapid decrease in low-MW DOC after 6 pm, and a concurrent increase in CO_2, due to the conversion of DOC by heterotrophs.

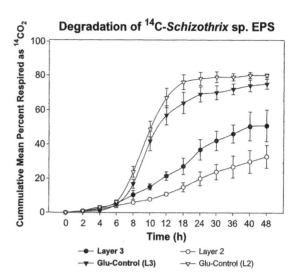

Fig. 7. Heterotrophic degradation over a 48-h period of ^{14}C-EPS into $^{14}CO_2$. Tracer concentrations of ^{14}C-EPS were injected into intact stromatolite mat layers, and the cumulative evolution of ^{14}C respired as $^{14}CO_2$ was followed. Glu-control represents the ^{14}C-glucose (control) injected into the mat.

uptake was higher upon the addition of acetate and ethanol (87–207 and 78–184 μmol ml^{-1} slurry h^{-1}, respectively) than when EPS, glucose, mannose, or xylose was added (26–78, 35–101, 68–93, and 42–107 μM ml^{-1} slurry h^{-1}, respectively). Substrate consumption started within 1 min upon addition, as was evident by decreasing oxygen concentration. An exception was EPS, which required a lag phase of 30–40 min before O_2 consumption increased over its endogenous rate.

Table 2
Aerobic respiration (O_2 uptake) after the addition of organic substrates in slurries prepared from individual mat layers of a stromatolite (NS.8), Highborne Cay, Bahamas

Carbon source	O_2 uptake rate (d[O_2]/dt) (μmol ml^{-1} h^{-1})		
	Layer 1	Layer 2	Layer 3
Acetate	207 (±11)	87 (±13)	172 (±13)
Ethanol	184 (±25)	78 (±11)	155 (±13)
Glucose	101 (±14)	35 (±4)	78 (±7)
Xylose	107 (±10)	42 (±10)	64 (±8)
Mannose	93 (±14)	n.d.	68 (±14)
EPS	78 (±8)	26 (±4)	51 (±4)
Endogenous rate	16 (±3)	8 (±1)	18 (±2)

Numbers are mean O_2 uptake rates (standard deviation is shown in parentheses). n.d.=not determined.

4. Discussion

The mat examined in our study at Highborne Cay is one example of stromatolites under current investigation. Reid and colleagues have examined a wide range of stromatolite mats and their associated microbial communities (Reid et al., 2000). They characterized three major types of surface communities, which occur along a continuum and exhibit varying degrees of precipitation. These range from rapidly-accreting, non-lithifying mats (Type I) to lithifying mats having a thin surface (and sometimes subsurface) crusty layer(s) (Type II) to strongly laminated (Type III) mats having thick laminae with fused grains. The current study examined sample 6/97 NS.8 (an early stage Type II mat) to determine how EPS and its microbial processes may be associated with lithification processes. This mat had a distinct surface crust present.

The rates at which microbial EPS are produced and degraded may represent an important varying factor influencing calcification within a stromatolite mat. In mats where there is high photosynthetic production, there may be a correspondingly high production of EPS. A high continued production of EPS would result in "spongy or gelatinous" stromatolite mats in which little or no lithification appears to be occurring. If EPS production exceeds EPS degradation (i.e., net accumulation of EPS), much of the Ca^{2+} ions would be bound (assuming that available ligands remain constant), making less Ca^{2+} available for precipitation. In contrast, a more rapid degradation of EPS and its associated heterotrophic microbial activities could release Ca^{2+} ions previously bound by EPS, making the ions potentially available for precipitation as $CaCO_3$. Concentrated heterotrophic activities (i.e., EPS degradation and CO_2 production) and the concurrent release of Ca^{2+} ions within a mat layer may result in the localized precipitation of $CaCO_3$ and the formation of lithified layers.

Our results show that the abundance, degradation rates, and production rates of EPS varied significantly within the different layers of microbial mats forming stromatolites in Exuma Cays, Bahamas. Layers which had a relatively high natural abundance of EPS could have resulted from the relatively high production and/ or slow degradation of EPS occurring in these layers. Layers that had a relatively lower abundance of EPS

generally exhibited more rapid EPS degradation rates. The observed differences in abundance, degradation, and production of EPS, and microbial composition, suggest that the Type II stromatolite mat is a very heterogeneous, vertically-layered system. This heterogeneity may contribute to the precipitation of distinct $CaCO_3$ micritic laminae associated with specific layers.

4.1. Bacterial cell counts

Direct cell counts of bacteria (Fig. 2) revealed that the highest population was associated with Layer 1. Also, Layers 1 and 3 contained higher biomass than Layers 2 and 4 did. The photosynthetic rates as measured by O_2 production during the light–dark and dark–light shifts with microelectrodes (Visscher et al., 1998) was the highest at 0.5–0.7 mm depth, which coincides with Layer 1. Thus, cyanobacterial production of oxygen and fixed carbon (including EPS and DOC) is associated with this depth interval, and therefore, the highest heterotrophic biomass can be sustained there. Aerobic respiration rates, also measured with microelectrodes, demonstrated that aerobic respiration peaked just below the depth of maximum O_2 production (Visscher et al., 1998). The relatively high biomass in Layer 3 is somewhat surprising, since O_2 penetrates only to the top of this layer and only during the peak of photosynthesis. Aerobic respiration could not be detected below Layer 2 (Visscher et al., 1998), but sulfate reduction and a population of sulfate reducers were associated with Layer 3 (Visscher et al., 1999). Therefore, the population of anaerobes, including sulfate-reducing bacteria, could account for the observed cell counts in this layer. Interestingly, the bacterial population of Layer 3 degraded ^{14}C-labeled EPS and glucose aerobically, and more rapidly so than did the microbes in Layer 2. Similarly, in the slurry experiments, the addition of monosaccharides and EPS stimulated O_2 uptake more so in the slurries of Layers 1 and 3, then the slurries of Layer 2. This indicates the presence of a viable aerobic heterotrophic population. In addition, sulfate reduction in the presence of O_2 has been reported in various microbial mats (Canfield and DesMarais, 1991; Fründ and Cohen, 1992; Visscher et al., 1992; Jørgensen, 1994), including Highborne Cay stromatolites (Visscher et al., 1999). The [14]CO_2 production can therefore be attributed to both aerobes and anaerobes.

4.2. Abundance of natural EPS material

Although the green layers (i.e., Layers 1, 3, and 5) contained high abundances of cells, EPS concentrations were significantly ($P=0.001$) lower (per gram dry sed) than the adjacent white layers did (i.e., Layers 2 and 4) (Fig. 2). Analyses of EPS concentrations using the phenol–sulfuric acid method (Dubois et al., 1956) showed that hexose sugars comprised a significantly ($P=0.0024$) smaller portion of EPS when compared with total EPS dry weights. The ratio of hexose-sugar abundance to total EPS dry weight varied depending on the layer (Table 1). This implies that a large portion (30–70% wt/wt) of the extractable EPS in the natural mat layers were composed of non-hexose sugar components. The variability in the C:N ratios of EPS extracted from the different layers also supported this idea. The highest C:N ratios (i.e., carbon-rich EPS) were observed in layers with higher EPS abundance (i.e., Layers 2 and 4) compared to layers with lower EPS abundance. Layers that showed the most rapid EPS degradation rates also showed lower C:N ratios (Table 1), when compared to adjacent layers. Together, these data suggest that over time, the carbohydrate (i.e., carbon) portion of stromatolite EPS is being partially or selectively degraded by heterotrophic microorganisms, leaving behind a more N-enriched residual polymer.

4.3. Diel "new production" of EPS material

The observed diel variation in the production of new EPS, as determined by the uptake of ^{14}C-bicarbonate by cells and its subsequent incorporation into EPS, showed that ^{14}C incorporation into EPS occurred very quickly (i.e., in less than 4 h). The highest production of EPS occurred during times when photosynthetic activities were also high (i.e., during daylight). A decrease in the total bicarbonate uptake was observed between 4 and 8 h incubation (i.e., 10:00 am to 2:00 pm), however (Fig. 4). This coincides with the period of highest solar irradiance and may represent a potential photoinhibition effect (Leverenz et al., 1990). Visscher et al. (1998)

observed that the highest rates of photosynthesis occurred before 2:00 pm. Decho et al. (in prep) observed the effects of varying light levels on the photosynthetic uptake of bicarbonate and its subsequent incorporation into EPS. They found that while EPS production occurred primarily during light conditions (i.e., associated with photosynthesis), the highest ambient levels of light reduced the production of EPS in the uppermost surface layer (0–0.2 mm depth). In the production experiments of the present study, this uppermost mat layer was not differentiated from Layer 1b. Therefore, any effects of photoinhibition on EPS production, if present, could not be discerned given our experimental design. The highest production of ^{14}C-labeled EPS occurred between 10:00 am and 2:00 pm (even though total bicarbonate uptake decreased during this same period), and the production of new EPS generally represented less than 3% to 4% of the total bicarbonate uptake by cells. Other studies of EPS production in phytoplankton (Mague et al., 1980; Wolter, 1982; Fogg, 1983; Myklestad, 1998) showed that a relatively high percentage (i.e., 40–75%) of photoassimilate was used in mucilage production, and often in response to environmental stressors such as nutrient limitations. In contrast, studies of benthic diatoms showed that EPS production was not always directly linked to immediate photosynthetic assimilation of CO_2 (Cooksey and Cooksey, 1986; Smith and Underwood, 1998) but rather occurred primarily during migratory rhythms and tidal immersion of diatoms (Taylor and Paterson, 1998; Underwood and Kromkamp, 1999; Taylor et al., 1999; Smith and Underwood, 2000; Staats et al., 2000; De Brouwer and Stal, 2002). EPS production was influenced by irradiance levels (Wolfstein and Stal, 2002; Wolfstein et al., 2002) and could be further influenced by UV irradiation (Underwood et al., 1999).

Our data, derived from ^{14}C-bicarbonate production experiments, suggest that a large portion of the newly-produced EPS was rapidly (i.e., within 12 h after secretion) degraded by heterotrophs. We observed that the cumulative mean production of ^{14}C-EPS decreased after 8 h (i.e., dark portion) of the diel cycle (Fig. 5). This pattern was consistent across all replicates ($n=3$) for each time period and reduced the net production of EPS to less than 2% of the initial bicarbonate uptake by cells. We propose that approximately 40% to 60% of newly-secreted EPS carbon is partially degraded and converted to CO_2 by heterotrophs. Our degradation studies, using ^{14}C-labeled EPS, support this (see below) and suggested that 30–50% of added EPS carbon was degraded over a 48-h period, leaving behind a more-refractory portion of the EPS carbon.

Low MW DOC was a major carbon pool produced during the diel experiments (Fig. 4). The sampling time during which maximum ^{14}C-bicarbonate was incorporated into organic cellular and extracellular pools occurred after 6:00 pm (Table 3). During this period, low-MW DOC represented 49.7±3.1% (mean±S.E.; $n=3$) of the total incorporated ^{14}C-bicarbonate. The subsequent loss of ^{14}C-label from the DOC pool during the next 4 h coincided with an increase in the fraction recovered as ^{14}CO_2 (Fig. 6). These data strongly suggested that rapid degradation of ^{14}C-DOC into ^{14}CO_2 by heterotrophic bacteria occurred during two different time periods, between 10:00 am and 2:00 pm and between 6:00 pm and 10:00 pm, just after the conclusion of daylight. This infers a very close temporal and spatial coupling of photosynthetic production and heterotrophic utilization in the upper layers of the stromatolite mat. Since a net decrease in ^{14}C-EPS was observed during 12-h to 24-h incubation, we assume that no major conversion by heterotrophic bacteria of ^{14}C-DOC into ^{14}C-EPS occurred. Most of the ^{14}C-DOC was mineralized into ^{14}CO_2. The rapid degradation of dissolved organic carbon, derived from planktonic sources such as diatom blooms and phytoplankton, has also been observed by a number of investigators

(Cho and Azam, 1988; Kirchman et al., 1991; Amon and Benner, 1994; Norman et al., 1995).

4.4. Degradation of EPS

The degradation of ^{14}C-labeled EPS, derived from the cyanobacterium *Schizothrix* sp., varied depending on the stromatolite layer. The cumulative evolution of ^{14}CO_2 resulting from the heterotrophic decomposition of ^{14}C-EPS was best described by a third-order curve (Hu and Bruggen, 1997) over the 48-h time period. Initial degradation of ^{14}C-EPS, after addition of label, was preceded by a 6-h lag period, in which there was no significant capture of ^{14}CO_2. This lag could have resulted from: (1) the lack of degradation (uptake and hydrolysis of substrate) for the initial 6–8 h. This is not likely, however, since a similar 6–8 h lag was also observed in D-glucose controls (Fig. 7), and glucose represents an easily hydrolyzable substrate for many bacteria. (2) A more likely scenario is that much of the initial lag occurred because of ^{14}CO_2 diffusion from the mat. (3) It is possible that isotopic dilution occurred over the 48-h time series (Sawyer and King, 1993; King, 1997). For this reason, it may be concluded that our experimental results provided a relatively conservative (i.e., slow) estimate of EPS degradation. Once ^{14}CO_2 evolution was detectable, a rapid cumulative increase occurred and continued until approximately 40 h, after which time there was no net gain in ^{14}CO_2. A mass balance of EPS turnover (to CO_2), as estimated directly from the ^{14}C-EPS degradation experiments, was ca. 30–50% loss of C (as ^{14}CO_2) over 48 h. Estimates determined indirectly from the ^{14}C-EPS production experiments over 24 h were ca. 35–60%. The relatively close agreement of these two estimates suggests that a large portion of newly-produced EPS was degraded and ultimately respired by heterotrophs and further supports the idea that EPS was turned over rapidly in certain layers (e.g., Layers 1 and 3) of the stromatolite mat. Studies by Anderson et al. (1987) of hot spring Cyanobacterial mats showed that the dark fermentation of polymeric organic matter leads to the production of organic acids which supply heterotrophs. The partial and rapid degradation of newly-produced EPS may have reduced net accumulations to less than ca. 1% per day, even in the most active layers of the mat. In studies of other non-mat systems, the degradation of

Table 3
Summary table showing characteristics of processes within the upper mat layers of stromatolite "NS.8.n 6/97" and EPS-related processes that (may) favor potential precipitation of $CaCO_3$

Mat layer	Color	EPS prod.	EPS degrad.	EPS C:N	Net EPS abund.	Potential for major precip.
LAYER						
1A	'White'	HIGH	n.d.	HIGH	HIGH	YES
1B	'Yellow–green'			Low	Low	
LAYER 2	'White'	Low	Slow	HIGH	HIGH	NO
LAYER 3	'Green'	Low	FAST	Low	Low	YES
LAYER 4	'White'	n.d.	n.d.	HIGH	HIGH	NO

n.d.=not determined.

EPS and other polymers in various forms have indicated generally slow turnover times (e.g., days to months) (Henrichs and Doyle, 1986; Weaver and Hicks, 1995; Hu and Bruggen, 1997).

In slurries of mat layers, the labile DOC was degraded at the highest rates. This is in agreement with the ^{14}C radiolabel experiments that revealed that the DOC pool was a major constituent of new production and that this carbon pool was turned-over rapidly. Interestingly, monosaccharides and EPS were oxidized equally fast by slurries as determined by O_2 uptake. From EPS:O_2 stoichiometry (data not shown), the O_2 consumption seems to level off before all EPS is completely degraded to CO_2, similar to the $^{14}CO_2$ recovery experiment (Fig. 7). Both radiolabeling and slurry experiments showed a lag phase prior to a rapid consumption. This indicates that in contrast to labile DOC, induction or de-repression of hydrolytic enzyme systems is required. The slurry experiments had a much shorter lag than did the radiolabel study. In contrast to experiments with ^{14}C-labeled glucose in intact samples, the slurry experiments which have well-mixed conditions and initially contain O_2 to its saturation provide a maximum potential respiration rate. However, if organic substrates are only partially degraded, this is detected by O_2 uptake and not necessarily by $^{14}CO_2$ production. This suggests that a partial degradation of EPS may occur that does not yield CO_2, perhaps the presence of relatively labile sidechains that degrade more rapidly. Alternatively, higher rates obtained in the slurry experiments could be due to the better mixing of the substrate EPS and/or the higher O_2 concentration present.

When calculating the specific O_2 uptake rates from the epifluorescence counts and slurry experiments, the rate for acetate is 1, 44, and 4 nM cell^{-1} h^{-1}, respectively, for Layers 1, 2, and 3. For glucose, the cell specific rates are 0.5, 18, and 1.3 nM cell^{-1} h^{-1}, and for EPS, 0.4, 13, and 1.3 nM cell^{-1} h^{-1} in Layers 1, 2, and 3, respectively. The highest specific rates are clearly associated with Layer 2, which population seems to prefer labile organic compounds (acetate) to monosaccharides and EPS. The much higher rate per cell in Layer 2 indicates that the bacteria in that layer could be starved with respect to organic substrates. The relatively higher rates in Layer 3, in comparison to Layer 1, could be attributed to carbon consumption by other respiratory processes such as the sulfate

reduction that obviously is unaccounted for in O_2 consumption. The sulfide produced during sulfate reduction requires O_2 for its oxidation, but this oxidation is incomplete, yielding polythionates and elemental sulfur (Visscher et al., 2002; Van den Ende and Van Gemerden, 1993). Visscher et al. (1998, 1999) showed in an earlier study that in Layer 3, sulfate reduction accounted for a significant sink for organic carbon.

4.5. Mass balance of EPS turnover

The extracellular secretion of EPS and the release of a wide array of low-MW metabolic products by cyanobacteria provide a supply of readily-utilizable organic molecules to stimulate utilization by heterotrophic bacteria. We found that the bacterial degradation of EPS to CO_2 was more rapid in dark–green cyanobacteria-rich layers, when compared to adjacent white layers. This may have occurred because the abundance of organic molecules stimulated bacterial heterotrophy in this layer.

The abundances of EPS were significantly ($P < 0.001$) higher in white layers because both EPS production was relatively high and heterotrophic degradation was slower than in adjacent green layers. The rapid conversion of low-MW DOC by heterotrophic bacteria to CO_2 may be particularly important in funneling C production. Bateson and Ward (1988) observed rapid transfer of secreted low-MW DOC, specifically glycolate, from photosynthetically-active *Synechococcus* cells to filamentous Chloroflexus-like organisms in hot-spring cyanobacterial mats.

A simple mass balance of total ^{14}C cycling within Layers 2 and 3 of the stromatolite mat was constructed in order to estimate if there was a net-accumulation or net-decrease in EPS material within the different layers of the stromatolite mat. The mass balance was calculated where:

$$\text{Net EPS Accumulation (per24h)} = \text{EPS Production} \times (24h) - \text{EPS Degradation (24h)}$$

The proportion of ^{14}C-bicarbonate that was taken up by autotrophic cells and incorporated into EPS during the incubation periods was determined. The degradation of newly-secreted EPS was estimated as the cumulative mean decrease in ^{14}C-EPS after its peak production and was subtracted from this value. Net production, therefore, was a function of EPS production and degradation over a 24-h period. Net EPS accumulation values were determined for stromatolite Layers 2 and 3. Our results (Table 2) suggest that microbial heterotrophic activities (i.e., degradation) may influence the turnover of C to different extents, depending on the mat layer, and that the turnover process was dependent on microbial production in the upper layers (i.e., photosynthetic C production). The relative roles that aerobic and anaerobic heterotrophy play in carbon cycling are currently under investigation. The EPS pools of stromatolites, which are secreted largely by cyanobacteria (Kawaguchi et al., 2003), are partially degraded soon after secretion. The EPS-remnant may be compositionally different and less labile to microbial degradation and may constitute an important structuring agent in longer-term persistence of stromatolites.

Our data and others (Kawaguchi and Decho, 2002a,b) showed that natural biofilm EPS represent a "more-refractory remnant" of the original polymer molecule(s) that is initially secreted by photosynthetic microorganisms. Our findings also indicate that layer-specific differences in EPS turnover were present within stromatolites. In green layers (e.g., Layers 1 and 3), there was a net loss of EPS (i.e., EPS Degradation>EPS Production). Although the top layer (Layer 1) of the stromatolite mat showed the highest production of EPS, a more rapid degradation of EPS (and DOC) by heterotrophic bacteria also occurred. This may fuel a diel pattern of production/respiration that creates favorable conditions for the precipitation of $CaCO_3$. In the white layers (e.g., Layer 2) of stromatolites, there was a high abundance and net gain in EPS (i.e., EPS Production>EPS degradation). Although Layer 2 had a lower production of EPS, there was a slower degradation of EPS. This resulted in a slight net gain of EPS (i.e., EPS Production>EPS Degradation) over time. The higher net gain of EPS presents conditions (Ca^{++} binding) that may reduce the capacity for major precipitation of micritic laminae.

Reid et al. (2000) presented a model of stromatolite formation. The model was derived from existing data and showed how microbial communities may develop through a series of successional stages (Types I and II), which may culminate in the formation of regularly-spaced thick laminae (Type III). These laminae contribute to the longer-term stability of modern stromatolites and characterize fossil stromatolites. The model also emphasizes how intermittent environmental perturbations (e.g., sedimentation rate and burial events) and perhaps endogenous interactions may influence the continued succession development, or cessation, of stromatolite growth (Stal, 2000). A central role of microbial EPS in this process was proposed in the model, since EPS may contribute to sediment trapping, Ca^{2+} ion binding, and inhibition of calcification. The present study represents the first estimate of EPS turnover in a natural system and within stromatolites.

Acknowledgements

We acknowledge Drs. Hans Paerl, Timothy Steppe (Marine Sciences Institute, UNC-Chapel Hill, Morehead City, NC), and James Pinckney (Marine Sciences Program, USC, Columbia, SC) for their constructive discussions and input during this work. Thanks are extended to crew of the R.V. Calanus (University of Miami) for providing an efficient working environment during our cruise, and to the staff of the Highborne Cay Marina (Exumas, Bahamas). This work was supported by NSF grants (OCE 95-30215, 96-19314, and 96-17738; EAR-BE 0221796) and represents RIBS (Research Initiative on Bahamian Stromatolites) Contribution #25.

References

Amon, R.M.W., Benner, R., 1994. Rapid cycling of high-molecular weight dissolved organic matter in the ocean. Nature (Lond.) 369, 549–552.

Anderson, K.A., Tayne, T.A., Ward, D.M., 1987. Formation and fate of fermentation products in hot spring Cyanobacterial mats. Appl. Environ. Microbiol. 53, 2343–2352.

Awramik, S.M., 1992. The history and significance of stromatolites. In: Schidlowski, M., et al. (Eds.), Early Organic Evolution: Implications for Mineral and Energy Resources. Springer-Verlag, Berlin, pp. 435–449.

Bateson, M.M., Ward, D.M., 1988. Photoexcretion and fate of glycolate in a hot spring Cyanobacterial mat. Appl. Environ. Microbiol. 54, 1738–1743.

Canfield, D.E., DesMarais, D.J., 1991. Aerobic sulfate reduction in microbial mats. Science 251, 1471–1473.

Cho, B.C., Azam, F., 1988. Major role of bacteria in biogeochemical fluxes in the ocean's interior. Nature 332, 441–443.

Cooksey, K.E., Cooksey, B., 1986. Adhesion of fouling diatoms to surfaces: some biochemistry. In: Evans, L.V., Hoagland, K.D. (Eds.), Algal Biofouling: Studies in Environmental Science, vol. 28. Elsevier.

Davis, G.R., 1970. Algal laminate sediments, Gladstone embayment, Shark Bay, Western Australia. Am. Assoc. Pet. Geol., Mem. 13, 85–168.

De Brouwer, J.F.C., Stal, L.J., 2002. Daily fluctuations of exopolymers in cultures of the benthic diatoms *Cylindrotheca closterium* and *Nitzschia* sp. (Bacillariophyceae). J. Phycol. 38, 464–472.

De Brouwer, J.F.C., Ruddy, G.K., Jones, T.E.R., Stal, L.J., 2002. Sorption of EPS to sediment particles and the effect on the rheology of sediment slurries. Biogeochemistry 61, 57–71.

Decho, A.W., 1990. Microbial exopolymer secretions in ocean environments: their role(s) in food webs and marine processes. Oceanogr. Mar. Biol. Annu. Rev. 28, 73–153.

Decho, A.W., 2000. Microbial biofilms in intertidal systems: an overview. Cont. Shelf Res. 20, 1257–1273.

Decho, A.W., Kawaguchi, T., 1999. Confocal imaging of natural in situ microbial communities and their extracellular polymeric secretions (EPS) using Nanoplast resin. BioTechniques 27, 1246–1251.

Des Marais, D.J., 1991. Microbial mats, stromatolites and the rise of oxygen in the Precambrian atmosphere. Palaeogeogr. Palaeoclimatol. Palaeoecol. 97, 93–96.

De Winder, B., Staats, N., Stal, L.J., Paterson, D.M., 1999. Carbohydrate secretion by phototrophic communities in tidal sediments. J. Sea Res. 42, 131–146.

Dill, R.F., Shinn, E.A., Jones, A.T., Kelly, K., Steinin, R.P., 1986. Giant subtidal stromatolites forming in normal saline waters. Nature 324, 55–58.

Dravis, J.J., 1983. Hardened subtidal stromatolites, Bahamas. Science 219, 385–386.

Dubois, M., Gilles, K.A., Hamilton, J.K., Rebers, P.A., Smith, F., 1956. Colorimetric methods for determination of sugars and related substances. Anal. Chem. 28, 350–356.

Fogg, G.E., 1983. The ecological significance of extracellular production of phytoplankton photosynthesis. Bot. Mar. 26, 3–14.

Fründ, C., Cohen, Y., 1992. Diurnal cycles of sulfate reduction under oxic conditions in microbial mats. Appl. Environ. Microbiol. 58, 77.

Grotzinger, J.P., 1990. Geochemical model for Proterozoic stromatolite decline. Am. Sci. 290, 80–103.

Grotzinger, J.P., Knoll, A.H., 1999. Stromatolites in Precambrian carbonates: evolutionary mileposts or environmental dipsticks? Annu. Rev. Earth Planet. Sci. 27, 313–358.

Henrichs, S.M., Doyle, A.P., 1986. Decomposition of ^{14}C-labeled organic substances in marine sediments. Limnol. Oceanogr. 31, 765–778.

Hoffman, P., 1976. Stromatolite morphogenesis in Shark Bay, Western Australia. In: Walter, M. (Ed.), Stromatolites. Elsevier Sci. Publ., Amsterdam, pp. 261–273.

Hu, S., Bruggen, A.H.C., 1997. Microbial dynamics associated with multiphasic decomposition of ^{14}C-labeled cellulose in soil. Microb. Ecol. 33, 134–143.

Jørgensen, B.B., 1994. Sulfate reduction and thiosulfate transformation in a Cyanobacterial mat during a diel oxygen cycle. FEMS Microbiol. Ecol. 13, 303–312.

Kawaguchi, T., Decho, A.W., 2000. Biochemical characterization of Cyanobacterial extracellular polymers from modern marine stromatolites. Prep. Biochem. Biotechnol. 30, 321–330.

Kawaguchi, T., Decho, A.W., 2002a. Isolation and biochemical characterization of extracellular polymeric secretions (EPS) from modern marine stromatolites and its inhibitory effect on $CaCO_3$ precipitation. Prep. Biochem. Biotechnol. 32, 51–63.

Kawaguchi, T., Decho, A.W., 2002b. A laboratory investigation of Cyanobacterial extracellular polymeric secretions (EPS) in influencing $CaCO_3$ polymorphism. J. Crystal Growth 240, 230–235.

Kawaguchi, T., Al Sayegh, H., Decho, A.W., 2003. Development of an indirect competitive enzyme-linked immunosorbent assay to detect extracellular polymeric substances (EPS) secreted by the marine stromatolite-forming cyanobacterium, *Schizothrix* sp. J. Immunoass. Immunochem. 24, 29–39.

King, G.M., 1997. Applications of ^{14}C and ^{3}H radiotracers for analysis of benthic organic matter transformations. In: Hurst, C.J., Hudson, G.R., McInerney, M.J., Stetzenbach, L.D., Walter, M.V. (Eds.), Manual of Environmental Microbiology. Amer Soc Microbiol Press, Washington, pp. 317–323.

Kirchman, D.L., Suzuki, Y., Garside, C., Ducklow, H.W., 1991. High turnover rates of dissolved organic carbon during a spring phytoplankton bloom. Nature (Lond.) 352, 612–614.

Knoll, A.H., 1992. The early evolution of eukaryotes: a geological perspective. Science 256, 622–627.

Leverenz, J.W., Falk, S., Pilstrom, C.-M., Samuelsson, G., 1990. The effects of photoinhibition on the photosynthetic light-response curve of green plant cells (*Chlamydomonas reinhardtii*). Planta 182, 161–168.

Liu, D., Wong, P.T.S., Dutka, P.J., 1973. Determination of carbohydrate in lake sediment by a modified phenol–sulfuric acid method. Water Res. 7, 741–746.

Logan, B.W., 1961. Cryptozoon and associated stromatolites from the Recent, Shark Bay, Western Australia. J. Geol. 69, 517–533.

MacIntyre, I.G., Reid, R.P., Stenek, R.S., 1996. Growth history of stromatolites in a Holocene fringing reef, Stocking Island, Bahamas. J. Sediment. Res. 66, 231–242.

MacIntyre, I.G., Prufert-Bebout, L., Reid, R.P., 2000. The role of endolithic Cyanobacteria in the formation of lithified laminae in Bahamian stromatolites. Sedimentology 47, 915–921.

Mague, T.H., Friberg, E., Hughes, D.J., Morris, I., 1980. Extracellular release of carbon by marine phytoplankton: a physiological approach. Limnol. Oceanogr. 28, 262–279.

Myklestad, S., 1998. Production, chemical structure, metabolism and biological function of the (1-3)-linked, ?-D-glucans in diatoms. Biol. Oceanogr. 6, 313–326.

Norman, B., Zweifel, U.L., Hopkinson, C.S., Fry, B., 1995. Production and utilization of dissolved organic carbon during an experimental diatom bloom. Limnol. Oceanogr. 40, 898–907.

Paerl, H.W., 1997. Primary productivity and producers. In: Hurst, C.J., Hudson, G.R., McInerney, M.J., Stetzenbach, L.D., Walter, M.V. (Eds.), Manual of Environmental Microbiology. Amer. Soc. Microbiol. Press, Washington, pp. 252–262.

Paerl, H.W., Steppe, T.F., Reid, R.P., 2001. Microbially mediated lithification in marine stromatolites: who's responsible? Environ. Microbiol. 3, 123–130.

Piegorsch, W.W., Bailer, A.J., 1997. Statistics for Environmental Biology and Toxicology. Chapman and Hall, NY. 579 pp.

Pinckney, J., Paerl, H.W., Reid, R.P., Bebout, B., 1994. Ecophysiology of stromatolitic mats, Stocking Island, Exuma Cays, Bahamas. Microb. Ecol. 29, 19–37.

Reid, R.P., Browne, K.M., 1991. Intertidal stromatolites in a fringing Holocene reef complex, Bahamas. Geology 19, 15–18.

Reid, R.P., MacIntyre, I.G., Browne, K.M., Steneck, R.S., Miller, T., 1995. Modern marine stromatolites in the Exuma Cays, Bahamas: uncommonly common. Facies 33, 1–18.

Reid, R.P., MacIntyre, I.G., Steneck, R.S., 1999. A microbialite/ algal ridge fringing reef complex, Highborne Cay, Bahamas. Atoll Res. Bull. 465, 1–18.

Reid, R.P., Visscher, P.T., Decho, A.W., Stolz, J., Bebout, B., MacIntyre, I.G., Dupraz, C., Pinckney, J., Paerl, H., Prufert-Bebout, L., Steppe, T., Des Marais, D., 2000. The role of microbes in accretion, lamination and early lithification of modern marine stromatolites. Nature (Lond.) 406, 989–992.

Rippka, R., Deruelles, J., Waterbury, J.B., Herdman, J., Stanier, R.Y., 1979. Generic assignments, strain histories and properties of pure cultures of Cyanobacteria. J. Gen. Microbiol. 111, 1–61.

SAS Institute, 1985. SAS user's guide: statistics, version 5 edition. SAS Institute, Cary, NC.

Sawyer, T.E., King, G.M., 1993. Glucose uptake in an intertidal marine sediment: metabolism and endproduct formation. Appl. Environ. Microbiol. 59, 120–128.

Schopf, J.W., 1996. Are the oldest fossils Cyanobacteria. In: Roberts, D.M., Sharp, P., Alderson, G., Collins, M. (Eds.), Evolution of Microbial Life. Cambridge Univ. Press, Washington, DC, pp. 23–62.

Smith, D.J., Underwood, G.J.C., 1998. Exopolymer production by intertidal epipelic diatoms. Limnol. Oceanogr. 43, 1578–1591.

Smith, D.J., Underwood, G.J.C., 2000. The production of extracellular carbohydrates by estuarine benthic diatoms: the effects of growth phase and light and dark treatment. J. Phycol. 36, 321–333.

Staats, N., Stal, L.J., de Winder, B., Mur, L.R., 2000. Oxygenic photosynthesis as a driving process in exopolysaccharide production in benthic diatoms. Mar. Ecol., Prog. Ser. 193, 261–269.

Stal, L.J., 1991. The metabolic diversity of the mat-building cyanobacterium *Microcoleus chthonoplastes* and *Oscillatoria limosa* and its ecological significance. Algol. Stud. 64, 453–467.

Stal, L.J., 1995. Physiological ecology of Cyanobacteria in microbial mats and other communities. New Phytol. 131, 1–32.

Stal, L.J., 2000. Cyanobacterial mats and stromatolites. In: Whitton, B.A., Potts, M. (Eds.), The Ecology of Cyanobacteria. Kluwer Acad. Press, Dordrecht, pp. 61–120.

Stolz, J.F., 2000. Structure of microbial mats and biofilms. In: Riding, R.E., Awramik, S.M. (Eds.), Microbial Sediments. Springer-Verlag, Berlin, pp. 1–8.

Taylor, I.S., Paterson, D.M., 1998. Micropatial variation in carbohydrate concentrations with depth in the upper millimetres of intertidal cohesive sediments. Estuar. Coast. Shelf Sci. 46, 359–370.

Taylor, I.S., Paterson, D.M., Mehlert, A., 1999. The quantitative variability and monosaccharide composition of sediment carbohydrates associated with intertidal diatom assemblages. Biogeochemistry 45, 303–327.

Underwood, G.J.C., Kromkamp, J., 1999. Primary production by phytoplankton and microphytobenthos in estuaries. Adv. Ecol. Res. 29, 93–153.

Underwood, G.J.C., Paterson, D.M., Parkes, R.J., 1995. The measurement of microbial carbohydrate exopolymers from intertidal sediments. Limnol. Oceanogr. 40, 1243–1253.

Underwood, G.J.C., Nilsson, C., Sundbäck, K., Wulff, A., 1999. Short-term effects of UVB irradiation on chlorophyll fluorescence, biomass, pigments and carbohydrate fractions in a benthic diatom mat. J. Phycol. 35, 656–666.

Van den Ende, F.P., Van Gemerden, H., 1993. Sulfide oxidation under oxygen-limitation by a *Thiobacillus thioparus* isolated from a marine microbial mat. FEMS Microbiol. Ecol. 13, 69–78.

Visscher, P.T., Prins, R.A., Van Gemerden, H., 1992. Rates of sulfate reduction and thiosulfate consumption in a marine microbial mat. FEMS Microbiol. Ecol. 86, 283–394.

Visscher, P.T., Reid, R.P., Bebout, B.M., Hoeft, S.E., MacIntyre, I.G., Thompson, J.A., 1998. Formation of lithified micritic laminae in modern marine stromatolites (Bahamas): the role of sulfur cycling. Am. Mineral. 83, 1482–1493.

Visscher, P.T., Gritzer, F.R., Leadbetter, E.R., 1999. Low-molecular weight sulfonates, a major substrate for sulfate reducers in marine microbial mats. Appl. Environ. Microbiol. 65, 3272–3278.

Visscher, P.T., Reid, R.P., Bebout, B.M., 2000. Microscale observations of sulfate reduction: correlation of microbial activity with lithified micritic laminae in modern marine stromatolites. Geology 28, 919–922.

Visscher, P.T., Surgeon, T.M., Hoeft, S.E., Bebout, B.M., Thompson Jr., J., Reid, R.P., 2002. Microelectrode studies in modern marine stromatolites: unraveling the Earth's past? In: Taillefert, M., Rozan, T. (Eds.), Environmental Electrochemistry: Analyses of Trace Element Biogeochemistry, Am. Chem. Soc. Symp. Ser., vol. 811. Oxford Univ Press, New York, NY, pp. 265–282.

Weaver, D.T., Hicks, R.E., 1995. Biodegradation of *Azotobacter vinelandii* exopolymer by Lake Superior microbes. Limnol. Oceanogr. 40, 1035–1041.

Wolfstein, K., Stal, L.J., 2002. The production of extracellular polymeric substances (EPS) by benthic diatoms: the effect of irradiance and temperature. Mar. Ecol., Prog. Ser. 245, 13–21.

Wolfstein, K., De Bouwer, J.F.C., Stal, L.J., 2002. Biochemical partitioning of photosynthetically fixed carbon by benthic diatoms during short-term incubations at different irradiances. Mar. Ecol., Prog. Ser. 245, 22–31.

Wolter, K., 1982. Bacterial incorporation of organic substances released by natural phytoplankton populations. Mar. Ecol., Prog. Ser. 7, 287–295.

Available online at www.sciencedirect.com

Palaeogeography, Palaeoclimatology, Palaeoecology 219 (2005) 87–100

ELSEVIER

PALAEO

www.elsevier.com/locate/palaeo

Microbial mats as bioreactors: populations, processes, and products

Pieter T. Visscher[a], John F. Stolz[b],*

[a]*Department of Marine Sciences, University of Connecticut, Groton, CT 06340, United States*
[b]*Department of Biological Sciences, Duquesne University, Pittsburgh, PA 15282, United States*

Received 10 May 2004; accepted 29 October 2004

Abstract

Microbial mats are dynamic and complex ecosystems exhibiting spatial and temporal heterogeneity. The physical/chemical environment is typified by steep gradients and distinct microenvironments. These microenvironments support a great diversity of species with a wide range of metabolic processes. These processes often result in coupled reactions and biogeochemical cycles, and produce important end products such as trace gases and mineral precipitates. The latter can impact the composition and character of the sediment, imparting a "biosignature." These biosignatures can be preserved in the rock record and are useful in the interpretation of fossil record on Earth and possibly as an indication of life on other planetary bodies. The modern marine stromatolites of the Exuma Cays, Bahamas, provide an ideal system for studying the populations, processes, and products in a microbial ecosystem using a multidisciplinary approach. In order to acquire redox energy, microbial populations need to carry out metabolic reactions at rates faster than the equivalent chemical (abiotic) reactions. As such, microbes can be viewed as bioreactors that preferably oxidize carbon to CO_2 to maximize the energy yield. The study of the microbial role in carbonate sedimentation and lithification in these stromatolites provides a picture of microbial mats as bioreactors producing a biosignature.
© 2004 Elsevier B.V. All rights reserved.

Keywords: Stromatolite; Microbial mats; Bioreactors; Carbonate precipitation; Lithification

1. Introduction

Microbial mat is a general term that is used to describe a variety of microbial communities, from the cyanobacterial slime that forms in a drainage ditch to the complex, multilayered microbial ecosystems commonly observed in salt marshes and estuaries (Krumbein et al., 1977; Stal et al., 1985; Nicholson et al., 1987). Their propensity for trapping, binding, and precipitating sediments results in the formation of sedimentary structures such as laminated sediments and stromatolites (Riding, 2000). These structures can be preserved in the rock record and it is from these remains that we reconstruct the early history of life on Earth (Schopf, 1983; Des Marais, 1990, 1997;

---

* Corresponding author. Tel.: +1 412 396 6333; fax: +1 412 396 5907.
  E-mail address: stolz@duq.edu (J.F. Stolz).

0031-0182/$ - see front matter © 2004 Elsevier B.V. All rights reserved.
doi:10.1016/j.palaeo.2004.10.016

Golubic, 1991; Grotzinger and Knoll, 1999). From an ecological perspective, microbial mats are ecosystems even though their geographic extent may be exceedingly small (on the order of a few meters). They contain the essential trophic groups (e.g., primary producers, consumers, and decomposers) and their populations are organized into specific communities interacting with each other and their environment (Stolz et al., 1988). These populations can be further grouped into specific guilds and assemblages, based on their metabolic properties (Visscher et al., 1992, 2002) and taxonomic affiliations (Ward et al., 1998).

The physical/chemical environment is typified by gradients (i.e., oxygen, sulfide, and light) that are often quite steep (i.e., the oxygen concentration may go from supersaturated to immeasurable within a few millimeters), resulting in distinct microenvironments. These microenvironments provide a plethora of habitats resulting in a community structure that exhibits both spatial and temporal heterogeneity. Thus, microbial mats are dynamic ecosystems that support a great diversity of species with a wide range of metabolic processes that take place in close proximity. These processes often result in coupled reactions (e.g., reduction and oxidation of an element such as C, S, N, etc., or, alternatively, stepwise oxidation or reduction of a compound) that support robust biogeochemical cycles. While these processes may be subject to the temporal environmental oscillations (i.e., diurnal and seasonal cycles; Visscher and van den Ende, 1994), the net result can be the formation of important endproducts such as trace gases and mineral precipitates (Fig. 1). As such,

microbial mats may be viewed from the materials science perspective as bioreactors. The modern marine stromatolites of Highborne Cay, Bahamas, provide an opportunity to explore this concept. This paper reviews net $CaCO_3$ precipitation from a microbial perspective, exploring the effect of metabolism on geochemical characteristics of Ca^{2+} and $CO_2/HCO_3^-/CO_3^{2-}$ activity.

The classic view of a microbial mat, with a surface community of oxygenic cyanobacteria underlain by subsequent layers of anoxygenic phototrophic bacteria and sulfate-reducing bacteria (SRB) (Krumbein, 1983; Cohen et al., 1984), has been dramatically revised. Instead of a layering that results from a sequence of metabolic reactions determined by gradients of light and redox potential (resulting in the sequence use of O_2, Fe(III)/Mn(IV), NO_3^-, SO_4^{2-}, etc., as electron acceptor), most types of metabolism are found in association with the cyanobacterial layer. For example, populations of SRB have now been found at the surface of mats (Fründ and Cohen, 1992; Visscher et al., 1992), and methanogenesis may peak there as well (Hoehler et al., 2002). Regardless of the vertical structure, marine microbial mats are composed of four major functional groups: oxygenic phototrophs (CYN), aerobic heterotrophic bacteria (HET), sulfate-reducing bacteria (SRB), and sulfide-oxidizing bacteria (SOB). The oxygenic phototrophs, which include primarily cyanobacteria, are the primary producers of the mat system, coupling light energy to CO_2 fixation. They may also provide assimilable nitrogen through nitrogen fixation. The aerobic heterotrophic bacteria oxidize a significant fraction of the fixed organic

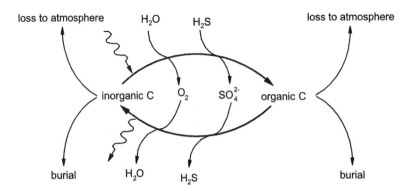

Fig. 1. Coupled cycles of carbon and oxygen and sulfur. Arrows on extreme left and right indicate the potential formation of biogenic signatures; downward arrow on the left is a means by which $CaCO_3$ precipitates.

carbon (e.g., photosynthate) in the system, gaining energy through respiration using O_2 as the terminal electron acceptor. The sulfate-reducing bacteria are the dominant anaerobic respiring organisms, producing sulfide in the process. The sulfide-oxidizing bacteria are chemolithotrophic organisms that oxidize reduced sulfur compounds using either O_2 or nitrate as terminal electron acceptors (Jørgensen et al., 1983; Stal et al., 1985; van Gemerden, 1993; Visscher et al., 1992, 1998). They recycle the reduced sulfur and may provide additional organic carbon through autotrophic CO_2 fixation. However, this outline emphasizes the "metabolic specialty" of the respective functional groups. A significant metabolic diversity and flexibility exists, and in light-driven ecosystems such as microbial mats, it is important to distinguish between daytime and nighttime processes (Table 1).

Microbial mat ecosystems are efficient in element cycling and, once developed, require little more than light to function. As such, they can be viewed as semiclosed systems, making it relatively easy to create mass balances and study element cycling. Compared to other benthic ecosystems, microbial mats have extremely high rates of oxygenic photosynthesis, aerobic respiration, sulfate reduction, and sulfide oxidation (e.g., Revsbech et al., 1986; Canfield and Des Marais, 1993). The metabolic rates of organisms comprising the mats are so high that the community production per unit mass rivals that of rain forests (Krumbein et al., 2003; Jørgensen, 2001).

2. Carbonate sedimentation and lithification in modern marine stromatolites

Highborne Cay is one of many localities in the Exuma Cays where stromatolites are found forming in open ocean waters of normal salinity (Dravis, 1983; Dill et al., 1986; Reid and Brown, 1991; Reid et al., 1995, 2000; Steneck et al., 1996, 1997). The term *stromatolite* is used here for organosedimentary structures formed by trapping and binding of sediment and net carbonate-precipitating activities of micro-organisms. These structures, which are characterized by alternating soft and hard layers, may be a few centimeters to over 2 m in height and form within a reef complex lying along the windward east coast of the Exuma Cays (Reid et al., 1995). The stromatolites are composed primarily of fine-grained carbonate sands that range in size from 125 to 250 μm and are stabilized by lithified layers, giving rise to a classic stromatolitic structure (Reid et al., 1995; Fig. 2). The microbial communities involved in the deposition of the Highborne Cay stromatolites have been under investigation for over a decade (Reid et al., 1995, 2000; Visscher et al., 1998, 2000, 2002; Stolz et al., 2001). Three different types of carbonate laminations have been discerned: thick (200 μm to several millimeters) layers composed of unconsolidated carbonate ooids (carbonate grains); thin (10–30 μm) micritic (microcrystalline $CaCO_3$ with diameter smaller than 4 μm) layers; and thicker (100–300

Table 1
Daytime and nighttime metabolic activities of key functional groups of a microbial mat community

Functional group	Daytime metabolic function	Nighttime metabolic function
Cyanobacteria	Carbon fixation (photosynthesis): $CO_2+H_2O \rightarrow CH_2O+O_2$	Fermentation (including H_2 production), N_2 fixation, glycogen degradation
Aerobic heterotrophs	Carbon oxidation (respiration): $CH_2O+O_2 \rightarrow CO_2+H_2O$	Fermentation, denitrification: $5CH_2O+2H_2O \rightarrow HCO_3^-+H^++4CH_3O$ and $5CH_2O+4NO_3^- \rightarrow 5HCO_3^-+H^++2N_2+H_2O$
Sulfide oxidizers	Sulfide oxidation: $H_2S+2O_2 \rightarrow SO_4^{2-}+2H^+$ (sometimes coupled to carbon fixation)	Denitrification, fermentation: $5HS^-+8NO_3^- \rightarrow 5SO_4^{2-}+4N_2+ H_2O+3OH^-$
Phototrophic sulfide oxidizers	Carbon fixation (anoxygenic photosynthesis coupled to sulfide oxidation): $2CO_2+H_2S+2H_2O \rightarrow 2CH_2O+SO_4^{2-}+2H^+$	Fermentation, synthesis of Bchl$a$, degradation of glycogen
Anaerobic heterotrophs-sulfate reducers	Carbon oxidation (sulfate respiration): $2CH_2O+SO_4^{2-} \rightarrow 2HCO_3^-+2H_2S$	Same as daytime
Anaerobic heterotrophs-methanogens	Carbonate respiration: $4H_2+CO_2 \rightarrow CH_4+2H_2O$ and $2CH_2O \rightarrow CH_4+CO_2$	Same as daytime

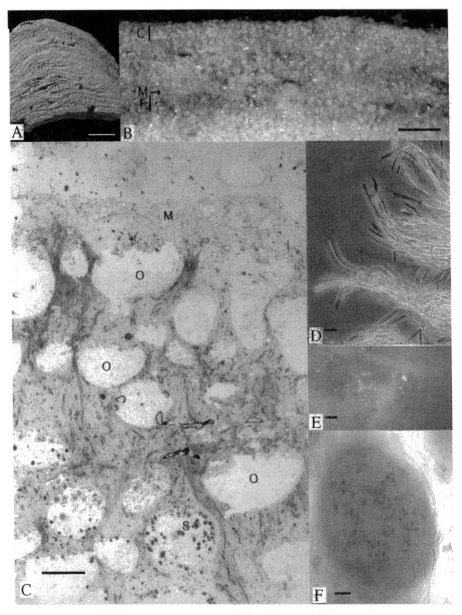

Fig. 2. The surface microbial communities of the modern marine stromatolites of Highborne Cay. (A) Vertical section through a stromatolite revealing the laminations; bar=2 cm. (B) Vertical section of the stromatolite surface showing a caramel-colored stage 1 community at the surface; bar=2 mm. (C) A subsurface micrite crust (arrow) and a subsurface *Solentia* sp.-infested layer (F); bar=100 μm. (C) Bright field light micrograph of an oriented thick section of the top 1.2 mm of stage 3 surface community showing the surface micritic layer (M), ooids (O), and *Solentia* sp.-infested ooids (S); bar=10 μm. (D) Phase contrast light micrograph of *Schizothrix gebeleinii* from the surface of a stage 1 community; bar=10 μm. (E) Fluorescence micrograph of the micritic crust. The bacteria have been stained with acridine orange and are fluorescing (bright spots); bar=10 μm. (F) Bright field light micrograph of an ooid colonized by *Solentia* sp. (the dark spots); bar=10 μm.

μm) hard laminations composed of fused ooid grains. Three different surface microbial communities have been identified, each representing a different growth stage, and each involved in the deposition of a particular form of carbonate (Reid et al., 2000; Fig. 2B and C).

The surface community involved primarily in the trapping of the carbonate ooids is dominated by the filamentous cyanobacterium *Schizothrix gebeleinii* and characterizes the stromatolite surface during periods of rapid sediment accretion (Fig. 2D). This cyanobacterial community, which is clearly visible as a caramel-green layer, provides the stromatolite with organic carbon during photosynthesis. The vertical profile of O_2 varies depending on the time of day, penetrating to less than 2 mm at daybreak and reaching a maximum of 5 mm with concentrations near twice saturation during late afternoon (Visscher et al., 1998, 2002). At depth, the caramel color bleaches with burial, resulting in white unlithified layers of unconsolidated ooids. The community associated with the development of a micritic crust is typified by a stromatolite surface populated by a community of heterotrophic bacteria embedded in amorphous extracellular polymeric secretions (EPS) underlain by filamentous cyanobacteria (Fig. 2E). In situ measurements indicate high rates (approximately 15 μM h^{-1}) of sulfate reduction (Visscher et al., 2000), which are one to three orders of magnitude higher than measured in continental margins (Ferdelman et al., 1999). The micritic crust, which forms during periods of little sediment accretion (i.e., few ooids are trapped), persists at depth in the stromatolite. Prolonged periods of low sedimentation lead to the further development of the community and the colonization of the near subsurface by the endolithic cyanobacterium *Solentia* sp. (Fig. 2F). This mat community shows the greatest species diversity with a variety of filamentous cyanobacteria, anoxyphototrophic bacteria, SRB, and SOB (Visscher et al., 1998, 2000; Stolz et al., 2001, Stolz, 2003). This community is also typified by high rates of sulfate reduction (ca. 20 μM h^{-1}; Visscher et al., 2000). The color of the layer remains for several centimeters below the surface before eventually becoming gray-green. The continued boring activity, characteristic for the endolithic cyanobacteria, and carbonate deposition within the bore holes of the ooids result in a fused-grain layer (Macintyre et al., 2000). In situ observation of the three different surface communities and their characteristic sedimentological products has led to a conceptual model in which the transformation from one type into another occurs in a cyclic fashion (Reid et al., 2000). The cyclical succession and subsequent burial of these surface communities result over time in the growth and laminated structure of the stromatolite. As much as the stromatolites of Highborne Cay are an example of the dynamic nature of microbial ecosystems, their physical structure represents the net result of carbonate precipitation and dissolution.

3. Carbonate precipitation and dissolution

The majority of the carbonates on the Earth's surface is biogenic and results from the precipitation of the CO_2 generated during microbial metabolism (e.g., Castanier et al., 1999, 2000). Due to the low solubility of carbonates, deposits are formed, especially in marine environments. Carbonates can be precipitated intracellularly or extracellularly. Extracellular carbonates (e.g., $CaCO_3$) have been linked to aerobic respiration, which increases the inorganic carbon concentration, resulting in an abiotic reaction with Ca^{2+} ions in the environment (Chafetz and Buczynski, 1992). Locally elevated CO_2 concentrations increase HCO_3^- and CO_3^{2-}, creating favorable conditions for $CaCO_3$ precipitation. Alternatively, HCO_3^- can dissociate under alkaline conditions; the proton can enter the cell while the CO_3^{2-} precipitates with Ca^{2+} (De Vrind-deJong and de Vrind, 1998). A similar scenario has been proposed for intracellular $CaCO_3$ precipitation. HCO_3^- is transported over the cell membrane via a bicarbonate transporter, then is cleaved intracellularly (by carbonic anhydrase type enzymes) to be used in CO_2 fixation (Robbins and Yates, 1998), while the other fraction precipitates with cations like Ca^{2+} and/or Mg^{2+}. It has been hypothesized that microorganisms benefit from $CaCO_3$ precipitation by production of a H^+, which results from the reaction of Ca^{2+} and HCO_3^- (see reaction equations). The precipitation of $CaCO_3$ occurs extracellularly, and the H^+ assists in the generation of a proton motive force (Δp), which provides a mechanism for energy generation, uptake of substrates, discharge of metabolites, and other cellular processes (McConnaughey and Whelan, 1997).

The precipitation depends on the saturation index SI, which is a function of the solubility product constant k_{SP} and the in situ pH and Ca^{2+}, or, SI=log (IAP/k_{SP}), where IAP is the ion activity product in the sample. By default, SI=0 at equilibrium. The solubility product constant $k_{SP}=Ca^{2+}\times CO_3^{2-}/CaCO_3=$ $10^{-8.42}$ for calcite and $10^{-8.22}$ for aragonite. If IAP>k_{SP}, then the solution is oversaturated. Laboratory experimentation has demonstrated that an SI≥0.8 is required before precipitation occurs (Kempe and Kazmierczak, 1994). Interestingly, preliminary investigations have demonstrated that the [Ca^{2+}] in Highborne Cay stromatolites is approximately two to five orders of magnitude higher (24–88 mM) within the surface mats than in the surrounding seawater (P. Visscher, unpublished data). In summary, the requirements for $CaCO_3$ precipitation are: (1) supersaturation with respect to $CaCO_3$ caused by high activities of Ca^{2+} or CO_3^{2-}; and (2) the onset of nucleation, which is the point at which the activation energy is overcome and precipitation of the critical nuclei begins. Calcium carbonates are also formed directly by organisms as surface structures of cells (e.g., calcifying algae) and some protozoan shells (foraminifera and coccolithophorids), as well as mollusks and bryozoa.

Clearly, many factors play a role in net precipitation of $CaCO_3$. The pH of seawater is relatively well-buffered and only a significant shift in the alkalinity (the total concentration of bases in the water) will result in precipitation or dissolution of $CaCO_3$. As is true for all the biochemical reactions below, the observed precipitation of $CaCO_3$ in cultures of aerobic heterotrophs (Castanier et al., 2000; Folk and Chafetz, 2000; Chafetz and Buczynski, 1992; Rivadeneyra et al., 1993, 1999) needs very careful evaluation before extrapolation to the field. Experiments with pure cultures (typically >10^9 cells/ml, or on plates in colonies) are not very relevant for the in situ precipitation, and molecular ecological studies (e.g., fluorescent in situ hybridization) and/or geochemical investigations (e.g., stable isotope ratios) are essential for an appropriate interpretation of laboratory data.

4. Microbes as biochemical reactors

In order to understand the role of microbes in precipitation and dissolution, their metabolic activity and associated biochemical reactions need to be evaluated. The combination of the biotic reactions and their abiotic (geochemical) counterparts deter-

mines the net effect that a microbial functional group will have on carbonate precipitation and dissolution.

As outlined above, there are four functional groups that are key players in microbial mats (i.e., CYN, HET, SRB, and SOB; van Gemerden, 1993; Visscher et al., 1998). However, before discussing the specifics of these four groups of microbes on carbonate precipitation in mats, a number of functional groups of general biogeochemical interest will be reviewed. The underlying consideration is the fact that most microbes engage in redox reactions, whether organic or inorganic, in order to generate and conserve energy. In doing so, they must be more efficient in mediating a certain reaction than the abiotic equivalent. Coupling of several redox reactions results in element cycling, which has multiple effects on the environment: (1) physical effects, including dissolution, precipitation, volatilization, and fixation of elements; (2) chemical processes, such as hydrolysis, condensation, biosynthesis, biotransformation, and biodegradation; and (3) spatial translocations, including transport driven by concentration gradients and physical processes. For this review, we will focus on the role microbes play in catalyzing chemical reactions that result in calcium carbonate precipitation or dissolution. Since ecosystems, including microbial ones, maximize the energy yield, we assume here that during respiration, organic carbon is transformed to the most oxidized form, CO_2, and electron donors are reduced to the most reduced forms. It should be noted that under in situ conditions, this may take place in a single metabolic reaction, or more likely in a sequence of aerobic and anaerobic conditions.

5. Coupled biotic–abiotic reactions that precipitate or dissolve $CaCO_3$

5.1. Photoautotrophy

During photosynthesis, microorganisms use light energy to generate ATP and reducing power to support carbon fixation. Oxygenic phototrophs (cyanobacteria) use predominantly H_2O as electron donor for photosynthetic electron transport, while anoxygenic phototrophs (purple and green anoxybacteria) primarily use reduced sulfur compounds as electron donors.

5.1.1. Oxygenic photoautotrophy

Cyanobacterial photosynthetic activity in microbial mats is typically very high (Revsbech, 1984; Jørgensen and Cohen, 1977; Visscher et al., 2002) and, as a result, the pH may increase to values higher than 10 (Visscher and van Gemerden, 1991). This increase in pH is the result of CO_2 production in a bicarbonate-buffered (marine) environment:

$$HCO_3^- \rightarrow CO_2 + OH^- \text{ (chemical)}$$

$$CO_2 + H_2O \rightarrow CH_2O + O_2 \text{ (microbial)}$$

$$Ca^{2+} + HCO_3^- \rightarrow CaCO_3 + H^+ \text{ (chemical)}$$

$$H^+ + OH^- \rightarrow H_2O \text{ (chemical)}$$

sum: $2HCO_3^- + Ca^{2+} \rightarrow CH_2O + O_2 + CaCO_3$

Thus, carbon fixation by oxygenic photoautotrophs yields 1 mol of $CaCO_3$ per mole of CO_2 consumed.

5.1.2. Anoxygenic photoautotrophy

Purple and green sulfur bacteria are commonly found in microbial mats and stratified water columns. Anoxygenic CO_2 fixation impacts the pH in a similar manner as the oxygenic counterpart, with the difference being that HS^- oxidation decreases alkalinity:

$$2HCO_3^- \rightarrow 2CO_2 + 2OH^- \text{ (chemical)}$$

$$HS^- + 2CO_2 + 2H_2O \rightarrow 2CH_2O + SO_4^{2-}$$
$$+H^+ \text{ (microbial)}$$

$$Ca^{2+} + HCO_3^- \rightarrow CaCO_3 + H^+ \text{ (chemical)}$$

$$2H^+ + 2OH^- \rightarrow 2H_2O \text{ (chemical)}$$

sum: $3HCO_3^- + Ca^{2+} + HS^- \rightarrow 2CH_2O + CaCO_3 + SO_4^{2-}$

The overall effect is that during anoxygenic photosynthesis, 0.5 mol of $CaCO_3$ precipitates per mole of CO_2 fixed.

5.2. Aerobic respiration (chemoorganoheterotrophy)

As outline above, heterotrophic microbes can only precipitate $CaCO_3$ if the system is very well-buffered

so that CO_2 production results in an increase in $[CO_3^{2-}]$ without a change in pH, and the environment contains sufficiently high $[Ca^{2+}]$. This has been shown in laboratory cultures (e.g., Chafetz and Buczynski, 1992; Rivadeneyra et al., 1993, 1999). Otherwise, the following respiration equations need to be considered:

$$CH_2O + O_2 \rightarrow HCO_3^- + H^+ \text{ (microbial)}$$

$$CaCO_3 + H^+ \rightarrow HCO_3^- + Ca^{2+} \text{ (chemical)}$$

sum: $CH_2O + CaCO_3 + O_2 \rightarrow 2HCO_3^- + Ca^{2+}$

The overall effect of aerobic heterotrophy is loss of one $CaCO_3$ per CH_2O oxidized.

5.2.1. Fermentation

In the absence of O_2 as terminal electron acceptor, many organisms are capable of fermentation of organic carbon, in which the same compound acts as electron donor and acceptor:

$$3CH_2O + H_2O \rightarrow HCO_3^- + H^+ + C_2H_6O \text{ (microbial)}$$

$$CaCO_3 + H^+ \rightarrow HCO_3^- + Ca^{2+} \text{ (chemical)}$$

sum: $3CH_2O + CaCO_3 + H_2O \rightarrow 2HCO_3^- + Ca^{2+}$
$$+C_2H_6O$$

Fermentation results in the loss of 1 mol of $CaCO_3$ per 5 mol of CH_2O used, and has only a minor impact on the dissolution process. It should be noted that the equation drafted above is derived from ethanol fermentation and that many other potential fermentative pathways may have other effects on the $CaCO_3$ budget.

5.3. Anaerobic respiration (chemoorganoheterotrophy)

In the absence of O_2, alternative terminal electron acceptors (TEAs) can be used in redox reactions. These TEA include: $FeOOH$, NO_3^-, SO_4^{2-}, and HCO_3^-.

5.3.1. Dissimilatory iron-reducing bacteria

Iron reduction may be important in freshwater and coastal marine environments (Nealson and Stahl, 1997; Lovley and Coates, 2000):

$$CH_2O + 4FeOOH + H_2O \rightarrow HCO_3^- + 4Fe^{2+}$$
$$+ 7OH^- \text{ (microbial)}$$

$$7Ca^{2+} + 7HCO_3^- \rightarrow 7CaCO_3 + 7H^+ \text{ (chemical)}$$

$$7H^+ + 7OH^- \rightarrow 7H_2O \text{ (chemical)}$$

$$sum: CH_2O + 4FeOOH + 7Ca^{2+} + 6HCO_3^-$$
$$\rightarrow 7CaCO_3 + 4Fe^{2+} + 6H_2O$$

Iron reduction precipitates $CaCO_3$ with a gain of 7 mol of $CaCO_3$ per mole of CH_2O oxidized.

5.3.2. Dissimilatory nitrate reduction

Denitrification, the dissimilatory reduction of nitrate to dinitrogen, may take place in a variety of sediments. However, in the marine environment, nitrogen is considered to be the limiting element and assimilatory nitrate reduction will compete with the dissimilatory use of NO_3^-:

$$5CH_2O + 4NO_3^- \rightarrow 5HCO_3^- + 2N_2 + 2H_2O$$
$$+ H^+ \text{ (microbial)}$$

$$CaCO_3 + H^+ \rightarrow HCO_3^- + Ca^{2+} \text{ (chemical)}$$

$$sum: 5CH_2O + 4NO_3^- + CaCO_3$$
$$\rightarrow 6HCO_3^- + 2N_2 + 2H_2O + Ca^{2+}$$

The net result is a minor loss (0.2 per CH_2O oxidized) of $CaCO_3$ during nitrate reduction.

5.3.3. Sulfate-reducing bacteria

The reduction of sulfate is a dominant respiratory pathway, especially in marine environments (Canfield et al., 1993): up to 80% of the carbon oxidation may proceed through sulfate reduction. In addition, the sulfide produced as a metabolic by-product is a sink for O_2 (see below):

$$2CH_2O + SO_4^{2-} + OH^- \rightarrow 2HCO_3^- + HS^-$$
$$+ H_2O \text{ (microbial)}$$

$$HCO_3^- + Ca^{2+} \rightarrow CaCO_3 + H^+ \text{ (chemical)}$$

$$H^+ + HCO_3^- \rightarrow CO_2 + H_2O \text{ (chemical)}$$

$$sum: 2CH_2O + SO_4^{2-} + OH^- + Ca^{2+}$$
$$\rightarrow CaCO_3 + CO_2 + 2H_2O + HS^-$$

Reduction of sulfate results in the precipitation of 1 mol of $CaCO_3$ per 2 mol of CH_2O oxidized. Some sulfate-reducing bacteria are capable of H_2 oxidation while fixing CO_2:

$$4H_2 + SO_4^{2-} \rightarrow HS^- + 3H_2O + OH^- \text{ (microbial)}$$

$$2CO_2 + 4H_2 + CoASH \rightarrow CH_3CO - SCoA$$
$$+ 3H_2O \text{ (microbial)}$$

$$2HCO_3^- \rightarrow 2CO_2 + 2OH^- \text{ (chemical)}$$

$$2Ca^{2+} + 2HCO_3^- \rightarrow 2CaCO_3 + 2H^+ \text{ (chemical)}$$

$$2H^+ + 2OH^- \rightarrow 2H_2O \text{ (chemical)}$$

$$sum: 8H_2 + SO_4^{2-} + 4HCO_3^- + 2Ca^{2+}$$
$$\rightarrow HS^- + CH_3CO - SCoA + 3H_2O + 2CaCO_3$$

This lithoautotrophic metabolism deploys the carbon monoxide dehydrogenase pathway for CO_2 fixation and is found in *Desulfobacterium autotrophicum* (Schauder et al., 1989). The net result of this type of sulfate reduction is 2 mol of $CaCO_3$ precipitated per 4 mol of H_2 oxidized or 1 mol of SO_4^{2-} reduced. Other pathways involve a modified TCA cycle (Spormann and Thauer, 1988; not further discussed here). The oxidation of H_2 coupled to CO_2 fixation is much less common than the heterotrophic oxidation of organic carbon.

In addition to the generic equation for heterotrophic sulfate reduction above, it is important to acknowledge the two different metabolic types of sulfate reduction that can be distinguished (Widdel, 1988): sulfate-reducing bacteria can either oxidize organic carbon (e.g., lactate) to CO_2 (complete oxidizers), or to acetate and CO_2 (incomplete oxidizers).

5.3.3.1. Sulfate reduction—complete oxidation

$$2C_2H_5O_3^- + 3SO_4^{2-} \rightarrow 3HS^- + 6HCO_3^-$$

$$+H^+ \text{ (microbial)}$$

$$3HCO_3^- + 3Ca^{2+} \rightarrow 3CaCO_3 + 3H^+ \text{ (chemical)}$$

$$3HCO_3^- + 3H^+ \rightarrow 3CO_2 + 3H_2O \text{ (chemical)}$$

sum: $2C_2H_5O_3^- + 3SO_4^{2-} + 3Ca^{2+}$

$$\rightarrow 3HS^- + 3CaCO_3 + 3CO_2 + 3H_2O$$

Complete oxidizers precipitate 1.5 mol of $CaCO_3$ per mole of lactate oxidized.

5.3.3.2. Sulfate reduction—incomplete oxidation

$$2C_2H_5O_3^- + SO_4^{2-} \rightarrow HS^- + 2HCO_3^- + H^+$$

$$+2C_2H_3O_2^- \text{ (microbial)}$$

$$HCO_3^- + Ca^{2+} \rightarrow CaCO_3 + H^+ \text{ (chemical)}$$

$$HCO_3^- + H^+ \rightarrow CO_2 + H_2O \text{ (chemical)}$$

sum: $2C_2H_5O_3^- + SO_4^{2-} + Ca^{2+}$

$$\rightarrow HS^- + CaCO_3 + CO_2 + H_2O + H^+ + H_2O$$

Incomplete oxidizers precipitate only 0.5 mol of $CaCO_3$ per mole of lactate oxidized. SRB have also been implicated in the dissolution of gypsum ($CaSO_4 \cdot 2H_2O$) while respiring organic C and liberating CO_2 and Ca^{2+} (Kah et al., 2001).

5.3.4. Methanogenesis

Methanogenesis is energetically the least favorable mode of respiration. However, methane production is prevalent in freshwater and marine environments, and certainly very important in marine microbial mats (Visscher and van Gemerden, 1991; Hoehler et al., 2002). In addition to simple organic molecules, many methanogens use H_2 and CO_2 during respiration:

$$CO_2 + 4H_2 \rightarrow CH_4 + H_2O \text{ (microbial)}$$

$$HCO_3^- \rightarrow CO_2 + OH^- \text{ (chemical)}$$

$$Ca^{2+} + HCO_3^- \rightarrow CaCO_3 + H^+ \text{ (chemical)}$$

$$H^+ + OH^- \rightarrow H_2O \text{ (chemical)}$$

sum: $2HCO_3^- + 4H_2 + Ca^{2+} \rightarrow CaCO_3 + CH_4 + 2H_2O$

Therefore, methanogens precipitate 0.5 mol of $CaCO_3$ per mole of HCO_3^- reduced. It should be noted that other types of methanogenesis (using methylated compounds and/or acetate) have different effects on the $CaCO_3$ budget.

5.4. Chemolithoautotrophy

The oxidation of H_2, CO, Fe^{2+}, NH_4^+, and HS^- is energetically very favorable. In marine environments, sulfide and ammonium oxidation are the most important modes of chemolithotrophy. The sulfur cycle in microbial mats is quite complicated (van Gemerden, 1993), with many intermediates between sulfate and sulfide. Most of these intermediates can either be oxidized, reduced, or fermented.

5.4.1. Aerobic sulfide oxidation

Oxidation of sulfide can take place with O_2 or NO_3^- as electron acceptor.

$$HS^- + 2O_2 \rightarrow SO_4^{2-} + H^+ \text{ (microbial)}$$

$$CaCO_3 + H^+ \rightarrow HCO_3^- + Ca^{2+} \text{ (chemical)}$$

sum: $HS^- + 2O_2 + CaCO_3 \rightarrow SO_4^{2-} + Ca^{2+} + HCO_3^-$

Aerobic sulfide oxidation results in dissolution of 0.5 mol of $CaCO_3$ per mole of HS^- oxidized.

5.4.2. Anaerobic sulfide oxidation

$$5HS^- + 8NO_3^- \rightarrow 5SO_4^{2-} + 4N_2 + H_2O$$

$$+3OH^- \text{ (microbial)}$$

$$3HCO_3^- + 3Ca^{2+} \rightarrow 3CaCO_3 + 3H^+ \text{ (chemical)}$$

$$3OH^- + 3H^+ \rightarrow 3H_2O \text{ (chemical)}$$

sum: $5HS^- + 8NO_3^- + 3HCO_3^- + 3Ca^2$

$$\rightarrow 3CaCO_3 + 4N_2 + 4H_2O + 5SO_4^{2-}$$

Under anaerobic conditions, sulfide oxidation using nitrate produces 3/5 mol of $CaCO_3$ per mole of HS^- oxidized. However, the nitrate (NO_3^-) concentration is typically low in microbial mats, including modern marine stromatolites (Visscher et al., 1998; Paerl et al., 2000). Furthermore, when sulfate reduction is maximal, O_2 availability is typically low. Sulfide-oxidizing bacteria have a preference for HS^- over other reduced sulfur compounds (Kuenen and Beudeker, 1982). Under oxygen limitation, microbial sulfide oxidation yields thiosulfates, polysulfides, and polythionates (van den Ende and van Gemerden, 1993). When oxygen is replenished, these compounds can be further oxidized to sulfate. This results in a two-step sulfide oxidation in which the two individual reactions can be separated in time and/or space.

5.4.3. Two-step sulfide oxidation

$$8HS^- + 6.5O_2 \rightarrow S_8O_6^{2-} + 6OH^- + H_2O \,(\text{microbial})$$

$$6HCO_3^- + 6Ca^{2+} + \rightarrow 6CaCO_3 + 6H^+ \,(\text{chemical})$$

$$6OH^- + 6H^+ \rightarrow 6H_2O \,(\text{chemical})$$

$$sum: 8HS^- + 6.5O_2 + 6HCO_3^- + 6Ca^{2+}$$
$$\rightarrow S_8O_6^{2-} + 6CaCO_3 + 7H_2O$$

During this first step, 1.75 mol of $CaCO_3$ precipitates per mole of HS^- oxidized. The following reactions:

$$S_8O_6^{2-} + 9.5O_2 + 7H_2O \rightarrow 8SO_4^{2-} + 14H^+ \,(\text{microbial})$$

$$14CaCO_3 + 14H^+ \rightarrow 14HCO_3^- + 14Ca^{2+} \,(\text{chemical})$$

$$sum: S_8O_6^{2-} + 9.5O_2 + 14CaCO_3 + 7H_2O$$
$$\rightarrow 8SO_4^{2-} + 14HCO_3^- + 14Ca^{2+}$$

result in dissolution of 1.75 mol of $CaCO_3$ per mole of S oxidized.

Another example of the potential significance of functional diversity on carbonate minerals can be found among thiobacilli. Two groups of thiobacilli can be distinguished based on the amount of energy required for autotrophic growth (Kelly, 1982, 1988). The first group, which includes *Thiobacillus neapolitanus*, *Thiobacillus thiooxidans*, and *Thiobacillus*

versutus, has a low growth yield (i.e., requires significant metabolic energy to fix CO_2) and oxidizes approximately 4 mol of HS^- per mole of CO_2 fixed, resulting in a net loss of 0.5 mol of $CaCO_3$ per mole of HS^- oxidized. The second group (e.g., *Thiobacillus thioparus*, *Thiobacillus aquaesulis*, and *Thiobacillus denitrificans*) has a high growth yield (i.e., needs much less metabolic energy to fix CO_2) and requires ~2 mol of HS^- to be oxidized per mole of CO_2 fixed, resulting in a dissolution potential of 0.5 mol of $CaCO_3$ per mole of HS^- oxidized. As a result, changes in the spatial or temporal distribution of the physiological activities mediated by these two groups could be expected to impact the fate of $CaCO_3$ in sediment communities.

5.4.4. Ammonium oxidation

Ammonium oxidation is another relatively common chemolithotrophic reaction in the marine environment. This oxidation is carried out by two different microbial species: the first is oxidizing ammonium to nitrite; the second species is oxidizing nitrite to nitrate (Capone, 2000):

$$NH_4^+ + 1.5O_2 \rightarrow NO_2^- + H_2O + 2H^+ \,(\text{microbial})$$

$$2CaCO_3 + 2H^+ \rightarrow 2HCO_3^- + 2Ca^{2+} \,(\text{chemical})$$

$$sum: NH_4^+ + 1.5O_2 + 2CaCO_3$$
$$\rightarrow NO_2^- + 2HCO_3^- + 2Ca^{2+}$$

During this metabolic reaction, 2 mol of $CaCO_3$ are lost per mole of NH_4^+ oxidized. However, these inorganic redox reactions are often coupled to CO_2 fixation. Most sulfide and ammonium oxidizers use the Calvin cycle to fix CO_2:

$$CO_2 + 2NADH + 3ATP \rightarrow CH_2O - P + 2NAD^+$$
$$+ 3ADP + 2Pi^- \,(\text{microbial})$$

$$HCO_3^- \rightarrow CO_2 + OH^- \,(\text{chemical})$$

$$Ca^{2+} + HCO_3^- \rightarrow CaCO_3 \,(\text{chemical})$$

$$H^+ + OH^- \rightarrow H_2O \,(\text{chemical})$$

$$sum: 2HCO_3^- + Ca^{2+} \rightarrow CH_2O + CaCO_3$$

The net effect of CO_2 fixation is the precipitation of 1 mol of $CaCO_3$ per mole of CO_2 and 9 ATP equivalents needed. Under standard conditions, sulfide oxidation yields $\Delta G'\,°=-798$ kJ/mol and ammonium oxidation $\Delta G'\,°=-287$ kJ/mol. Assuming that CO_2 fixation using the Calvin cycle requires 405 kJ when 100% efficient (1 ATP=45 kJ/mol), and that the actual efficiency of this process is 50% (Kelly, 1982) at best (requiring 810 kJ), the following net effects of chemolithoautotrophy on $CaCO_3$ can be expected: (1) complete aerobic sulfide oxidation coupled to CO_2 fixation produces 0.5 mol of $CaCO_3$ per mole of HS^-

oxidized; (2) incomplete aerobic sulfide oxidation to polythionates and CO_2 fixation generate approximately 1.2 mol of $CaCO_3$ per mole of HS^- oxidized; and (3) ammonium oxidation to nitrite coupled to CO_2 fixation yields a loss of ~2.7 mol of $CaCO_3$ per mole of NH_4^+ oxidized.

6. Concluding remarks

We have attempted in this brief review to assess the effects that metabolic processes typically associated

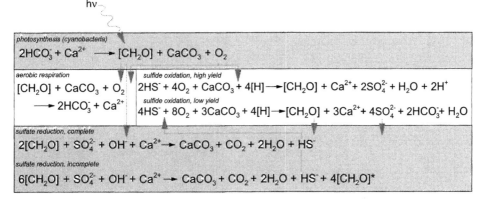

Examples:

Assuming that all C is oxidized through complete SR (SRB(c)) and HS^- is reoxidized by high yield SOB (SOB(hi); see text below), the effect on net $CaCO_3$ precipitation (in parentheses) is:

CYN	$3CO_2 + 3H_2O \rightarrow 3CH_2O + 3O_2$	(+3)
SRB(c)	$4CH_2O + 2SO_4^{2-} \rightarrow 4CO_2 + 2HS^- + 2H_2O + 2OH^-$	(+2)
SOB (hi)	$2HS^- + 4O_2 + CO_2 + 4[H] \rightarrow CH_2O + 2SO_4^{2-} + H_2O + 2H^+$	(-1)

Overall: $+4\ CaCO_3\ (1.33\ per\ CO_2\ fixed\ by\ CYN)$

Assuming that 50% of the C is oxidized through incomplete SR (SRB(i)) and HS^- is reoxidized By low yield SOB (SOB(lo)), the effect on net $CaCO_3$ precipitation (in parentheses) is:

CYN	$47CO_2 + 47H_2O \rightarrow 47CH_2O + 47O_2$	(+47)
HET	$34CH_2O + 24O_2 \rightarrow 24CO_2 + 24H_2O$	(-24)
SRB(i)	$24CH_2O + 4SO_4^{2-} \rightarrow 8CO_2 + 4HS^- + 4H_2O + 4OH^- + 16CH_2O^*$	(+2)
SOB (lo)	$4HS^- + 8O_2 + CO_2 + 4[H] \rightarrow CH_2O + 4SO_4^{2-} + H_2O + 4H^+$	(-3)

Overall: $+24\ CaCO_3\ (0.51\ per\ CO_2\ fixed\ by\ CYN)$

Fig. 3. Metabolic reactions of key functional groups coupled to geochemical precipitation/dissolution reactions in a microbial mat (equations include the effect on $CaCO_3$ balance; reactions assumed in a carbonate-buffered system). Dark boxes designate precipitation and light boxes dissolution. [CH_2O] denotes "generic" organic carbon; [CH_2O]* is a different C compound (i.e., [CH_2O] is lactate and [CH_2O]* is acetate in incomplete sulfate reduction). Example calculations show the overall effect on $CaCO_3$ precipitation of two hypothetical C flow scenarios.

with microbial mats have on carbonate deposition and dissolution. The lithifying marine stromatolites at Highborne Cay, Bahamas, were used as our model, but the general principles can be applied to other systems (e.g., organic-rich tidal flat, brackish estuary). We have illustrated that these effects can be chemically predicted in terms of net carbonate precipitation and, therefore, net impact on the geologic record (Visscher et al., 1998; Castanier et al., 1999; Vasconcelos and McKenzie, 1997). Processes such as photosynthesis and sulfate reduction result in net precipitation, while aerobic respiration and sulfide oxidation result in net dissolution (Fig. 3). Many studies have focused on photosynthesis and aerobic respiration as the key processes affecting precipitation (Krumbein et al., 1977; Chafetz and Buczynski, 1992; Pinckney et al., 1995). However, in microbial mats, photosynthesis and aerobic respiration are tightly coupled in space and time, and the coupling of these processes results in no net precipitation. In contrast, sulfate reduction and sulfide oxidation are more likely to be temporally and spatially separated in microbial mats due to the different environmental requirements of these processes (Fig. 3). Their combined activities should therefore result in "hot spots" of precipitation and dissolution (Visscher et al., 1998, 2002). Such a scenario has been observed in the Highborne Cay stromatolites where high rates of sulfate reduction near the surface coincide with microcrystalline $CaCO_3$ precipitation (Visscher et al., 2000) and the greatest abundance of both SRB and SOB (Visscher et al., 1998).

Precipitation outside the cell may contribute to the generation of a proton motive force (see above), but also seems deleterious for the individual cell, since it leads to entombment. As a collective action of a microbial community, however, precipitation of a microcrystalline $CaCO_3$ layer could be advantageous: modern marine stromatolites in Highborne Cay exist in a nutrient-poor environment and a surface $CaCO_3$ crust provides a seal, thereby limiting loss of nutrients to the environment. Furthermore, rather extreme hydrodynamic conditions (high, fluctuating flow in an intertidal/subtidal beach environment) require a significant mechanical strength, and the need to maintain a structure that is minimally impacted by burial due to sand movement requires that the stromatolite mats grow in an upwards direction. The alternating soft layers (bound and trapped ooids in a type 1 community) and hard $CaCO_3$ precipitates (type 2 and type 3 communities) provide exactly that. Finally, the hard surface of the stromatolites somewhat prevents the settlement of eukaryotic organisms.

Overall, our understanding of the effects of metabolic processes upon lithification is limited, and even less is known about how specific metabolic activities (e.g., differences of daytime and nighttime metabolism; see Table 1) and physiological diversity within functional groups can influence the overall metabolism of the community and the formation of mineral products (see Fig. 3; example calculations). Although oxygenic and anoxygenic photoautotrophic bacteria may be considered primary producers, their impact on net carbonate precipitation is different. While the sum total of the sulfate-reducing bacteria populations may be considered a common guild, the predominance of one physiological type over another (i.e., incomplete oxidizers vs. complete oxidizers) can also impact net carbonate precipitation. More information is needed as to the temporal and spatial relationships of metabolic processes and the impact that the coupling of these processes has on mineral production by microbial communities. This may provide insights into carbonate and mineral precipitation processes in general as well as shed light on mechanisms of microbe–mineral interactions.

Acknowledgements

The authors would like to thank the members of RIBS. This work is supported by the NSF Biocomplexity Program. This is RIBS contribution no. 26.

References

Canfield, D.E., Des Marais, D.J., 1993. Biogeochemical cycles of carbon, sulfur, and free oxygen in a microbial mat. Geochim. Cosmochim. Acta 57, 3971–3984.

Canfield, D.E., Jørgensen, B.B., Fossing, H., Glud, R., Gundersen, J., Ramsing, N.B., Thamdrup, B., Hansen, J.W., Nielsen, L.P., Hall, P.O.J., 1993. Pathways of organic carbon oxidation in three continental margin sediments. Mar. Geol. 113, 27–40.

Capone, D.G., 2000. The marine nitrogen cycle. In: Kirchman, D. (Ed.), Marine Microbial Ecology. J. Wiley, New York, pp. 455–493.

Castanier, S., Metayer-Levrel, G., Perthuisot, J.-P., 1999. Ca-carbonates precipitation and limestone genesis—the microbiologist's point of view. Sediment. Geol. 126, 9–23.

Castanier, S., Le Metayer-Levrel, G., Perthuisot, J.-P., 2000. Bacterial roles in the precipitation of carbonate minerals. In: Riding, R.E., Awramik, S.M. (Eds.), Microbial Sediments. Springer, Berlin, pp. 32–39.

Chafetz, H.S., Buczynski, C., 1992. Bacterially induced lithification of microbial mats. Palaios 7, 277–293.

Cohen, Y., Castenholz, R.W., Halvorson, H.O. (Eds.), Microbial mats: Stromatolites. Alan Liss, New York. 498 pp.

Des Marais, D.J., 1990. Microbial mats and the early evolution of life. Trends Ecol. Evol. 5, 140–144.

Des Marais, D.J., 1997. Long-term evolution of the biogeochemical carbon cycle. In: Banfield, J.F., Nealson, K.H. (Eds.), Geomicrobiology: Interactions between Microbes and Minerals, Mineralogical Society of America, Reviews in Mineralogy, vol. 35., pp. 427–448.

De Vrind-deJong, E.W., de Vrind, J.P.M., 1998. Algal deposits of carbonates and silicates. In: Banfield, J.F., Nealson, K.H. (Eds.), Geomicrobiology: Interactions between Microbes and Minerals, Mineralogical Society of America, Reviews in Mineralogy, vol. 35., pp. 267–307.

Dill, R.F., Shinn, E.A., Jones, A.T., Kelly, K., Steinen, R.P., 1986. Giant subtidal stromatolites forming in normal salinity water. Nature 324, 55–58.

Dravis, J.J., 1983. Hardened subtidal stromatolites, Bahamas. Science 219, 385–386.

Ferdelman, T.G., Fossing, H., Neumann, K., Schulz, H.D., 1999. Sulfate reduction in surface sediments of the southeast Atlantic continental margin between 15°38′ S and 27°57′ S (Angola and Namibia). Limnol. Oceanogr. 44, 650–661.

Folk, R.L., Chafetz, H.F., 2000. Bacterially induced microscale and nanoscale carbonate precipitates. In: Riding, R.E., Awramik, S.M. (Eds.), Microbial Sediments. Springer-Verlag, Berlin, pp. 40–49.

Fründ, C., Cohen, Y., 1992. Diurnal cycles of sulfate reduction under oxic conditions in microbial mats. Appl. Environ. Microbiol. 58, 70–77.

Golubic, S., 1991. Modern stromatolites—a review. In: Riding, R. (Ed.), Calcareous Algae and Stromatolites. Springer, Berlin, pp. 541–561.

Grotzinger, J.P., Knoll, A.H., 1999. Stromatolites in Precambrian carbonates: evolutionary mileposts or environmental dipsticks? Annu. Rev. Earth Planet. Sci. 27, 313–358.

Hoehler, T.M., Albert, D.B., Alperin, M.J., Bebout, B.M., Martens, C.S., Des Marais, D.J., 2002. Comparative ecology of H_2 cycling in sedimentary and phototrophic ecosystems. Antonie van Leeuwenhoek 81, 575–585.

Jørgensen, B.B., 2001. Space for hydrogen. Nature 412, 286–289.

Jørgensen, B.B., Cohen, Y., 1977. Solar Lake (Sinai): 5. The sulfur cycle of the benthic cyanobacterial mat. Limnol. Oceanogr. 22, 657–666.

Jørgensen, B.B., Revsbech, N.P., Cohen, Y., 1983. Photosynthesis and structure of benthic microbial mats: microelectrode and SEM studies of four cyanobacterial communities. Limnol. Oceanogr. 28, 1075–1093.

Kah, L.C., Lyons, T.W., Chesley, J.T., 2001. Geochemistry of a 1.2 Ga carbonate–evaporite succession, northern Baffin and Bylot Islands: implications for Mesoproterozoic marine evolution. Precambrian Res. 111, 203–234.

Kelly, D.P., 1982. Biochemistry of the chemolithotrophic oxidation of inorganic sulphur. Philos. Trans. R. Lond., B 298, 499–528.

Kelly, D.P., 1988. Oxidation of sulfur compounds. In: Cole, J.A., Ferguson, S.J. (Eds.), The Nitrogen and Sulphur Cycles. Cambridge Univ. Press, Cambridge, UK, pp. 65–98.

Kempe, S., Kazmierczak, J., 1994. The role of alkalinity in the evolution of ocean chemistry, organization of living systems and biocalcification processes. Bull. Inst. Oceanogr. (Monaco) 13, 61–117.

Krumbein, W.E., 1983. Stromatolites—the challenge of a term in space and time. Precambrian Res. 20, 493–531.

Krumbein, W.E., Cohen, Y., Shilo, M., 1977. Solar Lake (Sinai): 4. Stromatolitic cyanobacterial mats. Limnol. Oceanogr. 22, 635–656.

Krumbein, W.E., Paterson, D.M., Zavarzin, G. (Eds.), 2003. Fossil and Recent Biofilms: A Natural History of Life on Planet Earth. Kluwer Scientific Publishers, Dordrecht, The Netherlands. 482 pp.

Kuenen, J.G., Beudeker, R.F., 1982. Microbiology of thiobacilli and other sulphur oxidizing autotrophs, mixothrophs and heterotrophs. Philos. Trans. R. Soc. Lond., B 298, 473–497.

Lovley, D.R., Coates, J.D., 2000. Novel forms of anaerobic respiration of environmental relevance. Curr. Opin. Microbiol. 3, 252–256.

Macintyre, I.G., Prufert-Bebout, L., Reid, R.P., 2000. The role of endolithic cyanobacteria in the formation of lithified laminae in Bahamian stromatolites. Sedimentology 47, 915–921.

McConnaughey, T.A., Whelan, J.F., 1997. Calcification generates protons for nutrient and bicarbonate uptake. Earth-Sci. Rev. 42, 95–117.

Nealson, K.H., Stahl, D.A., 1997. Microorganisms and biogeochemical cycles: what can we learn from layered microbial communities? In: Banfield, J.F., Nealson, K.H. (Eds.), Geomicrobiology: Interactions between Microbes and Minerals, Mineralogical Society of America, Reviews in Mineralogy, vol. 35., pp. 5–34.

Nicholson, J.A.M., Stolz, J.F., Pierson, B.K., 1987. Structure of a microbial mat in a saltmarsh. FEMS Microbiol. Ecol. 45, 343–364.

Paerl, H.W., Pinckney, J.L., Steppe, T.F., 2000. Cyanobacterial–bacterial mat consortia: examining the functional unit of microbial survival and growth in extreme environments. Environ. Microbiol. 2, 11–26.

Pinckney, J., Paerl, H.W., Reid, R.P., Bebout, B.M., 1995. Ecophysiology of stromatolitic mats, Stocking Island, Exuma Cays, Bahamas. Microb. Ecol. 29, 19–37.

Reid, R.P., Brown, K.M., 1991. Intertidal stromatolites in a fringing Holocene reef complex, Bahamas. Geology 19, 15–18.

Reid, R.P., Macintyre, I.G., Steneck, R.S., Browne, K.M., Miller, T.E., 1995. Stromatolites in the Exuma Cays, Bahamas: uncommonly common. Facies 33, 1–18.

Reid, R.P., Visscher, P.T., Decho, A.W., Stolz, J.F., Bebout, B.M., Macintyre, I.G., Paerl, H.W., Pinckney, J.L., Prufert-Bebout, L.,

Steppe, T.F., DesMarais, D.J., 2000. The role of microbes in the accretion, lamination and early lithification of modern marine stromatolites. Nature 406, 989–992.

Revsbech, N.P., 1984. Analysis of microbial mats by use of electrochemical microsensors: recent advances. In: Stal, L.J., Caumette, P. (Eds.), Microbial Mats, Structure, Development, and Environmental Significance, NATO ASI Series. Series G : Ecological Sciences, vol. 3. Springer-Verlag, Berlin, pp. 135–148.

Revsbech, N.P., Madsen, B., Jørgensen, B.B., 1986. Oxygen production and consumption in sediments determined at high spatial resolution by computer simulation of oxygen microelectrode data. Limnol. Oceanogr. 31, 293–304.

Riding, R., 2000. Microbial carbonates: the geological record of calcified bacterial–algal mats and biofilms. Sedimentology 47, 179–214.

Rivadeneyra, M.-A., Delgado, R., Delgado, G., Ferrer, M., Del Moral, A., Ramos-Cormenzaqna, A., 1993. Carbonate precipitation by *Bacillus* sp. isolated from saline soils. Geomicrobiol. J. 11, 175–184.

Rivadeneyra, M.-A., Delgado, G., Soriano, M., Ramos-Cormenzaqna, A., Delgado, A., 1999. Biomineralization of carbonates by *Marinococcus albus* and *Marinococcus halophilus* isolated from Salar de Atacame (Chile). Curr. Microbiol. 39, 53–57.

Robbins, L.L., Yates, K.K., 1998. Production of carbonate sediments by a unicellular green alga. Am. Mineral. 83, 1503–1509.

Schauder, R., Preuß, A., Jetten, M., Fuchs, G., 1989. Oxidative and reductive acetyl CoA/carbon monoxide dehydrogenase pathway in *Desulfobacterium autotrophicum*: 2. Demonstration of the enzymes of the pathway and comparison of CO dehydrogenase. Arch. Microbiol. 151, 84–89.

Schopf, J.W. (Ed.), 1983. Earth's Earliest Biosphere, Its Origin and Evolution. Princeton University Press, Princeton, NJ.

Spormann, A.M., Thauer, R.K., 1988. Anaerobic acetate oxidation to CO_2 by *Desulfotomaculum acetoxidans*. Demonstration of enzymes required for the operation of an oxidative acetyl-CoA/carbon monoxide dehydrogenase pathway. Arch. Microbiol. 150, 374–380.

Stal, L.J., van Gemerden, H., Krumbein, W.E., 1985. Structure and development of a benthic marine microbial mat. FEMS Microbiol. Ecol. 31, 111–125.

Steneck, R.S., Miller, T.E., Reid, R.P., Macintyre, I.G., 1996. Ecological controls on stromatolite development in a modern reef environment: a test of the ecological refuge paradigm. Carbonates Evaporites 13, 48–65.

Steneck, R.S., Macintyre, I.G., Reid, R.P., 1997. Unique algal ridge systems of Exuma Cays, Bahamas. Coral Reefs 16, 29–37.

Stolz, J.F., 2003. Structure of marine biofilms, flat laminated mats and modern marine stromatolites. In: Krumbein, W.E., Paterson, D.M., Zavarzin, G. (Eds.), Fossil and Recent Biofilms: A Natural History of Life on Planet Earth. Kluwer Scientific Publishers, Dordrecht, The Netherlands, pp. 65–76.

Stolz, J.F., Botkin, D.B., Dastoor, M.N., 1988. The integral biosphere. In: Rambler, M.B., Margulis, L., Fester, R. (Eds.), Global Ecology: Towards a Science of the Biosphere. Academic Press, Boston, MA, pp. 31–50.

Stolz, J.F, Feinstein, T.N., Salsi, J., Visscher, P.T., Reid, R.P., 2001. Microbial role in sedimentation and lithification in a modern marine stromatolite: a TEM perspective. Am. Mineral. 86, 826–833.

van den Ende, F.P., van Gemerden, H., 1993. Sulfide oxidation under oxygen limitation by a *Thiobacillus thioparus* isolated from a marine microbial mat. FEMS Microbiol. Ecol. 13, 69–78.

van Gemerden, H., 1993. Microbial mats: a joint venture. Mar. Geol. 113, 3–25.

Vasconcelos, C., McKenzie, J.A., 1997. Microbial mediation of modern dolomite precipitation and diagenesis under anoxic conditions (Lagoa Vermelha, Rio de Janeiro, Brazil). J. Sediment. Res. 67, 378–390.

Visscher, P.T., van den Ende, F.P., 1994. Diel and spatial fluctuations of sulfur transformations. In: Stal, L.J., Caumette, P. (Eds.), Microbial Mats. Structure, Development and Environmental Significance. Springer-Verlag, Berlin, pp. 353–360.

Visscher, P.T., van Gemerden, H., 1991. Production and consumption of dimethylsulfoniopropionate in marine microbial mats. Appl. Environ. Microbiol. 57, 3237–3242.

Visscher, P.T., Prins, R.A., van Gemerden, H., 1992. Rates of sulfate reduction and thiosulfate consumption in a marine microbial mat. FEMS Microbiol. Ecol. 86, 283–294.

Visscher, P.T., Reid, R.P., Bebout, B.M., Hoeft, S.E., Macintyre, I.G., Thompson Jr., J.A., 1998. Formation of lithified micritic laminae in modern marine stromatolites (Bahamas): the role of sulfur cycling. Am. Mineral. 83, 1482–1494.

Visscher, P.T., Reid, R.P., Bebout, B.M., 2000. Microscale observation of sulfate reduction: evidence of microbial activity forming lithified micritic laminae in modern marine stromatolites. Geology 28, 919–922.

Visscher, P.T., Surgeon, T.M., Hoeft, S.E., Bebout, B.M., Thompson Jr., J.A., Reid, R.P., 2002. Microelectrode studies in modern marine stromatolites: unraveling the Earth's past? In: Taillefert, M., Rozan, T. (Eds.), ACS symposium series 220. Electrochemical Methods for the Environmental Analysis of Trace Metal Biogeochemistry. Cambridge Univ. Press, New York, NY, pp. 265–282.

Ward, D.M., Ferris, M.J., Nold, S.C., Bateson, M.M., 1998. A natural view of microbial diversity within the hotspring cyanobacterial mat communities. Microbiol. Mol. Biol. Rev. 62, 1353–1370.

Widdel, F., 1988. The microbiology and ecology of sulfate- and sulfur-reducing bacteria. In: Zehnder, A.J.B. (Ed.), Biology of Anaerobic Microorganisms. Wiley and Sons, New York, NY, pp. 469–585.

Available online at www.sciencedirect.com

ELSEVIER

Palaeogeography, Palaeoclimatology, Palaeoecology 219 (2005) 101–115

www.elsevier.com/locate/palaeo

Geobiology of microbial carbonates: metazoan and seawater saturation state influences on secular trends during the Phanerozoic

Robert Riding[a,*], Liyuan Liang[b,c]

[a]*School of Earth, Ocean and Planetary Sciences, Cardiff University, Cardiff CF10 3YE, United Kingdom*
[b]*School of Engineering, Cardiff University, Cardiff CF24 0YF, United Kingdom*
[c]*Oak Ridge National Laboratory, Oak Ridge, TN 37831-6038, USA*

Received 28 September 2004; accepted 22 November 2004

Abstract

Microbial carbonates are long-ranging, essentially bacterial, aquatic sediments. Their calcification is dependent on ambient water chemistry and their growth is influenced by competition with other organisms, such as metazoans. In this paper, these relationships are examined by comparing the geological record of microbial carbonates with metazoan history and secular variations in $CaCO_3$ saturation state of seawater. Marine abundance data show that microbial carbonates episodically declined during the Phanerozoic Eon (past 545 Myr) from a peak 500 Myr ago. This abundance trend is generally inverse to that of marine metazoan taxonomic diversity, supporting the view that metazoan competition has progressively limited the formation of microbial carbonates. Lack of empirical values concerning variables such as seawater ionic composition, atmospheric partial pressure of CO_2, and pH currently restricts calculation of $CaCO_3$ saturation state for the Phanerozoic as a whole to the use of modeled values. These data, together with palaeotemperature data from oxygen isotope analyses, allow calculation of seawater $CaCO_3$ saturation trends. Microbial carbonate abundance shows broad positive correspondence with calculated seawater saturation state for $CaCO_3$ minerals during the interval 150–545 Myr ago, consistent with the likelihood that seawater chemistry has influenced the calcification and therefore accretion and preservation of microbial carbonates. These comparisons suggest that both metazoan influence and seawater saturation state have combined to determine the broad pattern of marine microbial carbonate abundance throughout much of the Phanerozoic. In contrast, for the major part of the Precambrian it would seem reasonable to expect that seawater saturation state, together with microbial evolution, was the principal factor determining microbial carbonate development. Interrelationships such as these, with feedbacks influencing organisms, sediments, and the environment, are central to geobiology.

Keywords: metazoan diversity; microbial carbonates; Phanerozoic; saturation state; seawater chemistry

---

* Corresponding author. Tel.: +44 29 20874329; fax: +44 29 2087432.
  E-mail address: riding@cardiff.ac.uk (R. Riding).

0031-0182/$ - see front matter © 2004 Elsevier B.V. All rights reserved.
doi:10.1016/j.palaeo.2004.11.018

1. Introduction

Throughout the 545 Myr of the Phanerozoic Eon, marine organisms have extensively used the calcium carbonate minerals aragonite and calcite to create organic skeletons. From much earlier times additional deposits have resulted from algal and bacterial processes, such as photosynthesis and sulphate reduction, that promote carbonate precipitation in microenvironments adjacent to cells. In marine environments over geological timescales these processes of $CaCO_3$ biomineralization have had great significance for the fossil record. They have also contributed to the generation of large quantities of carbonate sediment. The resulting long-lived accumulations of limestones and dolostones constitute important sedimentary records in oceanic and crustal rocks. This carbon reservoir, dependent upon the presence of a hydrosphere on Earth, is much larger than that in modern biomass and fossil fuels (Rubey, 1951; Holland, 1978; Stumm and Morgan, 1996). Such large-scale sequestration of CO_2 in carbonate rocks dominates the long-term carbon cycle and is therefore likely to have been a key influence on Earth's climate (Walker et al., 1981; Berner et al., 1983; Kasting and Catling, 2003). It follows that the relationship between water chemistry and biomineralization in aquatic organisms is a major research area for geobiology at the interface between palaeobiology and carbonate sedimentology, with broad implications for Earth's surface environment.

Microbial carbonates have particular relevance in this field for two reasons. Firstly, microbial carbonates are very long ranging, with probably the longest geological record and most extensive facies distribution of any biogenic sediment (Riding, 2000). Secondly, microbial carbonates are very susceptible to environmental influence, both for the growth of the micro-organisms that localize them and for the precipitation processes that determine their sedimentary accretion and geological preservation. One of the most widely appreciated features of the secular distribution of microbial carbonates is the suggestion that they have declined in abundance from a peak in the Proterozoic. Phanerozoic stromatolite decline was noted by Fischer (1965) and Cloud and Semikhatov (1969) but recognition of its longer term pattern, and possible link to metazoan competition, resulted from

study of gastropod grazing on the Bahama Banks (Garrett, 1970) coupled with compilation of late Proterozoic reduction in stromatolite diversity (Awramik, 1971). Nonetheless, decline in microbial carbonates has also long been linked to changes in calcification (Fischer, 1965; Monty, 1973, 1977; Serebryakov and Semikhatov, 1974; Gebelein, 1976; Grotzinger, 1990), ultimately related to seawater chemistry.

Here we examine the Phanerozoic distribution of microbial carbonates in general and consider to what extent it may be possible to relate their secular abundance to competition with metazoans and to changes in seawater chemistry. This enquiry necessarily involves a number of assumptions, e.g., that metazoan diversity may be a proxy for metazoan abundance and, therefore, competition in its broadest sense, and also that currently available modeled estimates of Phanerozoic atmospheric and seawater composition may provide a window into changes in past seawater saturation state. Not the least aim of this study is to encourage further work that will lead to more robust information on which these interpretations can be based. Our approach to these questions is therefore essentially to compare secular variation in microbial carbonate abundance, metazoan abundance, and seawater saturation state for the Phanerozoic.

2. Microbial carbonates

Microbial carbonates can be regarded mainly as products of bacterial, and also algal, processes that promote the precipitation of $CaCO_3$ minerals such as aragonite and calcite, and trap sedimentary particles. Some microbial carbonates are difficult to confidently recognize in ancient carbonate sediments even though they are likely to have been volumetrically important. These may include micritic particles precipitated in the water column by pelagic microbes (Thompson, 2000) and those produced by the disintegration on the seafloor of calcified bacteria such as *Girvanella* (Pratt, 2001). In contrast, benthic microbial carbonates (Burne and Moore, 1987) produced by calcified microbial mats and biofilms are much more readily recognizable. These deposits include forms such as stromatolites and thrombolites that can be regarded as reef-builders along with skeletal algae and invertebrates.

Biocalcification in aquatic organisms ranges from biologically controlled, where organisms closely regulate their calcification, to biologically induced, where calcification is metabolically mediated but dependent on ambient water chemistry (Lowenstam, 1981; Leadbeater and Riding, 1986; Mann, 2001). In bacteria, a wide range of processes can lead to localized pH increase which affects carbonate speciation and favours $CaCO_3$ precipitation. In microbial mats these processes differ significantly, although over minute distances, between surface mat and sub-mat microenvironments. In oxic illuminated mat-surface environments, cyanobacterial photosynthetic uptake of CO_2 and/or HCO_3^- raises pH in ambient waters and in protective mucilaginous sheaths around the cyanobacterial cells, promoting calcification (Pentecost and Riding, 1986; Merz, 1992; Merz-Preiß, 2000; Arp et al., 2001). In submat environments, anaerobic degradation of photosynthetic mat material by organotrophic bacteria can result in pH rise and $CaCO_3$ precipitation. These latter energy yielding processes include ammonification, denitrification, and sulphate reduction (references in Riding, 2000). In all these cases, bacterial calcification is not obligate, and it is dependent on suitable environmental and micro-environmental conditions. It is therefore biologically induced. This limits it to environments, whether in open water bodies or in sediment pores, where ambient waters are significantly oversaturated with respect to $CaCO_3$ minerals.

3. Phanerozoic abundance of microbial carbonates

Until recently, data regarding microbial carbonate Phanerozoic abundance have been scarce. Awramik (1971) used measures of diversity, based on form–genera and form–species, to show marked Neoproterozoic decline in stromatolites. However, comparable data are sparse for the Phanerozoic although it is well-known that microbial carbonates continued to be prominent, especially in reefs (Pratt, 1982; Webb, 2001). Riding (1992, 1993) qualitatively assessed Phanerozoic variation in the abundance of marine calcified cyanobacteria and found that

their abundance can broadly coincide with that of ooids and marine cements, e.g., in the Cambrian–Early Ordovician, Late Devonian, and Permian–Triassic. More recently, improved data sets have been compiled for microbial carbonate abundance. On the basis of extensive literature review, Arp et al. (2001) plotted 864 reported occurrences of Phanerozoic calcified marine cyanobacteria, normalized to 10 Myr intervals. Kiessling and Flügel (2002) collected semiquantitive data from a literature search of 3050 global Phanerozoic reef sites to establish a Paleoreef Database (PaleoReefs). Using this data set, Kiessling (2002) reduced the fourteen reef-builder categories to nine, one of which is microbes, to show the 'number of reefs in which a particular reef-building group is dominant' and plotted these data against thirty-two 'supersequence' time-slices (Kiessling, 2002, Fig. 16). Riding (in press) replotted these microbial reef data using time ranges for each supersequence based on Golonka and Kiessling's, 2002) evaluation of absolute ages. Kiessling's (2002) reefal microbial

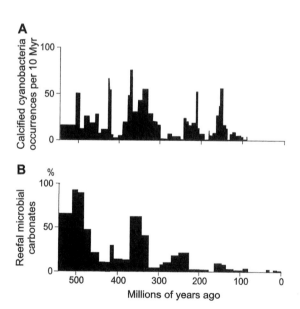

Fig. 1. Phanerozoic secular distribution of microbial carbonates. (A) Reported occurrences of calcified cyanobacteria per 10 Myr Phanerozoic (from Arp et al., 2001). (B) Relative abundance of reefal microbial carbonates replotted from Kiessling (2002, Fig. 16) using Golonka and Kiessling's (2002, Fig. 1, pp. 13–18) time intervals for each supersequence supersequences (see Riding, in press).

carbonate abundance data, together with occurrences of calcified cyanobacteria compiled by Arp et al. (2001), are shown in Fig. 1. Kiessling's (2002) data show a steep Early Palaeozoic decline in reefal microbial carbonate relative abundance that becomes a more gradual decline towards the present-day. However, marked oscillations are superimposed on this trend, with successively diminishing peaks of relative abundance at about 500, 350, 240, and 150 Myr ago (Riding, in press). In contrast, Arp et al.'s (2001) data show more occurrences of marine calcified cyanobacteria prior to ~300 Myr ago, fewer 300–100 Myr ago, and hardly any in the past 100 Myr (Fig. 1). There are peaks in occurrence about 500, 420, 370, 330, 210, and 150 Myr ago. These peaks are similar in size except near 420 and 370 Myr, when they are a little larger.

4. Metazoan diversity

Marine metazoan diversity, as compiled by Sepkoski (1997) shows exponential Early Cambrian to Late Ordovician increase, lesser overall change until marked Late Permian decline ~250 Myr ago, and then increase to the present-day, with the general diversity level of the Palaeozoic being exceeded by ~100 Myr ago during the Cretaceous (Fig. 2). This diversification resulted in the number of genera increasing from zero to ~2500 over the Phanerozoic as a whole.

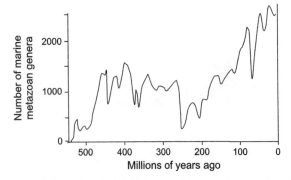

Fig. 2. Marine metazoan generic diversity data (from Sepkoski, 1997) showing exponential Early Cambrian to Late Ordovician increase, lesser overall change until marked Late Permian decline ~250 Myr ago, followed by increase to the present-day.

5. Seawater chemistry

Fundamental controls on marine carbonate precipitation through time can be expected to include saturation state of seawater with respect to aragonite and calcite (Kempe and Kazmierczak, 1990; Opdyke and Wilkinson, 1990; Webb, 2001) and biologically controlled biomineralization processes (Bosscher and Schlager, 1993). Mg^{2+}/Ca^{2+} ratio affects the kinetics of calcite precipitation, and is widely believed to have produced alternate calcite and aragonite seas through time (Sandberg, 1975, 1983; Stanley and Hardie, 1998). The relative importance of these controls on carbonate sedimentation for the Phanerozoic as a whole is still unclear (Opdyke and Wilkinson, 1990; Bosscher and Schlager, 1993; Stanley and Hardie, 1998; Leeder, 1999). Suggested broad stability of ocean water composition during the past 1000 Myr (Holland, 1978) does not preclude variations in Ca^{2+}, Mg^{2+}, carbonate alkalinity, seawater pH, and temperature (e.g., Berner et al., 1983; Sandberg, 1983; Karhu and Epstein, 1986; Morse and Mackenzie, 1990; Hardie, 1996; Berner and Kothavala, 2001; Stanley and Hardie, 1998; Veizer et al., 1999) that could significantly affect the precipitation of aragonite and calcite. In particular, such variations in seawater chemistry would influence saturation states with respect to these minerals (Arvidson et al., 2000; Riding and Liang, 2004, in press). Variation in marine abundances of aragonite and calcite has been recognized throughout the Phanerozoic (Sandberg, 1983; Burton and Walter, 1987; Wilkinson and Algeo, 1989; Mackenzie and Morse, 1992), but calculation of the saturation state of seawater for $CaCO_3$ minerals is more problematic, particularly prior to 100 Myr ago (Riding and Liang, 2004, in press).

6. Seawater saturation state

Saturation state with respect to $CaCO_3$ minerals, defined by $\Omega = \{Ca^{2+}\}\{CO_3^{2-}\}/K_{sp}$ (where Ω signifies saturation ratio, and $\{\ \}$ signifies activity), is determined by activity product of calcium and carbonate over the solubility constant at a given temperature and pressure condition (Stumm and Morgan, 1996). Arvidson et al. (2000, p. 2) calculated saturation for calcite and dolomite for the past 100 Myr using partial pressure of CO_2 (p_{CO2}) and seawater composition data of Berner et al. (1983) and showed that saturation state declined towards the present-day. Calculation of saturation state for periods prior to 100 Myr has been hindered by lack of estimates of oceanic carbonate speciation. In order to constrain the system it is necessary to know at least two of the four key parameters: carbonate alkalinity, p_{CO2}, dissolved inorganic carbon (DIC), and pH (Stumm and Morgan, 1996). Estimated values for p_{CO2} can be derived from the model named GEOCARB III (Berner and Kothavala, 2001), but additional information is needed to estimate carbonate species for the Phanerozoic. In the absence of pH data, Riding and Liang (2004, in press) determined carbonate concentration by estimating DIC from the correlation established for p_{CO2} and HCO_3^- for the last 100 Myr by Lasaga et al. (1985) who revised data originally obtained by Berner et al. (1983) using the so-called BLAG model.

Using this approach, together with published estimates of other major ions (Hardie, 1996; Stanley and Hardie, 1998), Riding and Liang (2004, in press) calculated $\Omega_{aragonite}$, $\Omega_{calcite}$, and $\Omega_{dolomite}$ for the Phanerozoic assuming chemical equilibrium. A constant temperature (15 °C) was used in the calculations. They found that the resulting calculated trends of $\Omega_{aragonite}$ and $\Omega_{calcite}$ exhibit broad positive covariation with carbonate platform accretion rate (see Bosscher and Schlager, 1993) and with periods of abundance of microbial and nonskeletal $CaCO_3$ precipitates (Riding and Liang, 2004, in press). Riding and Liang (2004, in press) concluded that these relationships suggest a primary control by seawater chemistry on limestone formation in general and microbial carbonates in particular. In addition, they found that higher values of saturation state and limestone accumulation correspond with 'calcite seas', and lower values with 'aragonite seas'.

7. Development of saturation ratio estimates

Previously, we calculated past seawater saturation state for major carbonate minerals ($\Omega_{aragonite}$, $\Omega_{calcite}$, and $\Omega_{dolomite}$) with a stoichiometric chemical equilibrium approach (Riding and Liang, 2004, in press). Numerical calculations were performed using the PHREEQC code, for phase reaction equilibria written in the C programming language (version 2, Parkhurst and Appelo, 1999). Estimation of activity coefficients in PHREEQC employs the extended Debye–Hückel limiting law (Debye and Hückel, 1923) for aqueous species. However, for concentrated solutions, including seawater and brines, Pitzer's (1991) approximation is deemed more appropriate to account for ionic interactions and to compute activity coefficients (Stumm and Morgan, 1996). Furthermore, Riding and Liang (2004, in press) simplified calculations by using a constant temperature of 15 °C, whereas global sea surface temperatures are likely to have fluctuated substantially during the Phanerozoic (Karhu and Epstein, 1986; Frakes et al., 1992). Stable oxygen isotope ($\delta^{18}O$) data from marine invertebrate shells provide a means of assessing Phanerozoic temperature variation (Veizer et al., 1999). However, such palaeotemperature estimation is complicated by a long-term rising $\delta^{18}O$ trend (Veizer et al., 1999), that might indicate a long-term control on seawater $\delta^{18}O$ by mid-ocean ridge and riverine fluxes, in addition to temperature (Veizer et al., 2000). Accordingly, Veizer et al. (2000) derived detrended values by subtracting the least-squares linear fit from the data.

Recognizing the limitations with regard to activity coefficients and temperature in the previous modeling approach, here we recalculate calcite saturation state ($\Omega_{calcite}$) for the Phanerozoic Eon. In these new calculations, the input chemistry data, such as Ca, Mg, K, SO_4^{2-}, p_{CO2}, DIC, etc. remain the same as in Riding and Liang (2004, in press). However, two

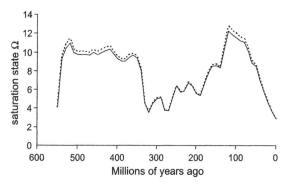

Fig. 3. Riding and Liang (2004, in press) $\Omega_{calcite}$ Phanerozoic trend (dashed line), compared with the same data recalculated using Pitzer equations (solid line). Use of Pitzer equations results in slightly (<5%) lower saturation ratio values.

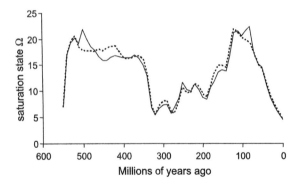

Fig. 4. Riding and Liang (2004, in press) $\Omega_{calcite}$ Phanerozoic trend recalculated at 25 °C (dashed line), compared with the same data incorporating Veizer et al's (2000, Fig. 3) temperature anomalies (solid line). The most conspicuous changes in saturation state resulting from incorporation of these temperature values are ~400–500 Myr ago and ~80 Myr ago.

additional factors are examined: (i) effect of ionic interaction, using an improved Pitzer formulations to calculate activity coefficients (Fig. 3); (ii) effect of temperature, using detrended temperature anomaly values in the range −4 °C and +6 °C, based on samples from low latitudes (Veizer et al., 2000, Fig. 3). The latter is implemented with a baseline of 25 °C, representing present-day mean tropical sea-surface temperature, to approximate temperatures during the Phanerozoic (Fig. 4).

8. Results

8.1. Saturation ratio trends

$\Omega_{calcite}$, recalculated using Pitzer equations from the input data of (Riding and Liang, 2004, in press), is compared with the original $\Omega_{calcite}$ trend in Fig. 3. Use of Pitzer equations results in slightly lower saturation ratio values, but the maximum deviation in calculated $\Omega_{calcite}$ is <5%. For example, calculated saturation ratio at its Phanerozoic peak 120 Myr ago is reduced from 12.8 to 12.3. The improvement by using Pitzer's formulation is small compared to the quality of the input data. For example, the error estimates on calculated GEO-CARB p_{CO2} CO$_2$ values are generally very large (Berner and Kothavala, 2001). For this reason, the remaining comparisons are made here using results obtained from extended Debye–Hückel limiting law.

The $\Omega_{calcite}$ Phanerozoic trend recalculated using Veizer et al's (2000, Fig. 3) temperature anomaly data is shown in Fig. 4, together with the $\Omega_{calcite}$ values recalculated at 25 °C. Calculated saturation ratio rises rapidly from 545 Myr ago to peaks ~500 Myr ago, declines to a high plateau between ~450 and 375 Myr, declines rapidly to low points ~320 and 280 Myr ago, rises to low peaks ~250 and 220 Myr ago, declines to ~190 Myr ago, and rises to high peaks ~100 Myr before steeply declining to the present day.

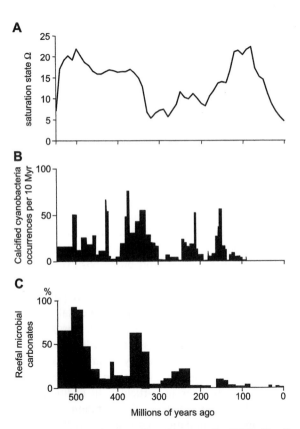

Fig. 5. (A) $\Omega_{calcite}$ incorporating Veizer et al's (2000, Fig. 3) temperature anomalies with 25 °C baseline (solid line in Fig. 4) compared with microbial carbonate abundance data of (B) Arp et al. (2001) and (C) Kiessling (2002). Periods of increased abundance of calcified cyanobacteria and reefal microbial carbonates prior to ~150 Myr ago broadly coincide with peaks of calculated saturation ratio. However, an anomaly occurs ~120–80 Myr ago when calculated saturation ratio is high and microbial carbonate abundance is low. This is suggested to reflect pelagic carbonate deposition by plankton. This removal, that would have reduced saturation state, is not incorporated into the saturation state values calculated here.

8.2. Saturation state and microbial carbonate abundance

The $\Omega_{calcite}$ trend (with temperature correction) is compared with microbial carbonate abundance data of Arp et al. (2001) and Kiessling (2002) in Fig. 5. Periods of increased abundance of calcified cyanobacteria and reefal microbial carbonates (e.g., ~500, 420, 370, 250–220, and 160 Myr ago) appear to coincide with peaks of calculated saturation ratio. However, whereas microbial carbonates decline in the past 150 Myr, calculated saturation ratio rises to marked peaks ~120–80 Myr ago.

8.3. Microbial carbonates and metazoan diversity

Phanerozoic abundance of microbial carbonates (Arp et al., 2001; Kiessling, 2002, Fig. 16) is compared with an inverse plot of marine metazoan generic diversity (data from Sepkoski, 1997) in Fig. 6. In general, microbial carbonates decline in abundance

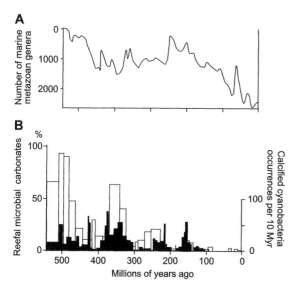

Fig. 6. (A) Inverse plot of marine metazoan generic diversity (data from Sepkoski, 1997). (B) Relative abundance of reefal microbial carbonates during the Phanerozoic. Open bar data from Riding (in press), replotted from Kiessling (2002, Fig. 16) using Golonka and Kiessling's, (2002, Fig. 1, pp. 13–18) time intervals for supersequences. Filled bars show reported occurrences of calcified cyanobacteria per 10 Myr Phanerozoic from Arp et al. (2001). The broadly inverse relationship supports the view that microbial carbonates declined through the Phanerozoic as metazoans diversified.

as metazoan diversity increases, and in some cases sharp declines (major extinctions) in diversity correspond with increases in microbial carbonates (e.g., ~370–360 and ~245 Myr ago).

9. Discussion

9.1. Constraints on saturation state estimates

As stated above (6, Seawater saturation state), calculation of $\Omega_{aragonite}$, $\Omega_{calcite}$, and $\Omega_{dolomite}$ is based on modeled values of atmospheric CO_2 (Berner and Kothavala, 2001), seawater major ion composition (Hardie, 1996; Stanley and Hardie, 1998), and a correlation established for p_{CO2} and HCO_3^- for the past 100 Myr (Lasaga et al., 1985). Many of the assumptions underlying these modeled values remain to be confirmed. For example, inference of seafloor spreading rates from changes in sealevel (Gaffin, 1987) is not supported for the past 180 Myr by estimates based on area–age distribution of oceanic crust (Parsons, 1982; Rowley, 2002). Nonetheless, Hardie's (1996) and Stanley and Hardie's (1998) modeled estimates of Phanerozoic Ca^{2+} and Mg^{2+} values are generally consistent with analyses of fluid inclusions preserved in marine halite (Lowenstein et al., 2001; Horita et al., 2002). Furthermore, model estimates of p_{CO2} from GEOCARB, with GEOCARB III (Berner and Kothavala, 2001) being the most recent version, for the Phanerozoic have broad support from p_{CO2} proxy data such as $\delta^{13}C$ values of palaeosols and fossil plant stomata (Ekart et al., 1999; Royer et al., 2001; Crowley and Berner, 2001). An additional difficulty concerns the lack of knowledge of variations in the carbonate system over time. In the absence of published data for either DIC, alkalinity, or pH for the Phanerozoic as a whole, Riding and Liang (2004, in press) estimated DIC using the correlation established for p_{CO2} and HCO_3^- for the last 100 Myr in the revised BLAG data (Lasaga et al., 1985) and extrapolated it to earlier periods from GEOCARB III p_{CO2} data. There are objections to this approach. In the first place, underlying modeling assumptions of Lasaga et al. (1985) differ from those of Hardie (1996) and Stanley and Hardie (1998), thus raising questions on consistency of the data. Furthermore, seawater chemistry of the earlier Phanerozoic is likely to differ from that of past 100 Myr, due to the varying inputs from continental weathering via rivers and inputs and outputs due to changes in mid-ocean ridge brine fluxes and reactions in sediments (Holland, 2004). Fundamentally, the relationship between p_{CO2} and DIC depends on ocean pH, and this is debatable even for the past 100 Myr (e.g., Pearson and Palmer, 2000; LeMarchand et al., 2000).

Given these uncertainties, how might we attempt to establish broad constraints on secular changes in this system? As mentioned earlier, if p_{CO2} is known the carbonate system can be constrained by either DIC, alkalinity or pH. Riding and Liang (2004, in press) used estimated DIC values to calculate saturation state. It is therefore relevant to compare the effects of (i) alkalinity and (ii) pH on carbonate saturation state.

(i) With the same input data but using present-day levels of seawater alkalinity (~2.3 mM), very different saturation states from those of Riding and Liang (2004, in press) are obtained. These new values are close to undersaturation in the Early Palaeozoic (~400 Myr ago; Fig. 7A) when modeled p_{CO2} was high. Holding alkalinity constant lowers seawater pH due to lack of buffering capacity. This appears unrealistic, since increased p_{CO2} can be expected to have increased subaerial weathering, supplying Ca^{2+}, Mg^{2+}, HCO_3^-, and other dissolved species, with the result that seawater alkalinity should increase and saturation state should rise. This is not seen where alkalinity is held constant (Fig. 7A).

(ii) The effect of holding pH constant is shown in Fig. 7B. Absolute saturation values differ markedly depending on the pH selected. For example, at constant pH 7.6, calculated saturation state becomes <1 at times of low p_{CO2}, whereas at constant pH 8.2 calculated saturation state reaches values >200 when p_{CO2} is very high. Despite these differences in absolute values, the saturation trends in both these cases are identical in shape, and although different in detail from that of Riding and Liang (2004, in press) remain similar in terms of most maxima and minima. When pH is held constant, alkalinity is allowed to increase to accommodate p_{CO2} increase and

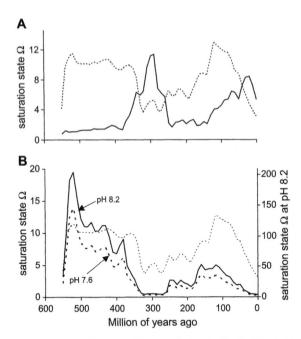

Fig. 7. (A) Saturation state ($\Omega_{calcite}$) calculated with alkalinity held constant at present-day level (~2.3 mM; solid line) compared with previous calculation (dashed line; Riding and Liang, 2004, in press). (B) Saturation state ($\Omega_{calcite}$) calculated with pH held constant at 7.6 (dark dashed line) and 8.2 (solid line), compared with previous calculation (thin dashed line; Riding and Liang, 2004, in press). Assumption of constant alkalinity (A) results in low saturation values at high p_{CO2}. This appears unrealistic, since increased p_{CO2} can be expected to increase subaerial weathering, supplying Ca^{2+}, Mg^{2+}, HCO_3^-, and other dissolved species. This in turn would result in increase in seawater alkalinity and corresponding increase in saturation state that is not seen in (A). In contrast, holding pH constant (B) permits alkalinity to increase with p_{CO2}. This is accompanied by increase in carbonate species and therefore results in higher saturation state.

carbonate species dominate the system. Thus, it is not surprising that the resulting saturation curve reflects the p_{CO2} curve. Of course, the extent of alkalinity increase at very high p_{CO2} could become unreasonable, as seen in the very high saturation state under these conditions. Nonetheless, this simple evaluation indicates that, compared with the constant alkalinity approach, the constant pH method is more reasonable because it permits the feedback mechanisms that increase alkalinity. We conclude that, although it is not possible to quantitatively constrain the carbonate system prior to 100 Myr ago, there is reason to consider that the revised BLAG relationship between p_{CO2} and HCO_3^- (Lasaga et al., 1985) used by Riding and Liang (2004, in press) may not be unrealistic.

Using this approach for the carbonate system, previous calculation yielded pH values in the range 7.6–8.2 (Riding and Liang, 2004, in press), which are within the ranges of sea-surface pH estimates for the past 300 Myr (Caldeira and Wickett, 2003; Pearson and Palmer, 2000; Zeebe, 2001). Furthermore, since change in alkalinity should be constrained by carbonate and silicate weathering, oceanic pH need not vary significantly even over longer timescales than the Phanerozoic (Grotzinger and Kasting, 1993). We conclude that, although the limitations on saturation state calculation noted above remain inescapable at present, it may be that absence of robust information regarding Phanerozoic DIC, alkalinity, or pH does not completely negate attempts to reconstruct secular variation in surface ocean saturation state.

9.2. Microbial carbonates compared with saturation trends incorporating temperature data

The recalculated $\Omega_{calcite}$ Phanerozoic trend, using Veizer et al.'s (2000, Fig. 3) temperature anomaly data (Fig. 4) is broadly similar to that of Riding and Liang (2004, in press) but with the following differences: a marked peak at ~500 Myr, an enhanced peak at ~420 Myr, a slight shift to a peak at ~370 Myr, and a marked peak at ~80 Myr ago. The earlier three of these peaks correspond with successive Palaeozoic peaks in calcified cyanobacteria abundance of Arp et al. (2001) and with Kiessling's (2002) Palaeozoic peaks of reefal microbial carbonate abundance (Fig. 5). Elevated sea-surface temperatures can be expected to increase saturation state, and therefore $CaCO_3$ precipitation, and also to stimulate microbial growth. It may not be coincidental, therefore, that intervals of higher temperatures in the Late Cambrian–Early Ordovician (~500 Myr), Late Devonian (~370 Myr), and Late Permian–Early Triassic (~250 Myr) recognized by Veizer et al. (2000, Fig. 3) from oxygen isotope analyses, broadly correspond with increased abundance of microbial carbonates (see also Riding, 1992).

However, steady decline in reefal microbial carbonate abundance peaks during the Phanerozoic to quite

low values during the past 200 Myr (Fig. 5C) creates marked discrepancy between low microbial carbonate abundance and elevated calculated saturation ratio ~80–120 Myr ago. Riding and Liang (2004, in press) noted a similar discrepancy with shallow-water limestone accumulation rate measured by Bosscher and Schlager (1993). They suggested that this might reflect the Late Cretaceous period of increased deposition of deep-sea carbonates in response to pelagic biomineralization. There is need, therefore, to incorporate biologically controlled $CaCO_3$ removal into geochemical models of global budgets. Notwithstanding uncertainties such as these, we conclude that, overall, there is a degree of coincidence between this newly calculated saturation state trend and microbial carbonate abundance measures prior to ~150 Myr ago that suggests a causal relationship.

9.3. Microbial carbonates compared with metazoan diversity

Comparisons between microbial carbonates and metazoan diversity rely on the assumption that taxonomic diversity may be a proxy for competitive interference, and this requires support. Nonetheless, the broadly inverse relationship between microbial carbonate abundance and marine metazoan generic diversity (Fig. 6) tends to support the view that microbial carbonates declined through the Phanerozoic as metazoans diversified and, directly or indirectly, provided competitive interference (Garrett, 1970; Awramik, 1971).

9.4. Dual control on microbial carbonates

Earlier assessments of the Phanerozoic history of microbial carbonates recognized a long-term decline (e.g., Maslov, 1959, Fischer, 1965) that was patterned by marked fluctuations in abundance (Riding, 1992, 1993), but there was a lack of stratigraphically resolved data to test this further. Data subsequently extracted from the literature for calcified cyanobacteria (Arp et al., 2001) and reefal microbial carbonates (Kiessling, 2002, Fig. 16) confirm a long-term pattern of fluctuating decline. Furthermore, comparisons between these measures of microbial carbonate abundance and both estimated saturation state and metazoan diversity data strengthen the likelihood that these have operated in conjunction as controlling factors.

In hindsight, therefore, it appears that suggestions concerning the importance of early lithification that go back many years (e.g., Walcott, 1914; Logan, 1961, p. 520; Fischer, 1965), and of interactions with eukaryotes in general and metazoans in particular (e.g., Fischer, 1965; Garrett, 1970; Awramik, 1971), including the related concept of stromatolites as 'disaster forms' resurging when metazoans are reduced (Schubert and Bottjer, 1992), have been on the right track. But it is also evident that neither factor should be considered in isolation. Riding (in press) concluded that microbial carbonates were most abundant when elevated saturation state coincided with low metazoan diversity (e.g., Cambrian–Early Ordovician and Late Silurian, and following Late Devonian, end-Permian, and end-Jurassic extinctions; Fig. 8) and were least abundant when reduced saturation state coincided with high metazoan diversity (e.g., Late Carboniferous, Early–Mid Jurassic, and Cenozoic). In other words, maxima and minima of microbial carbonate abundance depended on whether the effects of metazoan diversity and carbonate saturation state reinforced or countered one another.

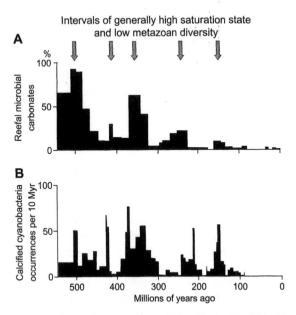

Fig. 8. Phanerozoic periods of microbial carbonate abundance (A, Kiessling, 2002; B, Arp et al., 2001) broadly coincide with intervals (arrows) when saturation state was elevated and metazoan diversity reduced.

9.5. Microbial carbonates and biocalcification

Confirmation that competition and early lithification have combined to determine the secular pattern of microbial carbonates observed in the Phanerozoic suggests a number of questions. For example, is it possible to judge whether – at different times – one of these factors was more important than another? Even more intriguingly, is it likely that these organisms – whether prokaryote or eukaryote – have not only competed for resources such as space and nutrients but also for the resources that promote biomineralization? In other words, could the link observed between metazoan increase and microbial carbonate decline reflect, inter alia, competition for $CaCO_3$? This focuses attention on the relationships between saturation state and biologically induced calcification on the one hand and biologically controlled calcification on the other. It could be expected that secular variation in seawater saturation state for $CaCO_3$ minerals should have influenced the long-term history of organisms that biologically induce, rather than closely control, their calcification. Preliminary comparisons of calculated seawater saturation state with patterns of marine calcified organisms during the Phanerozoic Eon suggest that the diversity of organisms with biologically induced calcification, such as chlorophytes, corals, and sponges, increased during periods of elevated saturation state and declined when saturation state was reduced. In contrast, organisms with relatively controlled calcification, such as molluscs, brachiopods, bryozoans, and echinoderms, appear to have been relatively unaffected by saturation state. This could indicate that $CaCO_3$ availability – governed by saturation state – has significantly influenced the diversity of organisms with biologically induced calcification. This could involve many tropical marine algae and invertebrates, especially those that are most involved in reef building. Thus, despite its apparent abundance, it may be that $CaCO_3$ has been so widely employed in biomineralization by aquatic organisms during the past ~550 million years that it has constituted a limiting resource, partitioned between organisms that biologically control and biologically induce their calcification, with the surplus being inorganically precipitated.

In contrast, for the major part of the Precambrian not only would metazoan interference have been reduced or absent (Awramik, 1971) but so would competition for $CaCO_3$, because controlled biomineralization was scarce prior to the Phanerozoic (Lowenstam, 1981). In these circumstances, seawater saturation state, together with microbial evolution, would have been the principal factors determining microbial carbonate development during the Precambrian (Grotzinger, 1994). The advent of biologically controlled calcification in the Palaeozoic is likely to have significantly affected this system. As a first approximation, we can envisage that marine carbonate sedimentation prior to the advent of controlled biocalcification would have reflected seawater $CaCO_3$ saturation state equilibrium. Other factors being equal, the evolutionary rise of organisms with controlled biocalcification (Lowenstam, 1981) must have reduced seawater saturation state and, thus, inorganic and biologically induced precipitation. During the Phanerozoic it seems likely that this switch from induced to controlled biocalcification progressively increased. Certainly, the record of microbial carbonates reflects long-term decline in microbially induced calcification. Biological influences on $CaCO_3$ precipitation would have been mediated by evolution and habitat availability, and by fluctuations in seawater saturation state-determined by global cycling—such as those that we endeavour to use here to estimate changes in saturation ratio. So long as most biocalcifiers were benthic, a prime control on habitat availability would have been the extent of shallow seas, and therefore dependent on sealevel fluctuations (Grotzinger, 1994). Cretaceous radiation of coccolithophore algae and globigerine foraminifers (Tappan and Loeblich, 1973) can be seen as part of the long-term rise in controlled biocalcification. But this substantial increase in oceanic planktic calcifiers also delivered large quantities of $CaCO_3$ to the deep-ocean rather than to shallow shelves (Hay, 1985; Opdyke and Wilkinson, 1988). This should have impacted negatively on shelf carbonate sedimentation (Wilkinson and Walker, 1989). Furthermore, because these pelagic carbonates were prone to dissolution in deep water prior to burial, they could be rapidly recycled in response to changes in global ocean chemistry. This has introduced a stabilizing factor into the marine carbonate system (Ridgewell et al., 2003) that continues to the present-day (Broecker and Clark, 2001). On the other hand, the rate at which buried deep-sea

carbonates are subducted could be quite irregular, and this may have been a factor contributing to low present-day p_{CO2} (Edmond and Huh, 2003). This in turn would have influenced current low seawater saturation state and reduced abundance of microbial carbonates. By this type of reasoning it should be possible to integrate the biological evolution of calcification, patterns of carbonate sedimentation, and variation in seawater chemistry over extensive timescales. This remains to be further elucidated.

10. Summary and conclusions

It has long been suggested that the geological distribution of microbial carbonates could reflect not only the rise of eukaryotes in general, and metazoans in particular, but also changes in seawater chemistry that determined the extent of microbial calcification (Fischer, 1965). Compilations of abundance data for calcified cyanobacteria (Arp et al., 2001) and reefal microbial carbonates (Kiessling, 2002) suggest that marine microbial carbonates in general have undergone long-term episodic decline since the Early Palaeozoic. A broadly inverse relationship between this pattern and that of metazoan diversity (Sepkoski, 1997) supports the view that microbial carbonates became less widespread as metazoans expanded (Riding, in press). Riding and Liang (2004, in press) calculated variation in saturation state for the Phanerozoic using published estimates of past seawater ionic composition and atmospheric CO_2 levels, and assuming constant temperature of 15 °C. They found broad positive coincidence between the secular trend of $\Omega_{calcite}$, shallow-water limestone accumulation rate (Bosscher and Schlager, 1993), and periods of increased abundance of microbial carbonates. These results suggest that microbial carbonate abundance responded to changes in the saturation state of seawater, as well as to changes in the presence of metazoans (Fig. 9), and that both of these factors need to be considered to understand the Phanerozoic history of marine microbial carbonates (Riding, in press), as Fischer (1965) anticipated.

Recognition of the likely importance of changes in seawater saturation state for understanding many aspects of limestone precipitation and biocalcification, including microbial carbonates (Kempe and Kaz-

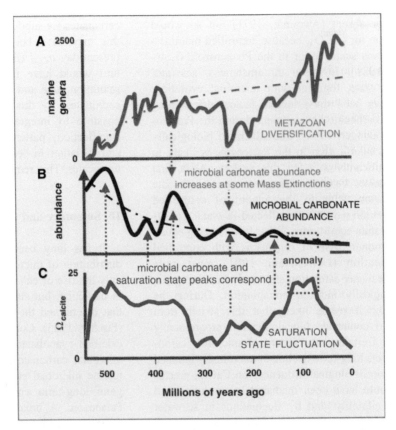

Fig. 9. Comparison of (A) marine metazoan generic diversity (Sepkoski, 1997), (B) microbial carbonate abundance generalized from Arp et al. (2001) and Kiessling (2002), and (C) calcite saturation ratio calculated in this paper. Microbial carbonates show oscillatory decline. The overall trend of microbial carbonate decline inversely mirrors that of metazoan diversification. Peaks of microbial carbonate abundance superimposed on the overall trend broadly coincide with periods of raised calcite saturation ratio (red arrows). An anomaly ~80–120 Myr ago occurs where high calculated saturation state does not correspond to increase in microbial carbonates. This could reflect reduction in saturation state due to Late Cretaceous increase in deep-sea carbonates; this has not been taken into account in calculating saturation state. Some peaks of microbial carbonate abundance also correspond with marked reductions (Mass Extinctions) in metazoan diversity (green arrows), although not all Mass Extinctions can be seen to have had this effect. This summary indicates that the Phanerozoic trend of microbial carbonate abundance broadly reflects the influence of both metazoan interference and seawater saturation state with respect to $CaCO_3$ minerals.

mierczak, 1990; Opdyke and Wilkinson, 1990; Webb, 2001), is not new. Unfortunately, calculation of past seawater saturation state is fraught with difficulty, particularly prior to 100 Myr ago. Comprehensive values of seawater ionic composition (e.g., Hardie, 1996; Stanley and Hardie, 1998) and atmospheric CO_2 levels (e.g., Berner and Kothavala, 2001) available for the Phanerozoic as a whole are modeled estimates. Interpretation of palaeotemperatures based on oxygen isotope ($\delta^{18}O$) data from marine invertebrate shells is complicated by an overall rising $\delta^{18}O$ trend (Veizer et al., 1999). Furthermore, there is a lack of information concerning seawater pH, alkalinity, or

DIC, one of which is required to constrain the carbonate system and calculate $\{CO_3^{2-}\}$.

Here we have endeavoured to improve estimation of Phanerozoic seawater saturation state by using Pitzer equations (Pitzer, 1991) to compute activity coefficients for seawater species, and by incorporating temperature anomaly data (Veizer et al., 2000, Fig. 3). We cannot overcome the lack of robust information concerning changes in Phanerozoic seawater pH. Nonetheless, we find that holding pH constant (e.g., at 7.6 and 8.2), while using the same ionic and p_{CO_2} input data, results in $\Omega_{aragonite}$ and $\Omega_{calcite}$ trends that are not very dissimilar from those of Riding and Liang

(2004, in press) who determined carbonate concentration by applying the correlation established for p_{CO2} and HCO_3^- for the last 100 Myr in the revised BLAG data (Lasaga et al., 1985) to the entire Phanerozoic. This calculation yielded pH variation between 7.6 and 8.2 (Riding and Liang, 2004, in press), with the change inversely linked to p_{CO2} increase, permitting feedback mechanisms to increase alkalinity to a reasonable level. Holding pH constant implies that as p_{CO2} increases, carbonate increases accordingly and controls seawater saturation state, resulting in similarity between the trends of saturation state and p_{CO2}. Similarities between the saturation ratio trend and the geological record therefore tend to substantiate GEOCARB III p_{CO2} values from which $\Omega_{calcite}$ was calculated. The $\Omega_{calcite}$ trend for the Phanerozoic, recalculated using temperature anomaly data from Veizer et al. (2000), shows correspondence with peaks of microbial carbonate abundance prior to ~150 Myr ago. The modeled estimates we have used are based primarily on global budgets of physico-chemical processes, and do not take account of biologically controlled removal of $CaCO_3$. Thus, contrast between (elevated) saturation state and (low) microbial carbonate abundance ~100 Myr ago could be due to the fact that calculated saturation ratio does not take account of $CaCO_3$ removed from the system by biologically controlled calcification such as pelagic plankton.

Long-term secular variation in seawater saturation state with respect to $CaCO_3$ minerals is of interest for a variety of reasons, not least for elucidating fundamental controls on $CaCO_3$ precipitation and recognizing their effects on geological patterns of sedimentation, as well as for its role in carbon sequestration. We conclude that similarities between calculated saturation state and the geological records of both limestone accumulation rate (Riding and Liang, 2004, in press) and microbial carbonate abundance provide evidence that secular variation in saturation state has significantly influenced marine carbonate formation. Improved estimates of seawater ionic composition (e.g., from fluid inclusion data), and of atmospheric p_{CO2} and past pH are needed for confident seawater saturation state calculation. Better measures of microbial and other non-skeletal carbonates through time are also required to further explore these questions.

Acknowledgements

We thank John Morse for advice and for provision of software to assist calculations incorporating Pitzer equations. Richard Whittecar kindly assisted with editing. We are grateful to Nora Noffke for helpful comments on the manuscript.

References

Arp, G., Reimer, A., Reitner, J., 2001. Photosynthesis-induced biofilm calcification and calcium concentrations in Phanerozoic oceans. Science 292, 1701–1704.

Arvidson, R.S., Mackenzie, F.T., Guidry, M.W., 2000. Ocean/atmosphere history and carbonate precipitation rates: a solution to the ãdolomite problemã. In: Glenn, C.R., Prévôt-Lucas, L., Lucas, J. (Eds.), Marine authigenesis: from global to microbial. SEPM Special Publication 66, Tulsa, OK, pp. 1–5.

Awramik, S.M., 1971. Precambrian columnar stromatolite diversity: reflection of metazoan appearance. Science 174, 825–827.

Berner, R.A., Kothavala, Z., 2001. GEOCARB III: a revised model of atmospheric CO2 over Phanerozoic time. Am. J. Sci. 301, 182–204.

Berner, R.A., Lasaga, A.C., Garrels, R.M., 1983. The carbonate-silicate geochemical cycle and its effect on atmospheric carbon-dioxide over the past 100 million years. Am. J. Sci. 283, 641–683.

Bosscher, H., Schlager, W., 1993. Accumulation rates of carbonate platforms. J. Geol. 101, 345–355.

Broecker, W.S., Clark, E., 2001. Glacial-to-Holocene redistribution of carbonate ion in the deep sea. Science 294, 2152–2154.

Burne, R.V., Moore, L.S., 1987. Microbialites: organosedimentary deposits of benthic microbial communities. Palaios 2, 241–254.

Burton, E.A., Walter, L.M., 1987. Relative precipitation rates of aragonite and Mg calcite from seawater: temperature or carbonate ion control? Geology 15, 111–114.

Caldeira, K., Wickett, M.E., 2003. Anthropogenic carbon and ocean pH. Nature 425, 365.

Cloud, P.E., Semikhatov, M.A., 1969. Proterozoic stromatolite zonation. Am. J. Sci. 267, 1017–1061.

Crowley, T.J., Berner, R.A., 2001. Paleoclimate-CO₂ and climate change. Science 292, 870–872.

Debye, P., Hückel, E., 1923. Zur Theorie der Elektrolyte. Phys. Z. 24, 185–206.

Edmond, J.M., Huh, Y., 2003. Non-steady state carbonate recycling and implications for the evolution of atmospheric P_{CO2}. Earth Planet. Sci. Lett. 216, 125–139.

Ekart, D.D., Cerling, T.E., Montanez, I.P., Tabor, N.J., 1999. A 400 million year carbon isotope record of pedogenic carbonate: implications for paleoatmospheric carbon dioxide. Am. J. Sci. 299, 805–827.

Fischer, A.G., 1965. Fossils, early life, and atmospheric history. Proc. Natl. Acad. Sci. 53, 1205–1215.

Frakes, L.A., Francis, J.E., Syktus, J.I., 1992. Climate Modes of the Phanerozoic: The History of the Earth's Climate Over rhe Past

600 Million Years. Cambridge University Press, Cambridge, England (274 pp.).

Gaffin, S., 1987. Ridge volume dependence on sea-floor generation rate and inversion using long-term sea-level change. Am. J. Sci. 287, 596–611.

Garrett, P., 1970. Phanerozoic stromatolites: noncompetitive ecologic restriction by grazing and burrowing animals. Science 169, 171–173.

Gebelein, C.D., 1976. The effects of the physical, chemical and biological evolution of the earth. In: Walter, M.R. (Ed.), Stromatolites, Developments in Sedimentology, vol. 20. Elsevier, Amsterdam, pp. 499–515.

Golonka, J., Kiessling, W., 2002. Phanerozoic time scale and definition of time slices. In: Kiessling, W., Flügel, E., Golonka, J. (Eds.), Phanerozoic Reef Patterns, SEPM Special Publication, vol. 72, pp. 11–20.

Grotzinger, J.P., 1990. Geochemical model for Proterozoic stromatolite decline. Am. J. Sci. 290-A, 80–103.

Grotzinger, J.P., 1994. Trends in Precambrian carbonate sediments and their implication for understanding evolution. In: Bengtson, S. (Ed.), Early Life on Earth, Nobel Symposium, vol. 84. Columbia University Press, New York, pp. 245–258.

Grotzinger, J.P., Kasting, J.F., 1993. New constraints on Precambrian ocean composition. J. Geol. 101, 235–243.

Hardie, L.A., 1996. Secular variation in seawater chemistry: an explanation for the coupled secular variation in the mineralogies of marine limestones and potash evaporites over the past 600 my. Geology 24, 279–283.

Hay, W.W., 1985. Potential errors in estimates of carbonate rock accumulating through geologic time. In: Sundquist, E.T., Broecker, W.S. (Eds.), The Carbon Cycle and Atmospheric CO_2: Natural Variations Archean to Present, Geophysical Monograph, vol. 32. American Geophysical Union, Washington, DC, pp. 563–573.

Holland, H.D., 1978. The Chemistry of the Atmosphere and Oceans. Wiley, New York (351 pp.).

Holland, H.D., 2004. The geologic history of seawater. In: Elderfield, H. (Ed.), The Oceans and Marine Geochemistry, pp. 583–625., Vol. 6. Treatise on Geochemistry (Holland, H.D., Turekian, K.K., Eds.), Elsevier-Pergamon, Oxford, 646 pp.

Horita, J., Zimmermann, H., Holland, H.D., 2002. Chemical evolution of seawater during the Phanerozoic: implications from the record of marine evaporates. Geochim. Cosmochim. Acta 66, 3733–3756.

Karhu, J., Epstein, S., 1986. The implication of the oxygen isotope records in coexisting cherts and phosphates. Geochim. Cosmochim. Acta 50, 1745–1756.

Kasting, J.F., Catling, D., 2003. Evolution of a habitable planet. Annu. Rev. Astron. Astrophys. 41, 429–463.

Kempe, S., Kazmierczak, J., 1990. Calcium carbonate supersaturation and the formation of in situ calcified stromatolites. In: Ittekot, V., Kempe, S., Michaelis, W., Spitzy, A. (Eds.), Facets of Modern Biogeochemistry. Springer, Berlin, pp. 255–278.

Kiessling, W., 2002. Secular variations in the Phanerozoic reef ecosystem. In: Kiessling, W., Flügel, E., Golonka, J. (Eds.), Phanerozoic reef patterns, SEPM Special Publication, vol. 72, pp. 625–690.

Kiessling, W., Flügel, E., 2002. Paleoreefs—a database on Phanerozoic reefs. In: Kiessling, W., Flügel, E., Golonka, J. (Eds.), Phanerozoic reef patterns, SEPM Special Publication, vol. 72, pp. 77–92.

Lasaga, A.C., Berner, R.A., Garrels, R.M., 1985. An improved geochemical model of atmospheric CO_2 fluctuations over the past 100 million years. In: Sundquist, E.T., Broecker, W.S. (Eds.), The Carbon Cycle and Atmospheric CO_2: Natural Variations Archean to Present, Geophysical Monograph, vol. 32. American Geophysical Union, Washington, DC, pp. 397–411.

Leadbeater, B.S.C., Riding, R. (Eds.), 1986. Biomineralization in Lower Plants and Animals. Special Volume, vol. 30. Systematics Association, Clarendon, Oxford (401 pp.).

Leeder, M., 1999. Sedimentology and Sedimentary Basins: From Turbulence to Tectonics. Blackwell, Oxford (592 pp.).

LeMarchand, D., Gaillardet, J., Lewin, E., Allegre, C.J., 2000. The influence of rivers on marine boron isotopes and implications for reconstructing past ocean pH. Nature 408, 951–954.

Logan, B.W., 1961. Cryptozoon and associated stromatolites from the Recent, Shark Bay, Western Australia. J. Geol. 69, 517–533.

Lowenstam, H.A., 1981. Minerals formed by organisms. Science 211, 1126–1131.

Lowenstein, T.K., Timofeeff, M.N., Brennan, S.T., Hardie, L.A., Demicco, R.V., 2001. Oscillations in Phanerozoic seawater chemistry: evidence from fluid inclusions. Science 294, 1086–1088.

Mackenzie, F.T., Morse, J.W., 1992. Sedimentary carbonates through Phanerozoic time. Geochim. Cosmochim. Acta 56, 3281–3295.

Mann, S., 2001. Biomineralization: Principles and Concepts in Bioinorganic Materials Chemistry. Oxford University Press, Oxford (240 pp.).

Maslov, V.P., 1959. Stromatolites and facies. Dokl. Akad. Nauk SSSR 125, 1085–1088. (in Russian).

Merz, M., 1992. The biology of carbonate precipitation by cyanobacteria. Facies 26, 81–102.

Merz-Preiß, M., 2000. Calcification in cyanobacteria. In: Riding, R., Awramik, S.M. (Eds.), Microbial sediments. Springer-Verlag, Berlin, pp. 50–56.

Monty, C.L.V., 1973. Precambrian background and Phanerozoic history of stromatolitic communities, an overview. Ann. Soc. Geol. Belg. 96, 585–624.

Monty, C.L.V., 1977. Evolving concepts on the nature and the ecological significance of stromatolites. In: Flügel, E. (Ed.), Fossil Algae, Recent Results and Developments. Springer-Verlag, Berlin, pp. 15–35.

Morse, J.W., Mackenzie, F.T., 1990. Geochemistry of sedimentary carbonates. Developments in sedimentology. Elsevier, Amsterdam (707 pp.).

Opdyke, B.N., Wilkinson, B.H., 1988. Surface area control of shallow cratonic to deep marine carbonate accumulation. Paleoceanography 3, 685–703.

Opdyke, B.N., Wilkinson, B.H., 1990. Paleolatitude distribution of Phanerozoic marine ooids and cements. Palaeogeogr. Palaeoclimatol. Palaeoecol. 78, 135–148.

Parkhurst, D.L., Appelo, C.A.J., 1999. User's guide to PHREEQC (version 2)—a computer program for speciation, batch-reaction, one-dimensional transport, and inverse geochemical calculations. Water-Resources Investigations Report 99-4259. US Geological Survey, Denver, Colorado, 326 pp.

Parsons, B., 1982. Causes and consequences of the relation between area and age of the ocean floor. J. Geophys. Res. 87, 289–302.

Pearson, P.N., Palmer, M.R., 2000. Atmospheric carbon dioxide concentrations over the past 60 million years. Nature 406, 695–699.

Pentecost, A., Riding, R., 1986. Calcification in cyanobacteria. In: Leadbeater, B.S.C., Riding, R. (Eds.), Biomineralization in Lower Plants and Animals, Systematics Association, Special Vol., vol. 30. Clarendon Press, Oxford, pp. 73–90.

Pitzer, K.S., 1991. Ion interaction approach: theory and data correlation. In: Pitzer, K.S. (Ed.), Activity Coefficients in Electrolyte Solutions. CRC Press, Boca Raton, Florida, pp. 75–153.

Pratt, B.R., 1982. Stromatolite decline—a reconsideration. Geology 10, 512–515.

Pratt, B.R., 2001. Calcification of cyanobacterial filaments; *Girvanella* and the origin of lower Paleozoic lime mud. Geology 29, 763–766.

Ridgewell, A.J., Kennedy, M.J., Caldeira, K., 2003. Carbonate deposition, climate stability, and Neoproterozoic ice ages. Science 302, 859–862.

Riding, R., 1992. Temporal variation in calcification in marine cyanobacteria. J. Geol. Soc. (Lond.) 149, 979–989.

Riding, R., 1993. Phanerozoic patterns of marine $CaCO_3$ precipitation. Naturwissenschaften 80, 513–516.

Riding, R., 2000. Microbial carbonates: the geological record of calcified bacterial–algal mats and biofilms. Sedimentology 47 (Suppl. 1), 179–214.

Riding, R., in press. Phanerozoic reefal microbial carbonate abundance: comparisons with metazoan diversity, mass extinction events, and seawater saturation state. Revista Española de Micropaleontología.

Riding, R., Liang, L., 2004. Marine limestone accumulation over the past 550 million years—control by seawater chemistry. Goldschmidt Conference 2004, Copenhagen, Conference Supplement. Geochim. Cosmochim. Acta, A354.

Riding, R., Liang, L., in press. Seawater chemistry control of marine limestone accumulation over the past 550 million years. Revista Española de Micropaleontología.

Rowley, D.B., 2002. Rate of plate creation and destruction: 180 Ma to present. Geol. Soc. Amer. Bull. 114, 927–933.

Royer, D.L., Wing, S.L., Beerling, D.J., Jolley, D.W., Koch, P.L., Hickey, L.J., Berner, R.A., 2001. Paleobotanical evidence for near present-day levels of atmospheric CO_2 during part of the Tertiary. Science 292, 2310–2313.

Rubey, W.W., 1951. Geologic history of seawater: an attempt to state the problem. Geol. Soc. Amer. Bull. 62, 1111–1147.

Sandberg, P.A., 1975. New interpretations of Great Salt Lake ooids and of ancient nonskeletal carbonate mineralogy. Sedimentology 22, 497–538.

Sandberg, P.A., 1983. An oscillating trend in Phanerozoic nonskeletal carbonate mineralogy. Nature 305, 19–22.

Schubert, J.K., Bottjer, D.J., 1992. Early Triassic stromatolites as post-mass extinction disaster forms. Geology 20, 883–886.

Sepkoski Jr., J.J., 1997. Biodiversity; past, present, and future. J. Paleontol. 71, 533–539.

Serebryakov, S.N., Semikhatov, M.A., 1974. Riphean and recent stromatolites: a comparison. Am. J. Sci. 274, 556–574.

Stanley, S.M., Hardie, L.A., 1998. Secular oscillations in the carbonate mineralogy of reef-building and sediment-producing organisms driven by tectonically forced shifts in seawater chemistry. Palaeogeogr. Palaeoclimatol. Palaeoecol. 144, 3–19.

Stumm, W., Morgan, J.J., 1996. Aquatic Chemistry: Chemical Equilibria and Rates in Natural Waters, 3rd ed. Wiley, New York (1022 pp.).

Tappan, H., Loeblich Jr., A.R., 1973. Evolution of the oceanic plankton. Earth-Sci. Rev. 9, 207–240.

Thompson, J.B., 2000. Microbial whitings. In: Riding, R., Awramik, S.M. (Eds.), Microbial Sediments. Springer-Verlag, Berlin, pp. 250–260.

Veizer, J., Ala, D., Azmy, K., Bruckschen, P., Buhl, D., Bruhn, F., Carden, G.A.F., Diener, A., Ebneth, S., Goddéris, Y., Jasper, T., Korte, C., Pawellek, F., Podlaha, O.G., Strauss, H., 1999. 87Sr/86Sr, δ13C and δ18O evolution of Phanerozoic seawater. Chem. Geol. 161, 59–88.

Veizer, J., Godderis, Y., Francois, L.M., 2000. Evidence for decoupling of atmospheric CO2 and global climate during the Phanerozoic eon. Nature 408, 698–701.

Walcott, C.D., 1914. Cambrian geology and paleontology: III. Precambrian Algonkian algal flora. Smithson. Misc. Collect. 64, 77–156.

Walker, J.C.G., Hays, P.B., Kasting, J.F., 1981. A negative feedback mechanism for the long-term stabilization of Earth's surface temperature. J. Geophys. Res. 86, 9776–9782.

Webb, G.E., 2001. Biologically induced carbonate precipitation in reefs through time. In: Stanley, G.D. (Ed.), The History and Sedimentology of Ancient Reef Systems. Kluwer Academic/Plenum Publishers, New York, pp. 159–203.

Wilkinson, B.H., Algeo, T.J., 1989. Sedimentary carbonate record of Ca–Mg cycling at the Earth's surface. Am. J. Sci. 289, 1158–1194.

Wilkinson, B.H., Walker, J.C.G., 1989. Phanerozoic cycling of sedimentary carbonate. Am. J. Sci. 289, 525–548.

Zeebe, R.E., 2001. Seawater pH and isotopic paleotemperatures of Cretaceous oceans. Palaeogeogr. Palaeoclimatol. Palaeoecol. 170, 49–57.

Available online at www.sciencedirect.com

Palaeogeography, Palaeoclimatology, Palaeoecology 219 (2005) 117–129

ELSEVIER

www.elsevier.com/locate/palaeo

The role of microorganisms and biofilms in the breakdown and dissolution of quartz and glass

Ulrike Brehm[a], Anna Gorbushina[a,*], Derek Mottershead[b,1]

[a]*Geomicrobiology, ICBM, Oldenburg University, P.O. Box 2503, D-26111 Oldenburg, Germany*
[b]*Department of Geography, Portsmouth University, Buckingham Building, Lion Terrace, Portsmouth, Hampshire PO1 3HE, UK*

Received 20 June 2003; accepted 29 October 2004

Abstract

Three different types of silica, (1) quartz sand, (2) crystalline (scepter) quartz, and (3) commercial glass, subjected to biological growth were investigated for evidence of biologically induced breakdown. The results of laboratory experiments with biofilms on quartz sand and glass were compared with material from the field. Microscopic (optical and SEM) analysis of quartz sand weathered in vitro by a microbial community of cyanobacteria, diatoms and heterotrophic bacteria demonstrate grain diminution of a sand fraction of North Sea sediment. The microbial community changes the minerals in the vicinity of the living cells, which leave their mark on the quartz surface in form of imprints, depressions and pits. Subsequently microbial growth on the surface brought about a general decrease of the grain size in the layer beneath a biofilm. Imprints of diatoms on glass surfaces colonized by diatoms and heterotrophic bacteria confirm the chemical etching activity of these organisms. Mixed cultures of diatoms and bacteria show depressions corresponding to the shape of individual cells. Our experiments confirm that diatoms, heterotrophic bacteria and cyanobacteria from natural biofilms can actively attack quartz and glass. Microscopic analysis of an idiomorphic scepter quartz crystal from a Tepui weathering environment reveals that the associated biofilms can create a local shift in the pH from 3.4 (pH of water on the Tepui) to evidently higher than 9 (necessary for quartz dissolution). The quartz covered with a biofilm is partially perforated to a depth of more than 4 mm. We conclude that biofilm growth in marine (sub-aquatic) and terrestrial (sub-aerial) conditions can significantly increase the breakdown of silica in the amorphous (glass), sub-crystalline (chert), crystalline and granular forms of quartz. Microbial growth may therefore substantially modify the processes of transformation of a major rock forming mineral of great chemical and physical resistance.
© 2004 Elsevier B.V. All rights reserved.

Keywords: Biocorrosion of quartz; Biologically accelerated weathering; Sub-aerial and sub-aquatic biofilms; Glass; Idiomorphic quartz

---

* Corresponding author. Tel.: +49 441 7983393; fax: +49 441 7983384.
  E-mail address: a.gorbushina@uni-oldenburg.de (A. Gorbushina).
[1] Tel.: +44 23 9284 2486; fax: +44 23 9284 2512.

0031-0182/$ - see front matter © 2004 Elsevier B.V. All rights reserved.
doi:10.1016/j.palaeo.2004.10.017

1. Introduction

All solid surfaces interfacing with the atmosphere and hydrosphere are susceptible to environmentally influenced alterations. Rock materials formed by high pressure and temperature within the Earth's crust are highly stable under the conditions of their formation. When they become exposed to the subaerial environment at the Earth's surface, however, they encounter a whole range of inorganic environmental influences such as water, atmospheric gases, and temperature fluctuations. Additionally and most significantly, living organisms also settle on their surfaces. These diverse environmental factors cause rocks to undergo alterations in mineralogical composition and material properties. The breakdown of rock forming minerals in this way is formally termed 'weathering'. During the weathering process rocks undergo sequential change in their chemical and physical properties, often including a loss of strength, mass, surface form and color. These alterations occur under the influence of physical (mechanical), chemical and biological processes, though subaerial breakdown generally involves interaction between all three types (White et al., 1992). As living organisms and especially microscopic ones are ubiquitous settlers on every solid surface on Earth, and frequently their action significantly accelerates the effects of chemical and physical factors (see for example, Papida et al., 2000), many weathering processes are closely affected by the presence of microbial growth. As a measure of the contribution of organic activity to rock weathering, it is estimated that 20–30% of stone weathering is the result of biological activity (Wakefield and Jones, 1998). The processes of mineral breakdown are also very important in biological fluxes on our planet since, as a result of microbial attack and consequent mineral breakdown, ions will be leached out into the surroundings and then translocated and used for growth and metabolism by primary producers such as plants and microorganisms (Lauwers and Heinen, 1974; Leyval and Berthelin, 1991; White et al., 1992).

As the environment, and especially biological growth, usually demonstrate a great degree of spatial heterogeneity, chemical and biological weathering processes often act not over the whole surface, but rather tend to concentrate at specific localities such as natural depressions, rock fissures and cleavage lines between mineral faces. This is partially a consequence of the structure of mineral crystals (Hochella and Banfield, 1995; Banfield and Hamers, 1997) and partially of the biological

growth response to surface irregularities (contact guidance or thigmotropism). This organismic response is expressed by following scratches, ridges and grooves, and by penetrating pores or tunnels (Hoffland et al., 2004). Certainly cases do exist whereby the organisms in question do not prefer any particular naturally favourable site, but rather simply penetrate the substrate in a manner congruent with their own inherent growth pattern. Microenvironments and biomorphogenic features created by the growth and metabolic activity of microorganisms are extremely important in the subsequent processes of chemical and physical breakdown. Microbial growth and accompanying biogenic products condition mineral surfaces and influence mineral breakdown by modifying the viscosity and chemical composition of adjacent water. Such biogenic conditioning can probably even influence internal weathering, through penetration of biogenic molecules into the internal space of a crystal where the flux of material to and from the primary/secondary interface is controlled by diffusion processes (Hochella and Banfield, 1995). The bonding between individual crystals of a rock will be strongly affected by chemically aggressive substances produced by the flora penetrating along the crystal boundaries (Krumbein and Dyer, 1985; Westall et al., 1990). Biochemical breakdown of rock forming minerals can result in microtopographic change of mineral surfaces through pitting and etching of their surfaces, mineral displacement reactions, widening of pores and mineral interfaces, and even complete dissolution of mineral grains (Ehrlich, 1998; Burford et al., 2003; Kumar and Kumar, 1991). Additionally, physical (mechanical) forces are exerted by microorganisms, as microscopic fungi spread within cracks and even through entire mineral bodies (Jongmans et al., 1997; Dornieden et al., 1997; Sterflinger, 2000).

Quartz is one of the most resistant of rock forming minerals (White and Brantley, 1995). It rates high on the Moh's Scale of Hardness with a value of 7. Its rate of dissolution is slow, about 10^{-17} mol cm^{-2} s^{-1} at 40 °C at near neutral pH in pure water, because the activation energy required to break Si–O bonds is high (Banfield and Hamers, 1997; Dove, 1995). As a component of rocks (e.g. granite, gneiss, sandstone) quartz crystals and grains have a higher resistance to weathering processes than many other common minerals such as feldspars and mica.

Quartz is also an abundant mineral on Earth—about 20% of the crust is composed of it (Nesbitt and Young, 1984). In terms of surface outcrop, quartz is estimated to occupy approximately 28% of the surface of the Earth's exposed (continental) crust (Leopold et al., 1964). It therefore contributes substantially to the surface geomorphology of the continents. Quartz is a major component of many building stones, where weathering processes may create issues of heritage conservation. Quartz is also present in the form of accumulating sediments in aqueous environments, particularly around continental margins. The abundance and ubiquity of quartz implies that any processes or organism contributing to the breakdown of this abundant mineral have potentially far reaching effects.

Quartz occurs as a primary mineral of rock formation in crystalline form, in which it may present as well formed geometric (euhedral or idiomorphic) crystals. Following rock breakdown resistant quartz crystals may be transported, abrading and perhaps fracturing as they travel, to be redeposited in secondary form as grains, either as an unconsolidated sedimentary mass or, subsequently, as consolidated sedimentary rock. Silica may also occur naturally in cryptocrystalline form as flint or chert, and in amorphous form as basalt or manufactured glass. Quartz and silica in all these forms may be subject to biologically influenced dissolution (Bennett and Siegel, 1987; Thorseth et al., 1995; Gorbushina and Palinska, 1999) and mechanical fracture (Krumbein et al., 1991).

In this paper we present examples of biologically induced breakdown of these highly resistant materials in various forms and environments. The forms investigated include natural idiomorphic quartz crystals, quartz sand, and window glass. The environments investigated include the major subaquatic and subaerial reservoirs of the mineral. Three different types of biofilm-quartz interaction were investigated:

■ A natural microbial community dwelling on the surface of an idiomorphic quartz crystal from a Tepui (tabular mountain) in Venezuela. In this case the microbial community was resident on and affecting the mineral surface for considerably longer than 10 years.

■ A natural intertidal microbial community cultivated under laboratory conditions on the surface of sterilized quartz sand from the North Sea. The well developed community was incubated on the sediment surface under controlled environmental conditions for a period of 9 months.

■ A selected sample from a natural community typical of intertidal sediment surfaces was inoculated on the surface of amorphous silica (glass). The organisms included diatoms and accompanying heterotrophic bacteria. The mesocosms were incubated under controlled environmental conditions in the laboratory, and within 9 months had developed a biofilm on the glass.

2. Materials and methods

2.1. Sources of experimental materials and organisms

The quartzitic tabular mountains or "Tepuis" of Venezuela form part of the Guyana shield. The Guyana shield has a continental erosion record spanning more then 60 million years. During these geological time periods, the climate changed from arid to extremely dry periods alternating with torrential rains. The current weathering environment is characterized by surface water with very low pH values (3.4±0.5) and high DOC concentrations (19±12 μg/ g) (Gorbushina et al., 2001). Several idiomorphic quartz crystals of 2–5 cm length and quartzite samples from the tepui surface which had experienced weathering in the field were collected. The quartz crystals were of the scepter type, in which the large prismatic head of the crystal is attached to a narrower stem beneath. The euhedral form of the head provides datum planes representing an initial surface, against which the depth of attack can be assessed. A predominantly cyanobacterial community is present on the quartz surface of the scepter crystal (Fig. 1), characteristic of the abundant carpet-like growth of cyanobacteria of the Tepui environment (Buedel et al., 1994; Gorbushina et al., 2001). Rock surfaces are therefore subject to the direct influence of microbial biofilms, which also penetrate into the substrate, and cause a significant degree of biological breakdown.

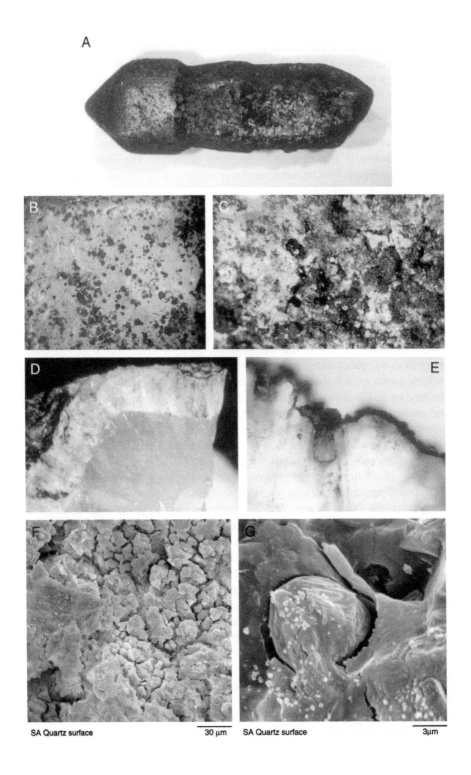

SA Quartz surface   30 μm   SA Quartz surface   3μm

2.2. Laboratory methods

In all mesocosm experiments natural microbial communities rather than pure cultures were deliberately used in order to simulate natural breakdown processes on quartz substrates.

In order to conduct controlled investigations of microbial communities on quartz and glass an in vitro laboratory system was used. A sample of a microbial mat community from surface sediments of the Wadden Sea (North Sea, Germany) was aseptically removed and placed on the surface of the model system. In the community various organisms were present, i.e. cyanobacteria (*Phormidium* sp., *Oscillatoria* sp., *Anabaena* sp. and *Synechocystis* sp.) as well as diatoms, green algae and chemotroph bacteria. The model substrates included sterilized natural sediments from the North Sea and commercial window glass. The grain size distribution in the North Sea sediment was 11.2% <0.315 mm, 43.8% 0.315–0.500 mm, 41.5% 0.5–1.0 mm and 3.5% >1.0 mm. The sediments were sterilized and placed in special glass containers, closed with a glass filter (0.2 μm porosity) at the bottom. This allowed the samples to moisten from below whilst leaving the surface wet but not water covered. The sediment surface was always higher than the water surface in the lower bowl, but capillary forces of the pore space in the sand ensured water supply throughout the whole sediment column. In this way the development of the growing biofilm at or near the sediment surface was supported by sufficient humidity but without any mechanical disturbance by water falling from above. The samples were wetted by natural North Sea water with a salinity of 35‰. The deteriorating influence of cyanobacteria on glass destruction is already known (Gorbushina and Palinska, 1999). Therefore in this experiment only diatom isolates with accompanying heterotrophic bacteria were used. Diatoms were selected for the glass experiment because they are (1) typical colonists

of sediment surfaces and (2) demand the presence of silica for metabolic processes and the production of a siliceous skeleton (Strelnikova, 2000). Additionally diatoms produce a significant amount of extracellular polymeric substances, which contribute to biofilm formation and may stimulate mineral material corrosion and growth of the accompanying heterotrophic bacteria. Pieces of ordinary window glass were inoculated with different diatoms from non axenic cultures (*Nitschia* sp., *Navicula* sp., *Cocconeis* sp.). Diatom strains were isolated from the Wadden Sea sediments by Cornelia Pfeiffer and kindly made accessible for our studies. The cultures included the accompanying heterotrophic bacteria and therefore represented a part of the natural microbial community.

2.3. Observation methods

Samples for scanning electron microscopy (SEM) were taken from the mesocosms or a natural sample, fixed with 2% glutaraldehyde in a phosphate buffer (pH 7.2), dehydrated by increasing the ethanol concentration, critical point dried (Balzers Union CPD 010), and coated with gold (Balzers Union SCD 030). They were examined using a Hitachi S-3200N scanning electron microscope with an accelerating voltage of 18–20 kV.

To observe the actual material changes on the glass surface under the biofilm, the colonized window glass samples were prepared by cleaning the surface with HCl to remove the microorganisms. For comparison, glass still covered with the biofilm was also analyzed.

Surface change in the idiomorphic crystal was observed with a dissecting microscope Zeiss DRC equipped with the photographic utensils MC63 and M35. Samples for thin sections were fixed with 2% glutaraldehyde in a phosphate buffer (pH 7.2), dehydrated by increasing ethanol concentration, and embedded in Spurr's resin (Spurr, 1969). In order to gain thin sections, embedded blocks were cut with a

Fig. 1. (A) Overview of scepter quartz crystal. The stem is 1.8 cm broad. (B) The head of the scepter quartz crystal has a smoother surface with unicellular cyanobacterial colonies loosely distributed on it. Single colonies measure 0.2 to 1.5 mm. (C) Pitted surface on the stem of the crystal, covered by abundant cyanobacteria. One can easily recognize considerable surface changes. (D) Cross-section through the crystal showing an increased porosity of the outermost layer of the crystal (a milky color reaction zone). (E) The distribution of organisms (red pigmented cyanobacteria) follows the pores and brings about distinguishable pitting. (F, E) Surface of an idiomorphic quartz crystal documented by scanning electron microscopy (SEM). (F) SEM micrograph of a quartz surface transformed into a sponge by the growth of unicellular cyanobacteria. (E) Contact of the cell with the quartz surface causes obvious surface pitting.

circular saw (Leitz 1600) and glued with Spurr resin to petrographic glass slides. The thin sections were examined using the microscope Axioplan (Zeiss).

In order to investigate the exclusively chemical interaction between the leaching activity of solutions with an elevated pH (>9) and sand size, in which total grain surface area is a measure of surface reactivity, NaOH solution with a pH of 11.34 was added to uncolonized quartz grains. Two grain size fractions (samples of 1 g) (i) smaller than 80 μm and (ii) 0.5–1.0 mm were used for this experiment to show the effect of increasing the reactive surface area on quartz mass dissolution rates. To assess the weight loss due to chemical solution, the samples were weighed before and after the experiment. The samples were leached in 500 ml NaOH for 1 week, washed, dried and weighed again.

3. Results

3.1. Idiomorphic crystal

The surface of the scepter quartz crystal covered by cyanobacterial biofilms exhibits a totally altered surface morphology as compared to freshly exposed scepter quartz from e.g. druses (Fig. 1A). The biofilm-induced morphology of the idiomorphic crystal shows partial destruction concentrated on certain zones. The head of the crystal has a smoother surface with unicellular cyanobacterial colonies loosely distributed on it (Fig. 1B). The stem of the crystal shows deeper pitting (Fig. 1C) and had probably been situated within the microbial mat in more humid environmental conditions, encouraging better growth of cyanobacteria. The weathering zone of the quartz crystal exactly corresponds to the crystal habitus and is evenly distributed on all sides of the crystal. The outermost layer of the crystal shows a milky reaction zone as seen in cross section (Fig. 1D), typical of the outer surface of many sub-aerially weathered desert chert samples (Krumbein, 1969). The milky semi-opaque appearance probably results from the increased porosity, reflecting the influence of environmental factors and morphologically expressing chemical and physical alteration of the silicate structure. The distribution of organisms in this opaque layer (red pigmented cyanobacteria, Fig. 1E) follows

the pores and is associated with recognizable pitting (Figs. 3 and 4). The stem is completely covered predominantly by filamentous cyanobacteria and shows very deep pitting (Fig. 1C), while the smoother surface of the crystal head demonstrates only a loose distribution of cyanobacterial colonies (Fig. 1B). These differences in surface changes between the tip and the stem of the crystal are closely correlated with the thickness and the species composition of the biofilm. The presence/coverage of a thicker carpet-like biofilm with dominating filamentous cyanobacteria at the stem causes more intensive surface changes and recognizable pitting patterns (Fig. 1C). The sparse biofilm of solitary colonies on the tip of the crystal leaves some surface areas completely uncovered, which is mirrored by the less developed pitting and partially unleached mineral surface. The outer milky weathered layer is also fragmentary or absent on the crystal head. These phenomena demonstrate the direct influence of cyanobacterial colonies on the intensity of the leaching and pitting processes on the crystal surface. Cyanobacterial growth accumulations at a single point on the surface produce deep pits around each single cell (Fig. 1F,G). The most likely cause of such leaching and pitting processes is an increase in pH in the vicinity of a photosynthesizing cell. However, extracellular mucilage and organic acids also can influence interactions with minerals. For instance, comparable breakdown and even perforation patterns (channels or tunnels) have been also described for bacterial cells on basaltic glass (Thorseth et al., 1995) and mycorrhizal fungi on feldspars (Jongmans et al., 1997). Destructive (pitting or drilling) patterns created by each single cell (e.g. Thorseth et al., 1995) and filament (Jongmans et al., 1997) can coalesce to form a continuous erosion front in the case of colony development (Fig. 1F). Such coalescence of individual boring patterns was described for limestone and chert in desert environments by Krumbein (1969) and Krumbein and Jens (1981). The influence of biological growth becomes even greater following multiple and massive colony and biofilm development. The biofilm in its complexity forms a continuous organic layer possessing sometimes very different properties, e.g. water storage capacity, heat conductivity and chemical composition including pH. The presence of a biofilm on the mineral substrate creates different geochemical conditions favoring acceleration or retardation of the weathering process. The accumulation and location of living colonies on the surface consequently influence the degree and the pattern of rock breakdown and change in mineral surface morphology. The resulting changes of mineral structure and porosity beneath the biofilm closely reflect the individual growth pattern and morphology of the biofilm. They can be used also for recognition of past (extinct) biological growth (Gorbushina et al., 2002).

3.2. Sterilized quartz sand

In the sterilized quartz sand experiment, rounded quartz grains served as a substrate for a cyanobacterial community with trichomes oriented and concentrated in bundles, in which several partially intertwined trichomes grow parallel to each other. This orientation ensures that the influence of cyanobacteria is concentrated in particular areas where the most biomass is present (Fig. 2A). Photosynthetic activity and exopolymer production by cyanobacteria and their accompanying heterotrophic bacteria lead to depressions adjacent to colonies vicinity (Fig. 2B). Further evolution of depressions in the areas of the quartz grains covered by these colonies, "belts" of trichomes results in more remarkable biogenic features (Fig. 2C–F). The ongoing breakdown and leaching under the influence of cyanobacterial filamentous belts result in progressive fragmentation of individual quartz grains (Fig. 2C). Thus even in the absence of any mechanical action by waves, the quartz grain fraction of the experimental set-up was reduced in size by fragmentation of large grains into smaller ones as a result of the locally concentrated growth of microorganisms (Fig. 2A,C). A consequence of the particle fragmentation is the increase in the overall grain surface area, which becomes susceptible to further chemical weathering (see geochemical calculations). Microenvironments created under the cells (filaments, colonies or biofilms) contribute to local biopitting effects by microbiota on quartz grain surfaces. By this localized dissolution the surface area is further enlarged and subsequently offers an even greater reactive surface for further interactions and weathering. From the Wadden Sea sand experiments biogenic grain reduction and cleavage of bigger quartz grains is clearly demonstrated.

3.3. Window glass

The leaching effects of diatoms could be very clearly demonstrated on the window glass. Diatoms and their accompanying bacteria were embedded in large amounts of extracellular polysaccharides covering the surface of the glass. After removal of the microorganisms the glass surfaces showed numerous conspicuous depressions or pitted zones, which were completely absent on the glass surfaces in their original condition (Fig. 3). Some of these depressions were aligned as if a number of diatoms had been living there (Fig. 3). It is, however, also possible that after having created a depression a diatom cell may move to a new place and resume the same etching process. The size of the diatoms correlated with the depressions on the cleaned glass surfaces following the experiment. The diatoms evidently derive silica from the glass surface for their metabolism and frustule construction. However, the associated heterotrophic bacteria are seemingly necessary in this ion transformation process, because the smaller moulds associated with heterotrophic bacteria are also visible on the glass surface (Fig. 3A). The combined etching activity of both diatoms and heterotrophic bacteria is clearly evidenced by our data. Since the cultivation of axenic diatoms is nearly impossible, the question of whether the diatoms, or the accompanying heterotrophic bacteria, are the main cause of the observed biopitting remains open. The successful interplay of diatoms and bacteria, however, seems to be beneficial for both. Diatoms produce polysaccharides useful for bacterial metabolism and survival; bacteria with their leaching activity provide Si ions for diatom frustule construction. This results in continuous leaching and degrading processes, which are part of the accelerated weathering induced by microorganisms. From the window glass experiment it is concluded that there is a synergistic interaction of individual microbiota which, in addition to the metabolic advantages, also increases the overall material breakdown.

3.4. Dissolution rates of quartz, grain size, and surface reactivity

In order to investigate the influence of grain size on quartz mass dissolution rates, an experiment was carried out in which two grades of sand were

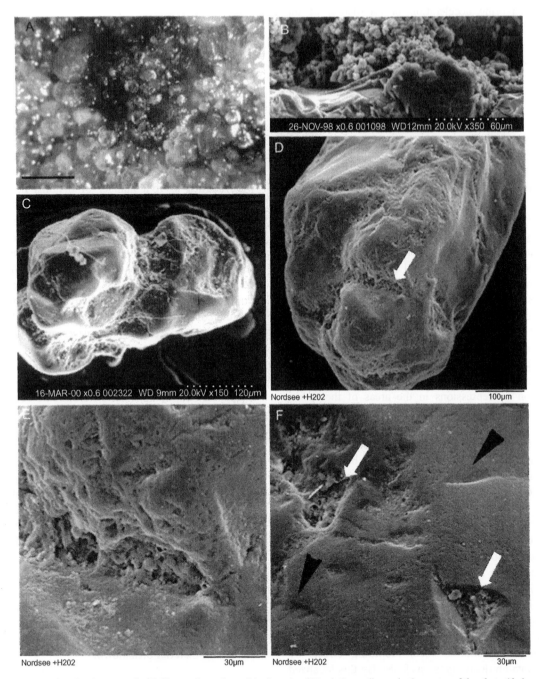

Fig. 2. (A) Abundant development of a biofilm on the surface of the inoculated North Sea sediment in the centre of the photo (darker area) is obviously associated with smaller (fragmented) grains. Scale bar is 800 μm. (B) bacterial colony growing in a depression on the quartz surface. (C) Fragmentation of quartz grains of a defined size in the areas of intensive cyanobacterial growth. If the process continues, splitting of the quartz grain by cyanobacterial filaments will soon become apparent. (D) Quartz grain, on which a biofilm has been removed by a peroxide treatment, exposes several regions with biogenic surface pitting. These regions have been covered by a biofilm and after their removal appear as furrowed etched areas. (E) Close-up of a biogenic pit with traces of two colonies of cyanobacteria. (F) Comparative micrograph of biogenic (white arrows) and typical mechanical/chemical pits (black arrowheads) on a quartz surface.

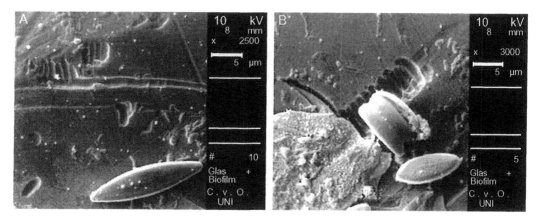

Fig. 3. (A) SEM micrographs of an etched window glass surface with recognizable traces and cells of bacteria and microscopic fungi or filamentous cyanobacteria. The remains of the biofilm including several bacteria and a diatom in the lower right corner are still present on the surface. (B) SEM micrograph of an etched window glass surface showing pits left by a diatom chain on the surface of glass. The pits are situated in a row characteristic of diatom colonies. Biofilm and several diatoms are clearly visible.

dissolved in a highly alkaline solution of NaOH. Uncolonized sand was used in order to simplify the solution process in the absence of potentially complicating biogenic effects.

Sand grains of two different size classes were used: (i) with a diameter 500 μm (=0.5 mm=0.05 cm) and (ii) with a diameter <100 μm (=0.1 mm=0.01 cm). Each sample weighed 1 g. For the calculations of surface area and volume of the grains the following formulae have been employed: $V_{sphere}=1/6\pi d^3$; $S_{sphere}=\pi d^2$.

The following assumptions are made: (i) sand grains are spherical; (ii) density of sand is 2.65 g cm^{-3}; (iii) solution loss is proportional to grain surface area.

From these calculations it follows (Table 1) that 1 g of sand grains of 100 μm grain diameter has a surface area (that is, a reactive weathering surface area) that is five times larger than the total grain surface area of 1 g of sand of 500 μm grain diameter. As the intensity of all weathering reactions is proportional to the total surface area of the grains, it can be assumed that the rate of weathering will be five times greater for the 100 μm grain fraction.

When all the 100 μm grains have been already completely dissolved, only about 0.2 g of the large grains (500 μm) are dissolved (proportion of the surface per gram, see Table 1). From this the following equation is derived:

$$\sum \Delta m = 0.2 \text{ g sand} \Rightarrow \sum \Delta V = 0.075 \text{ cm}^3$$

ΔV per sphere: 0.075 cm^3/g: 5765 spheres/ g=1.3×10^{-5} cm^3/sphere
The diameter of sand grain before leaching: $V_{sphere1}$=1/6π(0.05 cm)^3
After leaching: $V_{sphere2}=V_{sphere1}-1.3\times10^{-5}$ cm^3/ sphere
=6.5×10^{-5} cm^3/sphere−1.3×10^{-5} cm^3/sphere
=5.2×10^{-5} cm^3/sphere=1/6π($d_{sphere2}$)^3
$\Rightarrow d_{sphere2}$=463 μm
i.e. the 500 μm spheres are reduced by 37 μm in diameter.

This shows that in the time taken for the 100 μm fraction to dissolve completely, the grain diameter of the 500 μm fraction is reduced by only 8%. In a closed system the smaller grain fractions will be dissolved much more rapidly than the larger ones and are therefore much more susceptible to weathering (Atkins, 1994). Biogenic grain fragmentation as observed in the experiment with natural sediment surface communities (see the above section on

Table 1
Comparison of grain characteristics in two grain size fractions

Diameter of sphere (μm)	500	100
Volume of sphere (cm^3)	6.5×10^{-5}	5.2×10^{-7}
Mass of sphere (g)	1.7×10^{-4}	1.4×10^{-6}
Number of spheres per 1 g	5765	720702
Surface area of 1 g sand (cm^2)	45	226

Sterilized quartz sand) significantly accelerates the dissolution process and brings about faster dissolution of the smaller grains. This more rapid dissolution under the combined influence of chemical and biological factors indicates that smaller grains dissolve and finally disappear extremely quickly. The rapidity of this process implies that the fine fraction of sandy sediments may well become underrepresented in marine environments in which microbial activity is abundant. This is witnessed in most microbial mats of the North Sea intertidal zone where grain sizes of 150–300 μm prevail and smaller grains as well as larger grains are practically absent (Krumbein et al., 1994).

4. Discussion and conclusions

Living organisms strongly influence mineral and rock breakdown processes through both physical and chemical action. This applies even to the most resistant mineral materials, such as quartz, chert, glass and similarly resistant mineral forms. As demonstrated in this study, quartz surface morphology and grain size dramatically change under the influence of microbial communities (Figs. 1 and 2). The mechanisms involved in such processes are, however, diverse. Quartz solubility significantly increases in alkaline conditions of pH 9 or higher (Dove and Rimstidt, 1994; Schwoerbel, 1999). As homogeneous environments with such high alkaline pH values are rare, local variations in physicochemical conditions at the rock surface are most important in weathering processes. In most environments local changes can occur, for instance in the close proximity of microorganisms that conduct photosynthesis. Photosynthetic activity and associated CO_2 consumption locally raise pH in selected microenvironments around the cell (Fig. 4). In contrast declining photosynthetic activity (increase of CO_2 production) leads to the establishment of lower pH values. The pH value directly influences quartz degradation and at pH values lower than 3.5 no quartz dissolution is possible (Fig. 4), although mechanical processes of grain diminution may still continue. Above pH 9 the dissolution rate of quartz is at a maximum. Microscopic analysis of an idiomorphic scepter quartz crystal from a Tepui weathering environment (Fig.

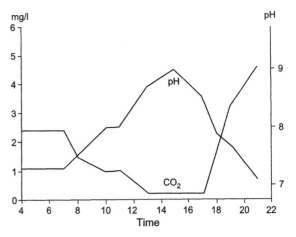

Fig. 4. Rates of quartz dissolution expressed as a function of pH and juxtaposed with CO_2 consumption associated with photosynthetic activity.

1) reveals that the associated biofilms can create a local shift in the pH from 3.4 (pH of water on the Tepui) to evidently higher than 9 (necessary for quartz dissolution). The quartz crystal covered with a biofilm is partially perforated to a depth of more than 4 mm, but the intensity of pit formation is closely correlated with the thickness and the species composition of the biofilm. Similar pit formation with following fragmentation of grains was also demonstrated for sand grains in the laboratory experiments (Fig. 2). The most likely cause of such leaching and pitting processes is an increase in pH in the vicinity of any photosynthesizing cell. However, at more advanced stages of weathering, enhanced access of fluids is probably accompanied by an increased impact of organic compounds, especially where microorganisms and their extracellular metabolites (e.g. mucilage) are present. Imprints of diatoms on glass surfaces colonized by diatoms and heterotrophic bacteria (Fig. 3) confirm the chemical etching activity of these organisms. Mixed cultures of diatoms and bacteria on glass (Fig. 3) and quartz (Fig. 2B) show depressions corresponding to the shape of individual cells. The presence of bacteria may play a significant role in this etching. It has been demonstrated that heterotrophic bacteria in such microbial communities are responsible for ion transport and thus are important in dissolution processes (Decho, 2000). Silica dissolution is significantly altered by the presence of silica-organic complexes (Bennett, 1991) and in NaCl solution (Xie and Walter, 1993) presumably because

these substances change the structure of water in a way that increases hydrogen bonding with H_4SiO_4 (Dove and Rimstidt, 1994). Mineral surfaces are very often affected by significant amounts of organic coatings (Jones and Uehara, 1973; Krumbein and Dyer, 1985; Westall et al., 1990) that are produced by bacteria, algae and fungi, as well as higher organisms that inhabit external and internal (fracture/fissure) surfaces of rock substrates. Important weathering interfaces may be covered by abundant biofilm products such as polysaccharides, their corresponding acids and other metabolic products which have a profound effect on weathering processes (Krumbein and Dyer, 1985; Westall et al., 1990; Welch and Vandevivere, 1994; Hochella and Banfield, 1995; Barker et al., 1997; Welch et al., 1999; Decho, 2000).

Mechanical water or wind induced weathering produces small pits, which evolve and grow larger by following the mineral symmetry and thus frequently approximate an ideal shape (e.g. Kempe et al., 2004; Fig. 2F). These patterns form a starting point for biological settlement and weathering, which superimpose the cell-shape driven morphology on the preceding pattern of predominant chemical and physical non-biological weathering (e.g. Fig. 2F).

From investigations of naturally weathered quartz and mesocosm experiments with quartz sand and amorphous silica, we conclude:

- Large idiomorphic quartz crystals (scepter quartz) from the quartzite-weathering environment of the Tepuis in Middle America, quartz grains from a recent marine sediment and window glass invariably show considerable morphological changes produced by different biofilm communities settling on and in them. Biopitting and grain dissection are the most characteristic effects.
- Micro-environmental photosynthesis-induced alkaline leaching of quartz is induced by biofilm growth even in a general pH environment between 3.5 and 8.5.
- Sedimentary quartz sand fractions can be reduced in grain size in situ rapidly without any water movement or mechanical action exclusively through the breakdown activity of biofilm communities.
- Some of the morphologies of the depressions (pit formation) and the fragmentation of large sand grains into smaller grains of about half the size

exploit mechanical weaknesses in the later stages of colonization (Fig. 2).
- Theoretical geochemical calculations using grain size and total reactive surface area are experimentally confirmed and reveal rapid biocorrosion or bioerosion processes.
- A synergistic association of diatoms with chemo-organotrophic bacteria may rapidly dissolve window glass thereby furnishing silica for frustule production under subaerial biofilm conditions.

In all the samples investigated the depressions in the quartz and glass surface corresponded exactly with the outline of microorganisms or biofilms (Figs. 1–3). This represents clear evidence of the capacity of microbial organisms directly to effect the dissolution and even the fragmentation of quartz.

Given the widespread distribution of quartz in almost all Earth surface environments, it is clearly implied that the microbial contribution to quartz, chert and (volcanic) glass breakdown is widely significant. The microbial attack on quartz has clear implications for the integrity of solid rock, both in natural outcrop and as cut stone in the built environment. The capacity to attack quartz sand in shallow marine environments, and in particular the disproportionate effect on the fine fraction implies that the evidenced microbial growth is capable of modifying the grain size distribution of marine sediments and, thereby, will make a significant contribution to the sediment transformation into solid rock, which can be designated as microbialites or siliciclastic stromatolites (Krumbein et al., 1994, 2003). Although not studied in these experiments many soils contain quartz sand, and microbial weathering takes place in the soil environment as well. The implications of the microbial attack on quartz extend more widely still, for if microbes can affect a tightly bonded mineral such as quartz, they will surely act more effectively on other less strongly bonded minerals with silicate sheet or chain frameworks, such as feldspars, amphiboles and pyroxenes, and micas and clay minerals. In addition to the 28% of the Earth's land surface formed by quartz an estimated 49% surface is occupied by other minerals with a silicate structure (Leopold et al., 1964). By extension, therefore, it can be inferred that the microbial contribution to mineral and rock weathering is extremely widespread, in respect of both mineral types and weathering environments.

Acknowledgements

This work was in part financially supported by DFG grant Kr 333/32 and EU-grant EVK4-CT-2002-00098 ("Inhibitors of biofilm damage on mineral materials"). The skilful microscopical and technical assistance of Renate Kort is deeply appreciated. Cultures of diatoms were kindly made accessible by Cornelia Pfeiffer and Katarzyna Palinska.

References

Atkins, P.W., 1994. Physical Chemistry. Freeman, New York.

Banfield, J.F., Hamers, R.J., 1997. Processes at minerals and surfaces with relevance to microorganisms and prebiotic synthesis. Rev. Miner. 35, 81–122.

Barker, W.W., Welch, S.A., Banfield, J.F., 1997. Biogeochemical weathering of silicate minerals. In: Banfield, J.F., Nealson, K.H. (Eds.), Geomicrobiology: Interactions Between Microbes and Minerals. Min. Society of America, Washington, pp. 391–428.

Bennett, P.C., 1991. Quartz dissolution in organic-rich aqueous systems. Geochim. Cosmochim. Acta 55, 1781–1797.

Bennett, P.C., Siegel, D.I., 1987. Increased solubility of quartz in water due to complexing by organic compounds. Nature (Lond.) 326, 684–686.

Buedel, B., Luettge, U., Stelzer, R., Huber, O., Medina, E., 1994. Cyanobacteria on rocks and soils of Orinoco lowlands and the Guayana Uplands, Venezuela. Bot. Acta 107, 422–431.

Burford, E.P., Fomina, M., Gadd, G.M., 2003. Fungal involvement in bioweathering and biotransformation of rocks and minerals. Mineral. Mag. 67, 1127–1155.

Decho, A., 2000. Exopolymer microdomains as a structuring agent for heterogeneity within microbial biofilms. In: Riding, R., Awramik, S.M. (Eds.), Microbial Sediments. Springer, Berlin, pp. 9–15.

Dornieden, Th., Gorbushina, A.A., Krumbein, W.E., 1997. Aenderungen der physikalischen Eigenschaften von Marmor durch Pilzbewuchs. Int. J. Restor. Build. Monum. 3, 441–456.

Dove, P.M., 1995. Kinetic and thermodynamic controls on silica reactivity in weathering environments. In: White, A.F., Brantley, S.L. (Eds.), Chemical Weathering Rates of Silicate Minerals, Bookcrafters, Michigan, Reviews in Mineralogy, vol. 31, pp. 235–290.

Dove, P.M., Rimstidt, J.D., 1994. Silica-water interaction. Rev. Miner. 29, 259–308.

Ehrlich, H.L., 1998. Geomicrobiology: its significance for geology. Earth-Sci. Rev. 45, 45–60.

Gorbushina, A.A., Palinska, K.A., 1999. Biodeteriorative processes on glass: experimental proof of the role of fungi and cyanobacteria. Aerobiologia 15, 183–191.

Gorbushina, A.A., Boettcher, M., Brumsack, H.-J., Krumbein, W.E., Vendrell-Saz, M., 2001. Biogenic forsterite and opal as a product of biodeterioration and Lichen Stromatolite formation in table mountain systems (Tepuis) of Venezuela. Geomicrobiol. J. 18, 117–132.

Gorbushina, A.A., Krumbein, W.E., Volkmann, M., 2002. Rock surfaces as life indicators: new ways to demonstrate life and traces of former life. Astrobiology 2 (2), 203–213.

Hochella Jr., M.F., Banfield, J.F., 1995. Chemical weathering of silicates in nature: a microscopic perspective with theoretical considerations. In: White, A.F., Brantley, S.L. (Eds.), Chemical Weathering Rates of Silicate Minerals, Bookcrafters, Michigan, Reviews in Mineralogy, vol. 31, pp. 354–406.

Hoffland, E., Kuyper, T.W., Wallander, H., Plassard, C., Gorbushina, A.A., Haselwandter, K., Holmström, S., Landeweert, R., Lundström, U.S., Rosling, A., Sen, R., Smits, M.M., van Hees, P.A.W., van Breemen, N., 2004. The role of fungi in weathering. Front. Ecol. 2 (5).

Jones, R.C., Uehara, G., 1973. Amorphous coatings on mineral surfaces. Proc.-Soil Sci. Soc. Am. 37, 792–798.

Jongmans, A., Van Breemen, N., Lundstrom, U., van Hees, P.A.W., Finlay, R.D., Srinivasan, M., Unestam, T., Giesler, R., Melkerud, P.A., Olsson, M., 1997. Rock-eating fungi. Nature 389, 682–683.

Kempe, A., Jamitzky, F., Altermann, W., Baisch, B., Markert, Th., Heckl, W.M., 2004. Discrimination of aqueous and aeolian paleoenvironments by atomic force microscopy—a database for the characterisation of martian sediments. Astrobiology 4 (1), 51–64.

Krumbein, W.E., 1969. Über den Einfluss der Mikroflora auf die exogene Dynamik (Verwitterung und Krustenbildung). Geol. Rundsch. 58, 333–363.

Krumbein, W.E., Dyer, B.D., 1985. This Planet is alive. Weathering and Biology—a multifaceted Problem. In: Drever, J.I. (Ed.), The Chemistry of Weathering. Reidel, Dordrecht, pp. 143–160 (324 pp.).

Krumbein, W.E., Jens, K., 1981. Biogenic rock varnishes of the Negev Desert (Israel): an ecological study of iron and manganese transformation by cyanobacteria and fungi. Oecologia 50, 25–38.

Krumbein, W.E., Urzi, C., Gehrmann, C., 1991. Biocorrosion and biodeterioration of antique and medieval glass. Geomicrobiol. J. 9, 139–160.

Krumbein, W.E., Paterson, D.W., Stal, L.J. (Eds.), 1994. Biostabilization of Sediments. BIS-Verlag, Oldenburg (526 pp.).

Kumar, R., Kumar, A.V., 1991. Biodeterioration of stone in tropical environments: an overview. The J. Paul Getty Trust, USA.

Lauwers, A.M., Heinen, W., 1974. Biodegradation and utilization of silica and quartz. Archiv. Mikrobiol. 95, 67–78.

Leopold, L.B., Wolman, M.G., Miller, J.P., 1964. Fluvial Processes in Geomorphology. W.H.Freeman & Co., San Francisco.

Leyval, C., Berthelin, J., 1991. Weathering of mica by roots and rhizospheric microorganisms of pine. Soil Sci. Soc. Am. J. 55, 1009–1016.

Nesbitt, H.W., Young, G.M., 1984. Prediction of some weathering trends of plutonic and volcanic rocks based on thermodynamic and kinetic considerations. Geochim. Cosmochim. Acta 48, 1523–1534.

Papida, S., Murphy, W., May, E., 2000. Enhancement of physical weathering of building stones by microbial populations. Int. Biodeterior. Biodegrad. 46, 305–317.

Schwoerbel, J., 1999. Einführung in Die Limnologie. Fischer Verlag, Stuttgart (465 pp.).

Spurr, A.R., 1969. A low-viscosity epoxy resin embedding medium for electron microscopy. J. Ultrastruct. Res. 26, 31–43.

Sterflinger, K., 2000. Fungi as geological agents. Geomicrobiol. J. 17, 97–124.

Strelnikova, N.I., 2000. Silicium as the basis of existence of the diatoms—one of the oldest group of algae. In: Witkowski, A., Sieminska, J. (Eds.), The Origin and Early Evolution of the Diatoms. Polish Academy of Sciences, Cracow, pp. 7–12.

Thorseth, I.H., Furnes, H., Tumyr, O., 1995. Textural and chemical effects of bacterial activity on basaltic glass: an experimental approach. Chem. Geol. 119, 139–160.

Wakefield, R.D., Jones, M.S., 1998. An introduction to stone colonising micro-organisms and biodeterioration of building stones. Q. J. Eng. Geol. 1998 (31), 301–313.

Welch, S.A., Vandevivere, P., 1994. Effect of microbial and other naturally occurring polymers on mineral dissolution. Geomicrobiol. J. 12, 227–238.

Welch, S.A., Barker, W.W., Banfield, J.F., 1999. Microbial extracellular polysaccharides and plagioclase dissolution. Geochim. Cosmochim. Acta 63, 1405–1419.

Westall, J.C., Brownawell, B.J., Chen, H., Collier, J.M., Hatfield, J., 1990. Adsorption of organic cations to soils and subsurface materials. USEPA Res. Dev. (EPA/600/S2-90/004).

White, A.F., Brantley, S.L., 1995. Chemical weathering rates of silicate minerals. Rev. Miner. 31, 1–22.

White, I.D., Mottershead, D.N., Harrison, S.J., 1992. Environmental Systems: An Introductory Text. Chapman & Hall, London, UK.

Xie, Z., Walter, J.V., 1993. Quartz solubilities in NaCl solutions with and without wollastonite at elevated temperatures and pressures. Geochim. Cosmochim. Acta 57, 1947–1955.

Available online at www.sciencedirect.com

Palaeogeography, Palaeoclimatology, Palaeoecology 219 (2005) 131–155

www.elsevier.com/locate/palaeo

Expanding frontiers in deep subsurface microbiology

Jan P. Amend[a],[*], Andreas Teske[b]

[a]*Department of Earth and Planetary Sciences, Washington University, St. Louis, MO 63130, United States*
[b]*Department of Marine Sciences, University of North Carolina, Chapel Hill, NC 27599, United States*

Received 16 June 2003; accepted 29 October 2004

Abstract

The subsurface biosphere on Earth appears to be far more expansive and physiologically and phylogenetically complex than previously thought. Here, several aspects of subsurface microbiology are discussed. Molecular and biogeochemical data, as well as characteristics from new isolates, suggest that ecosystems below deep-sea hydrothermal vents are inhabited primarily by thermophilic archaea and bacteria. The void spaces and conduits in basalt at mid-ocean ridges, and, even more so, at sediment-covered hydrothermal vent sites represent promising hunting grounds for novel chemosynthetic archaea and bacteria. As examples, we highlight the subsurface microbial communities within the basalt flanks of the Juan de Fuca Ridge off British Columbia, and the hydrothermal sediments at the Guaymas Basin in the Gulf of California and at Middle Valley on the Juan de Fuca Ridge. In the deep continental subsurface, microbial studies have primarily targeted aquifers in basaltic and granitic rock, with an almost exclusive emphasis on H_2-driven chemolithoautotrophy. For example, in the subsurface water from Lidy Hot Springs, Idaho (USA), H_2-consuming methanogenic archaea represent the most populous organisms, comprising >95% of all cells. Besides H_2, other potential sources of chemical energy (reducing power) are considered, including abiotically synthesized organic matter and reduced sulfur compounds. In addition, the well-established sequences of redox couples in microbial metabolism are investigated at elevated temperatures using a thermodynamic approach. Values of $p\varepsilon^o$ for 11 common reduction half-reactions are tabulated at 25, 50, 100, and 150 °C. Although many modes of metabolism may be employed by subsurface biota, sulfate reduction appears to be one of the more ubiquitous strategies. Recent studies of sulfate-reducing archaea and bacteria in marine and continental systems are reviewed.
© 2004 Elsevier B.V. All rights reserved.

Keywords: Subsurface biosphere; Marine hydrothermal systems; Deep continental biosphere; Sulfate-reducing prokaryotes; Redox potentials

1. Introduction

* Corresponding author. Tel.: +1 314 935 8651; fax: +1 314 935 7361.

E-mail addresses: amend@levee.wustl.edu (J.P. Amend), teske@email.unc.edu (A. Teske).

0031-0182/$ - see front matter © 2004 Elsevier B.V. All rights reserved.
doi:10.1016/j.palaeo.2004.10.018

Most of the biosphere that we see with our eyes or observe through a microscope is powered by solar radiation on the Earth's surface, either directly by photosynthesis, or indirectly by converting the products of photosynthesis—oxygen and primary biomass. However, microbial life has also colonized deep sediments and the Earth's crust, wherever available carbon and energy sources, and subsurface porosity permit. The deep subsurface microbial biosphere contributes significantly to overall biomass and biodiversity on Earth. In fact, a recent study reported that perhaps as much living carbon is scattered throughout the subsurface as is found in land plants (Whitman et al., 1998). As a telling indicator for the rapid growth of this field, several excellent reviews have appeared in recent years that cover the expanding data sets and new research directions in subsurface microbiology (e.g., Fredrickson and Onstott, 2001; Kieft and Phelps, 1997; Krumholz, 2000; Parkes et al., 2000; Parkes and Wellsbury, 2004; Pedersen, 1997; Pedersen, 2000a). Interest in the deep biosphere is further evidenced by the fact that culture collections specific to subsurface isolates have now been established (Balkwill et al., 1997). In the present overview, the focus is on selected topics in deep subsurface microbiology that have advanced significantly in the past few years: 1) the microbiology of the marine subsurface at hydrothermal vents; 2) the deep subsurface continental biosphere; 3) redox potentials of microbially mediated processes at elevated temperatures; 4) cultivations and molecular detection of sulfate-reducing microbes; and 5) contamination monitoring.

2. The hydrothermal biosphere in the marine subsurface

Hydrothermal vents on or near mid-ocean ridges are excellent candidates for biologically active hot spots in the Earth's crust, since they provide chemical energy sources that can be utilized by microorganisms (Karl, 1995). The deep hydrothermal subsurface includes the extensive zone where seawater is entrained into porous, freshly formed ocean crust and undergoes hydrothermal alteration. The steep thermal gradients there favor the presence of thermophilic (Table 1) microorganisms. At present, the temperature limit for growth of the two most hyperthermophilic microorganisms in pure culture stands at 113 and 121 °C; even higher temperatures may be tolerated during short-term survival (Blöchl et al.,

Table 1
Glossary of terms

Aerobe: an organism that uses oxygen (O_2) as its terminal electron acceptor.

Anaerobe: an organism that uses a terminal electron acceptor other than oxygen (O_2).

Archaea: one of the two prokaryotic and phylogenetic domains of life (the other being the Bacteria); it includes most of the extreme halophiles and hyperthermophiles and all of the methanogens.

Bacteria: one of the two prokaryotic and phylogenetic domains of life (the other being the Archaea); it includes cyanobacteria, Proteobacteria, Gram-positive bacteria, and other lineages, many with thermophilic representatives.

Chemolithoautotroph: an organism that obtains its metabolic energy from the oxidation of reduced inorganic compounds and its carbon for biosynthesis from CO_2 or CO.

Chemo(organo)heterotroph: an organism that obtains both its metabolic energy and its carbon for biosynthesis from organic compounds.

Domain: the highest taxonomic level in the classification of life; the three domains are the Archaea, Bacteria, and Eukarya.

Eukarya: one of the three domains of life, containing all non-prokaryotic organisms.

Exergonic: refers to a chemical reaction that is energy-yielding, i.e., for which $\Delta G_r < 0$.

Hyperthermophile: an organism with an optimum growth temperature ≥ 80 °C.

Mesophile: an organism that grows at ambient temperatures, generally between 15 and 45 °C.

Moderate thermophile: an organism with an optimum growth temperature between 45 and 80 °C.

Phenotype: the observable characteristics of an organism, including biochemical properties and metabolic pathways.

Phylotype: the phylogenetic marker sequence (commonly the 16S rRNA sequence) that defines the phylogenetic identity of an organism.

Prokaryote: single-celled microorganisms lacking a membrane-bound nucleus and other structural features of eukaryotic cells.

Terminal electron acceptor: the oxidant (e.g., O_2, NO_3^-, SO_4^{2-}, CO_2) in a biologically mediated overall redox process.

Thermophile: any organism that grows at high temperatures.

1997; Kashefi and Lovley, 2003). Interpreting these microbially permissible temperature maxima in the context of standard geothermal gradients in the Earth's crust suggest that the habitable zone in the subsurface may be several kilometers deep (Colwell, 2001). The extent of the habitable subsurface zone at hydrothermally active mid-ocean ridges is unknown, but must be limited by the depth of the magmatic heat source. At the fast-spreading East Pacific Rise (EPR), for example, the depth of the magmatic heat source is circa 1.6–2.4 km below the sea floor, but at the slow-spreading Mid-Atlantic Ridge (MAR), the geothermal heat source is located at 3–3.5 km below the rift valley floor (Alt, 1995). Independent of temperature constraints, low-permeability dikes that overlie the magmatic heat source probably limit the microbially accessible subsurface environment to the upper few hundred meters of permeable volcanic basalts and metal-sulfide deposits (Alt, 1995).

Although subsurface microorganisms have not been cultured from hydrothermal vent subsurface samples directly, it can be argued that these putative organisms are predominantly thermophilic, anaerobic archaea. This inference is based on the large number of archaea isolated and/or identified by 16S rRNA sequencing from hydrothermal vent water samples and from seafloor hydrothermal vent chimneys and sediments accessible to submarines. The porous hydrothermal vent mineral matrix harbors diverse archaeal populations; these archaea may colonize subsurface hydrothermal vent deposits of sufficient porosity if the temperature regime and other key environmental variables are compatible (Harmsen et al., 1997; Hoek et al., 2003; Huber et al., 2003; Schrenk et al., 2003; Stetter, 1999; Takai and Horikoshi, 1999; Takai et al., 2001). In addition, indirect evidence is accumulating that hyperthermophilic, anaerobic archaea reside in the hydrothermal vent subsurface, representing candidate organisms for a subsurface biosphere of unknown extent. Observations of elevated DNA content in superheated, sterile hydrothermal vent endmember fluids suggest that subsurface archaea growing in protected niches are entrained in dynamic and possibly meandering subsurface flow, and are subsequently detected as highly degraded remnant DNA in the vent effluent rather than as viable cells (Deming and Baross, 1993).

Thermophiles have been isolated directly from vent plumes and volcanic solids associated with undersea eruptions and diking events (Huber et al., 1990; Summit and Baross, 1998). These include, for example, sulfur-reducing archaea of the genera *Thermococcus*, *Pyrococcus*, and *Pyrodictium*, as well as sulfate reducers of the genus *Archaeoglobus*. Strains of the hyperthermophilic genera *Thermococcus* and *Methanococcus* were cultured from warm (3–30 °C) hydrothermal vent effluents that are, however, not hot enough to allow growth of these organisms. It follows that these organisms inhabited the higher temperature subsurface but were flushed out to the

surface by hydrothermal circulation (Holden et al., 1998). Hyperthermophilic archaea of the genera *Thermococcus* and *Pyrococcus* isolated from moderately hot vent fluids showed distinct 16S rRNA sequence motifs and intergenic spacer regions, indicating that these strains were distinct from previous isolates obtained from surficial vent samples (Summit and Baross, 2001). These findings are generally compatible with hydrothermal subsurface habitats harboring distinct strains and species of hyperthermophiles, but leave unresolved the specific biogeochemical regimes and depth extent of hydrothermal subsurface habitats.

2.1. Hydrothermal subsurface biosphere of oceanic crust at mid-ocean ridges and ridge flanks

Direct evidence for microorganisms in the basaltic rock and metal sulfides of the deep hydrothermal vent subsurface is limited. Complex mixtures of volcanic breccia, sulfide minerals, and basalts underlie hydrothermal vents, and provide sufficient pore space for microorganisms. However, drilling into a hydrothermally active sulfide mound on the Mid-Atlantic Ridge (TAG site, ODP leg 158) has not yielded any evidence of subsurface hyperthermophiles, at least at this location (Reysenbach et al., 1998). Due to extremely steep temperature gradients, the "hot spots" at mid-ocean ridges may not harbor an extensive deep subsurface biosphere; shallow subsurface environments are more likely where convective cooling and seawater entrainment create suitable conditions for microorganisms.

Temperate sites at a suitable distance from active vents may prove to be better candidates for deep hydrothermal habitats. Currently, the best evidence for a deep subsurface biosphere in mid-ocean ridge basalts comes from the eastern flanks of the Juan de Fuca Ridge offshore Washington and Vancouver Island. During the Ocean Drilling Program (ODP) leg 168, a transect (ODP sites 1023–1032) across the eastern flank of the Juan de Fuca Ridge allowed a detailed study of hydrothermal fluid flow patterns and geochemical evolution of crustal fluids in the ridge flank subsurface (Davis et al., 1997). Seawater enters the hydrothermal plumbing system near the non-sedimented ridge crest and through outcrops on the sedimented ridge flank. The thick sediment cover confines this subsurface fluid to the oceanic crust basement basalt; discharge on the sedimented ridge flanks appears to be guided by the small number of seamount outcrops that rise above the sediment cover (Fisher et al., 2003). During subsurface passage, the fluid heats up gradually (from circa 10 °C to 60 °C) and evolves geochemically from its original seawater composition towards a composition more characteristic of axial hydrothermal vent fluid (Elderfield et al., 1999).

A detailed microbiological study was performed at ODP hole 1026B. This borehole penetrated through the sediment cover into the ocean crust basement basalt, and was sheathed down to the basalt basement by an impervious steel liner that prevented sediment porewater influx. The borehole was further capped with a Circulation Obviation Retrofit Kit (CORK) to the effect that hot (65 °C) water from the basaltic crust accumulated within the capped borehole. An overpressure spigot on the CORK device prevented influx and mixing of marine bottom water into the borehole fluid. The microbial community growing within the bore hole fluid was sampled with a filtration column (BioColumn) attached to the CORK spigot (Cowen et al., 2003). Comparable to an in situ enrichment, the BioColumn filters collected a diverse, thermophilic microbial community that was growing in the bore hole. Based on 16S rRNA analysis, it included deltaproteobacterial and gram-positive sulfate-reducing bacteria, as well as relatives of the thermophilic, ammonia-producing chemolithoautotrophic bacterium *Ammonifex degensii*. Archaeal clones were related to crenarchaeotal phylotypes from hot springs in Yellowstone National Park and to the sulfate-reducing, thermophilic genus *Archaeoglobus* (Cowen et al., 2003). Thus, the moderately hot fluid circulating through ridge flank basalts supported thermophilic microbial communities, including diverse sulfatereducing bacteria and archaea. These sulfate-reducing populations probably contributed to decreasing sulfate concentrations and to biogenic sulfur isotopic signatures in subsurface fluids across the Juan de Fuca Ridge flank (Bach and Edwards (2003) and references therein). Energy sources of ridge flank microbial communities were modeled based on the gradual oxidation of ferrous iron and iron sulfides within ridge flank basalts of increasing age. Iron- and sulfuroxidizing chemolithoautotrophs that inhabit ridge

flank basalt crust potentially produce microbial biomass in a similar order of magnitude as anaerobic heterotrophs that degrade organic carbon of sea water origin (Bach and Edwards, 2003).

These pioneering studies suggest that the older and cooler basalt crust at mid-ocean ridge flanks, and its temperate hydrothermal fluids, may support a more active and extensive hydrothermal subsurface biosphere than the very hot hydrothermal fluids and basalt crust on active mid-ocean ridge crests.

2.2. The hydrothermal sediment biosphere

Sediment-covered hydrothermal vent sites represent unique natural laboratories where diverse chemical and microbial processes occur. In contrast to their unsedimented counterparts, where hydrothermal endmember fluids with variable admixtures of seawater emerge directly into the water column, fluids in sediment-covered vents rise through the accumulated sediment layers and undergo complex chemical transformations in tandem with the hydrothermally affected sediments (Von Damm, 1995; Von Damm, 1990; Von Damm et al., 1985). These extensive, up to several hundred meter thick hydrothermal sediments may be home to a large diversity of anaerobic and thermophilic microbial life. Further, in these sediments, organic and inorganic electron donors are abundant, as are carbonate, sulfate, and other terminal electron acceptors (TEAs). Sedimented vents have been and continue to be an inexhaustible source for novel microorganisms (Karl, 1995).

The subsurface of three sedimented deep-sea hydrothermal vent sites has been studied extensively: Guaymas Basin on the northernmost extension of the EPR in the Gulf of California, Escanaba Trough on the Gorda Ridge spreading center offshore California, and Middle Valley on the Juan de Fuca Ridge spreading center offshore Washington. These hydrothermal sediments have been explored by the ODP and the Deep Sea Drilling Program (DSDP). During DSDP leg 64, Guaymas Basin sediments were recovered at sites 477, 478, and 481 (Curray et al., 1982). ODP leg 139 at Middle Valley included sites 857 and 858 (Davis et al., 1992), and ODP leg 169 drilled sites 1037 and 1038 at the Escanaba Trough and sites 1035 and 1036 in Middle Valley (Fouquet et al., 1998). Depending on local sedimentation rate, primary production in the water column, and accumulation of allochtonous organic matter, hydrothermal sediments may contain a significant amount of organic matter, which may undergo accelerated diagenesis and thermal degradation to hydrothermal petroleum (Rushdi and Simoneit, 2002). These degradation processes increase the level of dissolved organic carbon (DOC) in interstitial waters (Simoneit and Sparrow, 2002). Thermal degradation products include methane (Welhan, 1988), C1–C8 short-chain hydrocarbons (Whelan et al., 1988), ammonia (Von Damm et al., 1985; You et al., 1994), short-chain fatty acids (Martens, 1990), alkanes (Rushdi and Simoneit, 2002), and alkenes (Von Damm et al., 1985). In addition, hydrothermal petroleum and tar with polycyclic aromatic compounds and alkanes tend to accumulate. Their carbon-isotopic signatures, commonly in the range of −25‰ to −30‰, are typical for organic matter fixed by Rubisco (Simoneit, 2002; Simoneit and Fetzer, 1996; Simoneit et al., 1997).

2.2.1. Guaymas Basin

Due to a very high sedimentation rates (1–2 mm/year) and high biological productivity in surface waters, the Guaymas Basin vents are covered with on average 100 m, and up to 500 m, thick layers of organic-rich, diatomaceous sediments (Fig. 1) (Calvert, 1966; Schrader, 1982). The organic carbon content of hydrothermally unaltered sediments is ~2%, which is significantly higher than that at Middle Valley and Escanaba Trough with ~0.5% (Simoneit and Fetzer, 1996).

Chemical interactions between organic-rich sediment and hydrothermal vent fluids accelerate high temperature pyrolysis of organic material (Kawka and Simoneit, 1987). Diffuse venting through the sediments at temperatures up to 200 °C releases vast quantities of petroleum hydrocarbons, short-chain organic acids, and ammonia via pyrolysis of complex organic substrates (Bazylinski et al., 1988; Martens, 1990; Von Damm, 1995; Whelan et al., 1988). The $\delta^{13}C$ carbon-isotopic composition of methane in the Guaymas Basin sediments (−43‰ to −51‰) indicates a thermocatalytic origin from pyrolysis of buried organic matter (Welhan, 1988). The methane content in the Guaymas vent fluids is 12–16 mM, approximately two orders of magnitude higher than that at most bare-lava vent sites (Welhan, 1988). In contrast to bare-lava sites, extensive precipitation of metal sulfides in the Guaymas hydrothermal sediments

Fig. 1. Hydrothermal sediments of Guaymas Basin. The sediment surface is covered with thick microbial mats, mostly sulfide-oxidizing *Beggiatoa* species, that intercept upwelling vent fluids. For scale, the in situ incubation device in the foreground is circa 0.5 m high (photo: Holger W. Jannasch).

reduces the metal content of the vent fluids (Kawka and Simoneit, 1987)–and thus their microbial toxicity–by several orders of magnitude (Von Damm, 1990; Von Damm et al., 1985).

Microbiologically, Guaymas is by far the best-studied sedimented vent site. However, the baseline studies of deep subsurface sediments at Guaymas (DSDP leg 64) did not include microbiology. So far, microbiological work has focused on the surficial sediments, where diverse bacterial and archaeal populations utilize the products of subsurface processes. For example, aerobic bacteria at Guaymas specialize in the complete oxidation of thermogenic petroleum compounds, including alkanes and aromatic compounds (Goetz and Jannasch, 1993); some sulfate-reducing bacteria have similar substrate preferences (Rueter et al., 1994). Dense mats of sulfur-oxidizing bacteria of the genus *Beggiatoa* cover hydrothermally active sediments; these mats intercept sulfide and potentially carbon substrates that originate in the sediments (Jannasch et al., 1989; Nelson et al., 1989). Steep temperature gradients, from circa 2–4 °C in seawater to >100 °C at 10–25 cm sediment depth, favor microaerophilic or anaerobic thermophiles. Many new thermophilic species and genera were discovered at Guaymas. They include new species of the microaerophilic, chemolithoautotrophic bacterial genus *Persephonella* (Götz et al., 2002); the sulfate-reducing bacterium *Thermodesulfobacterium* (Jeanthon et al., 2002); the hyperthermophilic archaeaon *Archaeoglobus* (Burggraf et al., 1990); the heterotrophic, sulfur-reducing archaeal genera *Thermococcus* and *Pyrococcus* (Canganella et al., 1998; Huber et al., 1995; Jannasch et al., 1992); the hyperthermophilic methanogenic genera *Methanocaldococcus* (Jeanthon et al., 1999; Jones et al., 1983; Jones et al., 1989) and *Methanopyrus* (Huber et al., 1989; Kurr et al., 1991); and the autotrophic, obligately Fe(III)-reducing, hyperthermophilic archaeon *Geoglobus ahangari* (Kashefi et al., 2002). Molecular studies of bacterial and archaeal populations of Guaymas Basin based on 16S rRNA genes (Edgcomb et al., 2002; Teske et al., 2002) and key functional genes (Dhillon et al., 2003; Teske et al., 2003) indicate that multiple phylogenetic lineages and novel organisms remain to be cultured. These novel organisms include dense populations of anaerobic methanotrophic archaea that assimilate methane produced by thermogenic decom-

position of organic matter in the Guaymas sediments (Teske et al., 2002).

Microbial activities in hydrothermally active sediments at Guaymas are controlled by temperature. The steep temperature gradients there confine microbial life to the near-surface sediment layers. PCR-amplifyable DNA for bacteria, archaea, and eukaryotes (Dhillon et al., 2003; Edgcomb et al., 2002; Teske et al., 2002), diagnostic lipid biomarkers (Schouten et al., 2003; Teske et al., 2002), numbers of active prokaryotic cells (Guezennec et al., 1996), and sulfate reduction rates (Elsgaard et al., 1994; Jørgensen et al., 1990; Weber and Jørgensen, 2002) all decrease drastically with depth within the upper 3–30 cm of geothermally active sediment. This trend also holds for processes with a very broad temperature spectrum, such as sulfate reduction (circa 0–90 °C; Weber and Jørgensen (2002) and references therein), or for conservative biomarkers with high preservation potential, such as bacterial and archaeal neutral lipids (Schouten et al., 2003). Consistent with microbial population and activity peaks at the sediment surface, concentration profiles of thermogenic microbial carbon substrates (acetate and propionate) show clear substrate depletion in the surface centimeters, but very high concentrations just below the microbially active surficial sediments (Martens, 1990).

As a working hypothesis, the abundant and diversified microbial life in the Guaymas sediments is the surface expression of subsurface processes (hydrothermal petroleum generation, upwelling of hydrothermal fluid and thermogenic substrates). The dense microbial populations at the sediment surface intercept products of subsurface processes, including methane, sulfide and organic acids. The downward extent of the Guaymas sediment microbiota is limited by temperature and electron acceptor availability (Fig. 2). To examine the depth range of the Guaymas microbial communities, hydrothermal sediments with moderate thermal gradients and hydrothermal flow should be targeted, where the temperature range from near 0 °C to at least 120 °C extends over tens or hundreds of meters into the sediment.

2.2.2. Middle Valley

Middle Valley is a sedimented ridge valley on the northern Juan de Fuca Ridge off the coast of British Columbia, and was sampled during ODP cruises 139

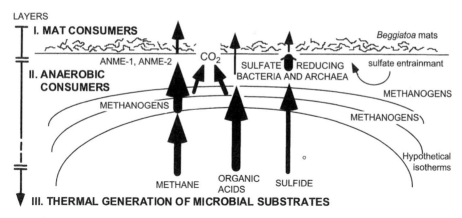

Fig. 2. Schematic cross-section through a hydrothermally active hot spot in Guaymas sediments, showing two-layer model for microbial alteration of thermogenic substrates of subsurface origin. I) Anaerobic consumers (sulfate reducers, anaerobic methanotrophs) intercept substrates of thermogenic subsurface origin (CH$_4$, organic acids) near the sediment surface. II) Their metabolic end products (CO$_2$, sulfide, simple organic compounds) are intercepted by the surface *Beggiatoa* mats. Arrows indicate hypothesized fluxes of chemical species.

(Davis et al., 1992) and 169 (Fouquet et al., 1998). At Middle Valley, young hot oceanic crust is covered by 200–1000 m of hemipelagic silty clays with occasional turbidite sequences. Hydrothermally heated and altered seawater penetrates up through the sediments, discharging locally with temperatures around 270 °C (Butterfield et al., 1994). Extended thermal gradients in Middle Valley sediments provided an opportunity to sample the entire microbially compatible temperature range on a scale of 10–100 m (Shipboard Scientific Party, 1998).

Microbial cells in hydrothermal sediments of Middle Valley were quantified by direct counts. In general, microbial cells were found throughout the upper sediment layers, from the mesophilic to the hyperthermophilic temperature range. Distribution patterns of microbial cells were generally controlled by in situ temperature, as shown by two contrasting examples (Cragg and Parkes, 1994; Cragg et al., 2000). A formerly active hydrothermal sulfide mound, the Bent Hill Massive Sulfide site, consisting mostly of hemipelagic and turbiditic sediments (ODP site 1035, hole 1035A), had a gradual temperature profile of ~0.7 °C increase per meter (Fouquet et al., 1998). The cell counts at this site showed a distinct zonation in mesophilic, thermophilic, and hyperthermophilic layers, with stepwise reduction in cell abundances compared to the standard abundance curve in non-hydrothermal sediments (Cragg et al., 2000). These standard cell abundances are based on the averaged cell counts from a wide range of marine subsurface

sediments (Parkes et al., 2000). The cell density profile in the mesophilic, upper part of the ODP site 1035 sediment column (~0–40 °C, 0–30 m) resembled the standard cell distribution for near-surface sediments. Cell densities in the thermophilic portion of the sediment column (~40–80 °C, 30–70 m) corresponded to ~10% of the standard cell densities. Cell densities in the hyperthermophilic sediment column (~80–110 °C, 70–170 m) decreased by 1–2 orders of magnitude in comparison to standard cell densities, dropping below the limit for statistically significant counts. Moderately thermophilic and hyperthermophilic prokaryotes persisted in the deeper sediment column, where it was too hot for mesophiles; their numbers, however, were 1–2 orders of magnitude lower than those of typical mesophilic populations in non-hydrothermal sediments (Cragg et al., 2000). This dataset shows, on the one hand, the gradual decrease in cell densities with increasing temperature, and, on the other, the persistent occurrence of subsurface microorganisms over the entire range of microbial growth temperatures.

Hydrothermal sediments with generally steeper temperature gradients were sampled over a lateral transect outward from a hot, active vent (ODP leg 139, Holes 858 A, B, C, D; ODP leg 169, Holes 1036A, B, and C). Here, very steep thermal gradients (3–12 °C per m) were observed, and two distinct trends in the cell counts became apparent. First, with increasing lateral distance from the center of active venting (9, 37, 71 m; holes 1036 A, B, and C), cell counts for the upper 15–20 m of the sediment column increased significantly (Cragg et al., 2000). Apparently, the environmental stress factors (toxic metal sulfide precipitates, high temperature) that limited microbial densities near the active venting "hot spot" lost some of their effect on a scale of <100 m lateral distance. Second, cell counts in all holes decreased rapidly with increasing depth and became statistically insignificant at 20–25 m below the sea floor (mbsf), most likely due to temperature constraints. Similar results were observed from a sample set (Holes 858A, B, C, D) obtained 5 years earlier at the same location (Cragg and Parkes, 1994).

Relatively high cell counts were reported from deep sediment layers where apparent temperatures were extremely high (155–185 °C), well above the generally accepted temperature limit of life. This apparent hyper-hyperthermophilic cell maximum was observed consistently in the same locations on leg 139 (Holes 858B and D) and again on leg 169 (Holes 1036A and 1036B). If accurate, these data push the temperature maximum for life considerably beyond the current limit of 121 °C for growth of pure cultures (Kashefi and Lovley, 2003). However, in situ temperatures in these ODP holes were not measured directly, but were modeled based on heat flow and lithological characteristics. Chlorinity profiles from the near-by sites obtained during ODP leg 139 (holes 858B and 858D) indicate that lateral seawater intrusion moderated these microbially incompatible in situ temperatures and created an "island of life" in a narrow sediment layer where seawater was mixed with hydrothermal fluid (Cragg et al., 2000). Seawater recharge, subsurface mixing, and lateral flow have to be taken into account at all sites (Stein et al., 1998). These studies emphasize the importance of high-resolution vertical and lateral sampling in hydrothermally influenced sediments to account for the highly dynamic and steep geochemical and thermal gradients, and to identify sudden permeability changes and lithological discontinuities in the heterogeneous hydrothermal sedimentary environment.

To analyze this community structure in different vent sediments in more detail, and to correlate characteristic community components with geochemical regime, molecular surveys based on rRNA genes and phylogenetically informative functional genes should be complemented by physiological studies on enrichments and pure culture isolates. So far, cultures from deep subsurface hydrothermal vent samples do not exist, but working hypotheses can be formulated to guide future research. For example, anaerobes that are abundant in near-surface sediments could, in principle, persist in hot, anoxic subsurface sediments. The energy-yielding reductive reactions that form the metabolic basis for thermophilic anaerobes (sulfur and sulfate reducers, methanogens) become thermodynamically more advantageous with increasing temperature and reduction potential, and give anaerobes a competitive edge over oxygen-respiring organisms, such as chemosynthetic sulfur oxidizers (McCollom and Shock, 1997). Some information on their community composition is given by phospolipid fatty acid (PLFA) profiles from Middle Valley and Guaymas sediments. The PLFA pool of all hydrothermal sediment samples tested (moderately thermophilic and hyperthermophilic samples from holes 1036A, B, and C at Middle Valley) was dominated by diverse bacterial membrane lipids; it did not include detectable amounts of archaeal diether or tetraether lipids (Summit et al., 2000). Parallel analyses of Guaymas Basin and Middle Valley sediment samples showed that the pools differed considerably in their relative content of saturated, terminally branched, and mono-unsaturated PLFAs, indicating major differences in microbial community structure (Summit et al., 2000).

Combinations of culture-dependent, nucleic acid-based, and organic biomarker approaches, together with microbial rate measurements, should allow functional and phylogenetic identification of bacteria, archaea, and eukaryotes throughout the microbially inhabited sediment column. Methodologically, recent ODP legs (e.g., 201) demonstrated the power of multidisciplinary study of deep subsurface ecosystems (D'Hondt et al., 2003). A polyphasic approach of this type should also clarify the question to what extent microorganisms leave their imprint on geochemical processes in the deep, hydrothermally active sediment subsurface, as they appear to do in cold marine subsurface sediments and in crustal continental rocks.

3. The deep subsurface continental biosphere

Deep *marine* hydrothermal vents as windows to the subsurface biosphere have garnered much attention in our quest to explore and understand the limits of life on Earth (e.g., Deming and Baross, 1993). However, the deep *continental* biosphere has not gone unnoticed. Numerous studies have targeted flood basalts, granites, sedimentary deposits, and mine shafts to find signs of past or present microbial life (e.g., Fredrickson and Onstott, 2001; Gold, 1992; Gold, 1999; Kotelnikova and Pedersen, 1998; Pedersen, 1997; Pedersen, 2000a,b; Pedersen, 2001; Stevens, 1997; Stevens and McKinley, 1995; Whitman et al., 1998). A recurring fundamental theme in these studies is the source of metabolic energy. Generally, the focus is on endogenous supplies of electron donors and acceptors, sources that are completely decoupled from surface processes. In other words, the deep subsurface biosphere considered in this and many other studies is obligately photosynthesis-independent. However, how can one be sure that a microbial process is truly decoupled from photosynthesis? For example, are sulfate-reducing prokaryotes (SRPs) photosynthesis-independent? The rise in seawater sulfate concentrations from <200 μM prior to 2.3 Ga to approximately present day levels by 1.8 Ga (Habicht et al., 2002; Lyons et al., 2004) is attributed to the preceding rise in molecular oxygen, which in turn is attributed to the emergence of cyanobacteria at or before ~2.7 Ga (Brocks et al., 1999; Summons et al., 1999). Further, can one readily differentiate a microbial community feeding on abiotically synthesized organic matter (terrestrial or extraterrestrial) from one that metabolizes photosynthesis-derived reduced carbon compounds? To minimize the ambiguity, previous studies have categorically dismissed chemoorganoheterotrophy as a surface-independent lifestyle and hence considered only chemolithoautotrophy. More specifically still, molecular hydrogen (H_2) is generally the sole electron donor considered in these investigations. In this section, we briefly summarize several recent studies of H_2-based microbial communities in the continental subsurface, remind the reader of the possible relevance to extraterrestrial ecosystems, discuss potential electron donors other than H_2 and organic matter, and propose the possibility of novel (undiscovered) redox strategies.

3.1. Microbes deep in granite and basalt

The aquifers of the Fennoscandian Shield (Swedish Baltic coast, northern Europe) represent arguably the best-studied natural laboratory of deep subsurface microbiology in granitic rock. Since 1987, the Deep Biosphere Laboratory at the University of Göteborg (Sweden) has explored this igneous province for signs of life. Comprehensive accounts of the geology, hydrology, geochemistry, and microbiology have been published (e.g., Pedersen, 1997); we highlight only the most salient points regarding the microbial diversity, activity, and metabolism. The microbial cell density at circa a dozen different sites, determined by epifluorescence microscopy, was generally 10^5–10^6 cells/mL of groundwater (Pedersen, 2001). However, the viable counts, determined as the number of colony-forming units, tended to be considerably lower, commonly 0.1–10% of cell counts (Pedersen and Ekendahl, 1990). Small subunit ribosomal RNA gene sequencing and culturing methods have identified a wide variety of aerobic and anaerobic, autotrophic and heterotrophic microbes, including both autotrophic and aceticlastic methanogens, homoacetogens, methanotrophs, SRPs, and Fe(III)-reducers (Kotelnikova et al., 1998; Kotelnikova and Pedersen, 1997; Kotelnikova and Pedersen, 1998; Pedersen, 2001).

Pedersen (1993) hypothesized an H_2-driven biosphere in these deep granitic aquifers. In his model, autotrophic methanogens and homoacetogens form the base of the food chain (Pedersen, 1997), consuming H_2 and CO_2 in net reactions represented, respectively, by

$$4H_2 + CO_2 \rightarrow CH_4 + 2H_2O \qquad (1)$$

and

$$4H_2 + 2CO_2 \rightarrow CH_3COOH + 2H_2O. \qquad (2)$$

Disproportionation and anaerobic respiration of acetic acid (CH_3COOH) may then support aceticlastic methanogens, SRPs, and Fe(III)-reducers in accord with, for example,

$$CH_3COOH \rightarrow CH_4 + CO_2, \qquad (3)$$

$$CH_3COOH + SO_4^{2-} + 2H^+ \rightarrow 2CO_2 + H_2S + 2H_2O, \qquad (4)$$

and

$$CH_3COOH + 8FeOOH + 16H^+ \rightarrow 2CO_2 + 8Fe^{2+} \\ + 14H_2O, \qquad (5)$$

where FeOOH denotes just one of several possible Fe(III)-bearing compounds. Biomass synthesized by this putative community of organisms may then serve as the carbon and energy sources for anaerobic heterotrophs. Indeed, at the Äspö Hard Rock Laboratory (HRL) in Sweden, homoacetogens and autotrophic and aceticlastic methanogens dominate the deep groundwater (Kotelnikova et al., 1998), directly supporting the H_2-driven biosphere model. Based on culture-dependent and culture-independent studies, Fe(III)-reducers, including *Shewanella putrefaciens* and relatives of *Pseudomonas*, were also identified at the Äspö HRL (Pedersen et al., 1996); organic compounds serve as the electron donors in the reduction of Fe(III) (Banwart et al., 1994; Banwart et al., 1996). In addition, SRPs have been documented at the Äspö HRL (Pedersen et al., 1996), and data on sulfide precipitation, ground water chemistry, and isotopic fractionations of C and S are suggestive of deep microbial sulfate reduction at Äspö (Laaksoharju et al., 1995; Tullborg, 2000), though alternate interpretations cannot be entirely ruled out.

A basalt equivalent of the H_2-based deep granite biosphere was proposed by Stevens and McKinley (1995) and later expanded by Stevens (1997) and McKinley et al. (2000). In the Columbia River flood basalts of western North America, deep aquifers contained up to 60 μM aqueous H_2. Metabolically diverse microorganisms were present, with acetogens and methanogens the most common; Fe(III)-reducers, SRPs, and fermenters were also found, but only in relatively low numbers. It was concluded that H_2, produced in reactions of anoxic water with freshly exposed basalt, served as the primary energy source for a photosynthesis-independent microbial ecosystem. It was further suggested that both mafic (high Mg+Fe) as well as some relatively felsic (high SiO_2) rocks can yield sufficient H_2 from water–rock reactions to support extensive chemolithoautotrophy in the subsurface. The authors also noted that the presence of heterotrophs in the deep basalts is nevertheless entirely consistent with an H_2-based biosphere, where in situ produced organic matter can support at least moderate rates of fermentation and anaerobic respiration.

In 1998, the evidence for an H_2-based ecosystem deep in the Columbia River basalt was called into question (Anderson et al., 1998). New experimental data were reported that showed that far too little H_2 is produced by reactions of basalt with water at "environmentally relevant, alkaline pH" (~8) to support an autotrophic subsurface microbial community. Anderson and colleagues stated that at a slightly acidic pH (~6), low concentrations of H_2 could be generated from water–basalt interactions, but this small amount of H_2 would not suffice as the primary energy source for chemolithoautotrophs. They reinterpreted the results published by Stevens and McKinley, suggesting that DOC was a more likely source of energy (i.e., reducing power) driving microbial metabolism in the Columbia River basalt. They further noted that carbon isotope data support their reinterpretation. Later, Fry et al. (1997) applied culturing-independent, molecular techniques to water samples from two different deep Columbia River basalt aquifers and found that the community structure at both sites was dominated by bacteria (92% and 64%), not archaea; in fact, archaea, the domain that accounts for all known methanogenic organisms, comprised <3% in each sample. This finding appears to contradict the assertion by Stevens and McKinley (1995) that methanogens are most populous in the Columbia River basalt aquifers. Anderson et al. (1998) conclude that while H_2 may not fuel the deep biosphere in the Columbia River basalt, reduced gases may still drive deep subsurface ecosystems elsewhere.

Recently, Chapelle et al. (2002) found just such an environment in subsurface water from Lidy Hot Springs, Idaho (USA). They employed several molecular biology techniques on water samples retrieved from a deep (200 m) well that taps anoxic hydrothermal (58.5 °C) fluid. Based on a method that couples the polymerase chain reaction with a most probable number analysis (PCR-MPN), archaea comprised up to 99% of all cells. A second method, quantitative PCR, yielded >95% archaea, and, in a third quantification, ~99% of all prokaryotic cells hybridized to a 16S rDNA probe for archaea. Based on 16S rDNA sequences of 10 clones and screening of 55 more clones with restriction fragment length polymorphism analysis, 95% of the archaeal sequences are closely related to autotrophic methanogens. These data, combining a multitude of molecular approaches confirm that methanogenic archaea dominate the microbial community at Lidy Hot Springs.

These results are also consistent with cell counts and in situ geochemistry. Chapelle et al. (2002) report that the hydrothermal groundwater at Lidy Hot Springs contains 2.8×10^5 microbial cells/mL, its H_2 concentration (13.0 nM) is sufficient to support active methanogens in nature, and the DOC is too low (<0.27 mg/l) to serve as the primary electron donor for a significant subsurface microbial community. The authors conclude that the archaea-dominated, specifically the methanogen-dominated biosphere glimpsed at Lidy Hot Springs is H_2-based and completely independent of photosynthesis; this ecosystem serves as an ideal window to the deep continental subsurface and represents perhaps the best current analog of what life may be like, or have been like, on extraterrestrial planets in our solar system, where the requirements for life (e.g., liquid water, redox gradients, carbon) are more likely to be encountered in the subsurface than at the surface.

Whether on extraterrestrial planets or on Earth, H_2-based subsurface microbial communities must be provided with a steady supply of H_2. For terrestrial systems, Pedersen (2000b) suggests three possible sources of photosynthesis-independent H_2: radiolysis of water driven by in situ-produced alpha radiation, interaction of anoxic water with iron-bearing igneous rocks, and reduced deep volcanic gases. The first (radiolysis of water) requires further study; the second (water–rock interaction) is generally supported by experimental and theoretical investigation; and the third (magmatic gases) appears to be ubiquitous in both continental and marine volcanic systems characterized by substantial fracturing and faulting. In a recent experimental study, it was shown that the concentrations of radiolytic H_2 in the subsurface can be high enough to support H_2-consuming microbes and inhibit H_2-producing ones (fermenters) (Lin et al., in press).

3.2. Alternate sources of reducing power

As noted above, photosynthesis-independent biosphere hypotheses rely on H_2 as the sole electron donor. It is worth questioning whether this is an absolute requirement. We now consider electron donors other than H_2 as possible sources of reducing power for chemotrophs in the deep subsurface. Based on evidence from cultured microorganisms, chemotrophs that require neither H_2 nor photosynthetically

produced organic carbon as electron donors are rare. However, several examples of such organisms are known. For example, the marine bacteria *Desulfocapsa sulfoexigens* and *Desulfocapsa thiozymogenes* catalyze the disproportionation of elemental sulfur (S^0), sulfite (SO_3^{2-}), or thiosulfate ($S_2O_3^{2-}$) in accord with

$$4S^0 + 4H_2O \rightarrow SO_4^{2-} + 3H_2S + 2H^+, \qquad (6)$$

$$4SO_3^{2-} + 2H^+ \rightarrow 3SO_4^{2-} + H_2S, \qquad (7)$$

and

$$S_2O_3^{2-} + H_2O \rightarrow SO_4^{2-} + H_2S, \qquad (8)$$

respectively (Finster et al., 1998; Jannsen et al., 1996). A number of mesophilic and thermophilic bacteria and archaea, including *Thermothrix*, *Pyrobaculum*, *Aquifex*, *Thioploca*, and *Ferroglobus*, mediate the oxidation of $S_2O_3^{2-}$ or S^0 with NO_3^- as the TEA (Brannan and Caldwell, 1980; Caldwell et al., 1976; Fossing et al., 1995; Hafenbradl et al., 1996; Huber et al., 1992; Völkl et al., 1993). Several Fe-reducers and Fe-oxidizers are also noteworthy. For example, members of *Sulfobacillus* and *Thiobacillus* catalyze the oxidation of $S_4O_6^{2-}$ or S^0 with Fe(III) as the TEA (Bridge and Johnson, 1998; Brock and Gustafson, 1976), and *Ferroglobus placidus* makes a living by oxidizing Fe^{2+} with NO_3^- (Hafenbradl et al., 1996). To what extent sulfur disproportionation or anaerobic respiration of Fe(II) and sulfur, or even NH_4^+, Mn(II), U(IV), and As(III) can support a subsurface biosphere is unknown. Based on thermodynamic modeling studies, reactions such as these can be exergonic under certain geochemical conditions (Amend et al., 2003; Amend and Shock, 2001). Consequently, the electron donors mentioned above–and perhaps others–may provide the reducing power required by organisms that constitute the base of a deep subsurface ecosystem.

It has been argued that if the base of the food chain in the deep subsurface is occupied by heterotrophs, the ecosystem must be linked to photosynthesis at the surface (Stevens, 1997). However, this theory ignores the possibility that under certain geochemical and geophysical conditions, organic compounds may be synthesized abiotically. Heterotrophic organisms that feed on such compounds would be photosynthesis-independent primary producers of biomass. How likely is the abiotic synthesis of organic compounds? Controversial theories regarding the abiotic origin of petroleum notwithstanding (Gold and Soter, 1982), it has been shown repeatedly that the synthesis of methane, carboxylic acids, amino acids, and other simple organic compounds from CO_2 (or CO) and H_2 can be thermodynamically favorable under reducing conditions, which may obtain in a variety of near-surface and deep subsurface environments (Shock, 1990; Shock and Schulte, 1998). In fact, Fischer–Tropsch-type reaction pathways are commonly invoked (Holm and Charlou, 2001; McCollom et al., 1999). Since organic synthesis of this type is not biologically mediated, the H_2 used as the electron donor does not represent the energy source at the base of the hypothesized food chain. Nevertheless, H_2 is essential to the emergence, propagation, and persistence of that food chain. In this postulated ecosystem, heterotrophic microbes that ferment or respire simple abiotically synthesized organic compounds, and not chemolithoautotrophs, would be the primary producers.

It is widely accepted that of the microbial species present in natural environments, only a few percent have been isolated, cultured, and at least partially characterized (Amann, 1995). Although some of the unknown majority may rely on well-known metabolic strategies (e.g., aerobic respiration, fermentation, methanogenesis, sulfur-reduction), it seems reasonable, in fact highly likely, that non-standard redox processes serve as energy-yielding reactions for yet to be discovered deep subsurface microorganisms. We cannot predict which novel redox reactions may prove most successful, but it should be noted that a number of non-standard redox reactions may be exergonic in deep subsurface environments, and hence, they represent plausible metabolisms (Amend et al., 2003). These include, for example, the oxidation of H_2S with NO_3^- as the TEA, yielding SO_4^{2-} or S^0 and NH_4^+; this mode of metabolism is known for members of *Beggiatoa* and *Thioploca* that occur at the sediment–water interface (McHatten et al., 1986; Otte et al., 1999), but it is unknown at depth. In addition, anaerobic methane oxidation by archaea with SO_4^{2-} as the oxidant was recently described in marine sediments (Boetius et al., 2000; Hinrichs et al., 1999; Orphan et al., 2001; Otte et al., 1999), but methane oxidation with Fe(III) or NO_3^- as the TEA can also be exergonic. In fact, Fe(III) might serve as the oxidant

to chemolithoautotrophs that use S^0 or H_2S as the electron donor.

4. Sequences of redox reactions as a function of temperature

Many geologic environments are redox stratified, including freshwater and marine sediments, closed marine basins, wetlands, soils, and the basaltic ocean crust. Above, we provided an overview of anaerobic organisms and the geochemical conditions that enable the progression of redox reactions in subsurface ecosystems. Based on chemical analyses of redox sensitive compounds and thermodynamic modeling of oxidation and reduction half reactions, a general sequence of these reactions with increasing depth has been established. In broad terms, the progression of TEAs begins with O_2 near the surface, followed, with increasing depth, by NO_3^-, NO_2^-, Mn(IV), Fe(III), SO_4^{2-}, and finally CO_2. This thermodynamically predicted progression, determined from values of the standard Gibbs free energies of reaction (ΔG_r^0), is typically paralleled by a microbial metabolic succession (Stumm and Morgan, 1996). In oxic systems, aerobic respiration tends to dominate, followed, with increasing depth and decreasing oxygen concentration, by denitrification, dissimilatory Mn- and Fe-reduction, sulfate-reduction, and lastly autotrophic methanogenesis; fermentation and acticlastic methanogenesis also occur at very low redox potentials. It is worth noting that the predicted succession of reactions is: a) based on thermodynamic properties under *standard state* conditions and not conditions that exist in the natural environment and b) calculated only at 25 °C and 1 bar and not at in situ temperatures and pressures. the in situ energy-yields from a reaction depend on temperature, pressure, and chemical composition, and are therefore site- or sample-specific.

It is beyond the scope of this communication to review the energetics of reactions in a host of redox stratified locales. However, as examples for the interested reader, the energetics of many chemolithoautotrophic reactions in the shallow marine hydrothermal system of Vulcano Island (Italy) or in continental hot springs at Obsidian Pool (Yellowstone National Park) are tabulated in Amend et al. (2003) and Shock et al. (2005), respectively. For the energetics at

Table 2
Values of $p\varepsilon^{\circ}$ as a function of temperature at P_{SAT} for 11 reduction half-reactions in microbial metabolism

Reaction		$p\varepsilon^{\circ}$			
		25 °C	50 °C	100 °C	150 °C
I	$O_2(aq)+4H^++4e^-\rightarrow2H_2O$	21.50	19.86	17.24	15.23
II	$NO_3^-+6H^++5e^-\rightarrow1/2N_2(aq)+3H_2O$	20.72	19.22	16.88	15.16
III	$NO_2^-+8H^++6e^-\rightarrow NH_4^++2H_2O$	15.23	14.14	12.45	11.20
IV	$NO_3^-+2H^++2e^-\rightarrow NO_2^-+H_2O$	13.88	12.90	11.36	10.20
V	$Fe_2O_3(s)+6H^++2e^-\rightarrow2Fe^{2+}+3H_2O$	13.06	11.88	10.01	8.60
VI	$SO_4^{2-}+10H^++8e^-\rightarrow H_2S(aq)+4H_2O$	5.08	4.90	4.67	4.57
VII	$CO_2(aq)+8H^++8e^-\rightarrow CH_4(aq)+2H_2O$	2.69	2.55	2.32	2.17
VIII	$S^0(s)+2H^++2e^-\rightarrow H_2S(aq)$	2.45	2.47	2.53	2.63
IX	$2CO_2(aq)+8H^++8e^-\rightarrow CH_3COOH(aq)+2H_2O$	2.17	2.06	1.82	1.66
X	$6CO_2(aq)+24H^++24e^-\rightarrow C_6H_{12}O_6(aq)+6H_2O$	0.17	0.18	0.11	0.07
XI	$H^++e^-\rightarrow1/2H_2(aq)$	-1.55	-1.30	-0.85	-0.44

elevated temperatures and pressures of several redox reactions in a deep-sea vent system and associated basaltic ridge flank, the reader is referred to McCollom and Shock (1997) and Bach and Edwards (2003). In the latter, the potential biomass production by chemolithoautotrophs within aging ridge flanks was thermodynamically evaluated. Bach and Edwards (2003) showed that aerobic and anerobic oxidation of ferrous iron and sulfide in basaltic ocean crust could fuel annual biomass production of $\sim5\times10^{11}$ g of cellular carbon. The reduction with H_2 of O_2, NO_3^-, Fe(III), SO_4^{2-}, and CO_2 may support even greater biomass production in the ridge flanks. The authors conclude that the total amount of biomass carbon produced annually by chemolithoautotrophs in these systems likely exceeds that derived from DOC assimilated by chemoorganoheterotrophs. They further note that while primary biomass synthesis in bare rock systems may be substantial, further theoretical studies and sampling efforts are required to more tightly constrain these microbial processes. To begin to address this shortcoming, we considered here the energetics at elevated temperatures for a number of ubiquitous redox reactions catalyzed by chemolithoautotrophs.

We computed values of the standard redox potentials (E_h°) of 11 half-reactions as a function of temperature at P_{SAT}[1] using the relation

$$E_h^{\circ} = \frac{-\Delta G_r^{\circ}}{nF}, \qquad (9)$$

---

[1] P_{SAT} is used in this study to denote saturation pressure for H_2O, in other words, the P–T boiling curve for water.

where ΔG_r° is as defined above at the temperature and pressure of interest, n represents the number of moles of electrons in reaction r, and F denotes the Faraday constant. As is commonly done, we converted E_h° to the negative logarithm of the electron activity ($p\varepsilon^{\circ}$) in accord with

$$p\varepsilon^{\circ} = \frac{FE_h^{\circ}}{2.303RT}, \qquad (10)$$

where R and T denote the gas constant and temperature in Kelvin, respectively. Values of $p\varepsilon^{\circ}$ can be further modified with the expression

$$p\varepsilon^{\circ\prime} = p\varepsilon^{\circ} + \frac{n_H}{2}\log K_W, \qquad (11)$$

where $p\varepsilon^{\circ\prime}$ denotes $p\varepsilon^{\circ}$ at neutral pH (the biological standard state), n_H stands for the number of moles of protons per mole of electrons in the half-reaction, and K_W represents the equilibrium constant at the temperature and pressure of interest for the water dissociation reaction written as

$$H_2O\rightarrow H^+ + OH^-. \qquad (12)$$

It should be reiterated that the biological standard state (given the prime ($\prime$) notation) refers to neutral pH and not pH 7. It is, after all, only at 25 °C and 1 bar that pH 7 represents neutrality; for example, at P_{SAT} and 100 °C, neutral pH is 6.13, and at 200 °C, it drops to 5.64.

Values of $p\varepsilon^{\circ}$ and $p\varepsilon^{\circ\prime}$ at P_{SAT} and 25, 50, 100, and 150 °C are given in Tables 2 and 3 for 11 common reduction half-reactions known to be mediated by microorganisms. These values were computed with

Table 3
Values of $p\varepsilon^\circ$ as a function of temperature at P_{SAT} for 11 reduction half-reactions in microbial metabolism

Reaction		$p\varepsilon^{\circ\prime}$			
		25 °C	50 °C	100 °C	150 °C
I	$O_2(aq)+4H^++4e^-\rightarrow 2H_2O$	14.50	13.23	11.12	9.42
II	$NO_3^-+6H^++5e^-\rightarrow 1/2N_2(aq)+3H_2O$	12.32	11.26	9.53	8.18
III	$NO_2^-+8H^++6e^-\rightarrow NH_4^++2H_2O$	5.90	5.29	4.28	3.45
IV	$NO_3^-+2H^++2e^-\rightarrow NO_2^-+H_2O$	6.88	6.27	5.24	4.39
V	$Fe_2O_3(s)+6H^++2e^-\rightarrow 2Fe^{2+}+3H_2O$	−7.94	−8.03	−8.37	−8.85
VI	$SO_4^{2-}+10H^++8e^-\rightarrow H_2S(aq)+4H_2O$	−3.67	−3.39	−2.99	−2.70
VII	$CO_2(aq)+8H^++8e^-\rightarrow CH_4(aq)+2H_2O$	−4.31	−4.09	−3.81	−3.65
VIII	$S^0(s)+2H^++2e^-\rightarrow H_2S(aq)$	−4.55	−4.17	−3.60	−3.19
IX	$2CO_2(aq)+8H^++8e^-\rightarrow CH_3COOH(aq)+2H_2O$	−4.83	−4.58	−4.31	−4.16
X	$6CO_2(aq)+24H^++24e^-\rightarrow C_6H_{12}O_6(aq)+6H_2O$	−6.83	−6.46	−6.02	−5.75
XI	$H^++e^-\rightarrow 1/2H_2(aq)$	−8.55	−7.94	−6.98	−6.26

Eqs. (9), (10), (11) using values of ΔG_r° calculated with the standard Gibbs free energies of reactants and products (ΔG_i°) given in Amend and Shock (2001) in accord with the relation

$$\Delta G_r^\circ = \sum_i v_{i,r}\Delta G_i^\circ, \tag{13}$$

where $v_{i,r}$ stands for the stoichiometric reaction coefficient of species i in reaction r. It can be seen in Tables 2 and 3 that values of $p\varepsilon^\circ$ and $p\varepsilon^{\circ\prime}$ for several of the reactions are quite sensitive to temperature. For example, $p\varepsilon^\circ$ for the reduction of O_2 (reaction I) decreases by 29% from 21.50 at 25 °C to 15.23 at 150 °C, and $p\varepsilon^\circ$ for the reduction of Fe(III) in hematite (reaction V) decreases by 34% from 13.06 to 8.60 over this temperature range. The $p\varepsilon^\circ$ of several other reactions, including sulfate-reduction (reaction VI), autotrophic methanogenesis (reaction VII), and homo-acetogenesis (reaction IX), decreases only slightly with increasing temperature. Interestingly, values of $p\varepsilon^\circ$ for S^0 and H^+ reduction (reactions VIII and XI) show modest increases between 25 and 150 °C.

Values of $p\varepsilon^{\circ\prime}$ at P_{SAT} and 25, 50, 100, and 150 °C for the 11 redox couples are plotted in the form of "electron towers" in Figs. 3–6. Although the general sequence of redox couples observed at 25 °C (discussed above) remains the same even at high temperatures, slight rearrangements of the order at low $p\varepsilon^{\circ\prime}$ values are seen with increasing temperature. For example, $p\varepsilon^{\circ\prime}$ of the CO_2/CH_4 couple drops below that of the S^0/H_2S couple between 50 and 100 °C. In addition, $p\varepsilon^{\circ\prime}$ of the Fe_2O_3/Fe^{2+} and the H^+/H_2 couples swap positions on the electron tower as temperature

increases from 25 to 50 °C. It can also be seen in Figs. 3–6 that while values of $p\varepsilon^{\circ\prime}$ for reactions (V–XI) stay relatively constant between −3 and −9, those of

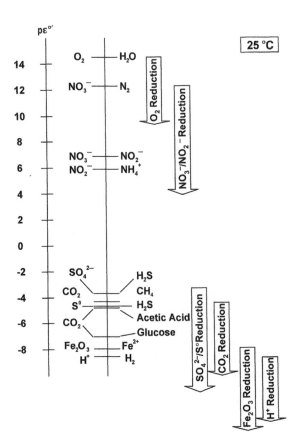

Fig. 3. Sequence of microbially mediated reduction half-reactions determined from values of $p\varepsilon^{\circ\prime}$ at 25 °C and 1 bar.

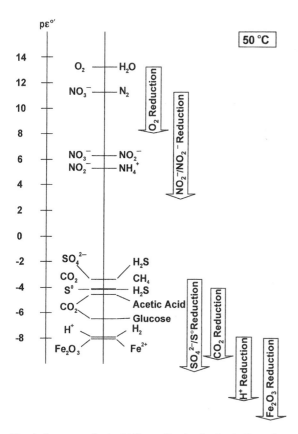

Fig. 4. Sequence of microbially mediated reduction half-reactions determined from values of $p\varepsilon^{o'}$ at 50 °C and 1 bar.

oxidized metals, bicarbonate or CO_2, even fermentable organic matter can serve as TEAs in redox processes. Anaerobic bacteria and archaea that utilize these oxidants, in particular sulfate reducers, metal reducers, and methanogens, are ubiquitous in the continental and marine deep subsurface. Different species and genera with distinct physiological adaptations to their subsurface habitat have been isolated, and novel phylogenetic lineages have been detected with 16S rRNA and functional gene sequencing. Here, we focus on a specific functional and phylogenetic group–the sulfate reducers–in order to illustrate the diversity, habitat range, and physiological adaptability of subsurface microorganisms. In addition, we evaluate the impact of molecular methods on subsurface microbiology.

Sulfate is the most important TEA in marine sediments (Jørgensen, 1982). Sulfate profiles of deep marine sediments indicate pervasive activity of SRPs,

reactions (I–IV) drop significantly, from 6–15 at 25 °C to 3–9 at 150 °C. In other words, if the activities of the reactants and products remain unchanged, the differences in energy-yield between two modes of metabolism may show a substantial decrease with increasing temperature. Consequently, thermophilic aerobes may still outcompete thermophilic SRPs, but the difference in available energy may be significantly less.

5. Sulfate-reducing bacteria and archaea

Above, we discussed some of the geochemical parameters that shape microbial life in the deep, anoxic subsurface. There, any oxygen is quickly consumed, and other TEAs are reduced during anaerobic metabolisms. As noted, sulfate and other oxidized sulfur compounds, ferric iron and other

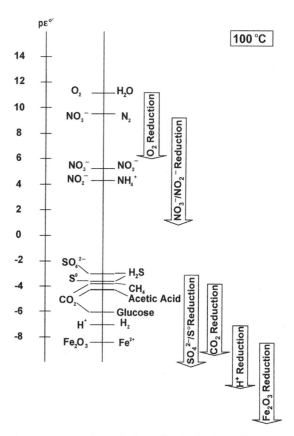

Fig. 5. Sequence of microbially mediated reduction half-reactions determined from values of $p\varepsilon^{o'}$ at 100 °C and 1 bar.

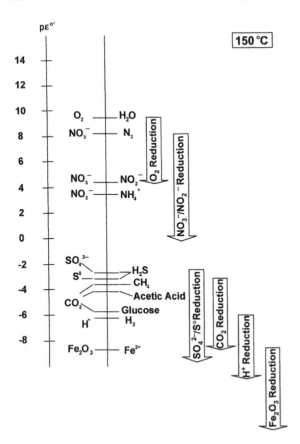

Fig. 6. Sequence of microbially mediated reduction half-reactions determined from values of $p\varepsilon^{o\prime}$ at 150 °C and 1 bar.

predominantly in near-shore and continental slope carbon-rich subsurface sediments (D'Hondt et al., 2002). In the sediments of a shallow marine hydro-thermal system at Vulcano (Italy), sulfate-reduction was the dominant microbial metabolism at 90 °C (Tor et al., 2003). Mesophilic, moderately thermophilic, and hyperthermophilic sulfate reducers have been isolated from the deep subsurface in marine and continental environments. A great phylogenetic and physiological diversity of mesophilic SRPs has been established, but mesophilic subsurface isolates are thus far represented only by the mutually related genera *Desulfovibrio* and *Desulfomicrobium*. These organisms convert organic substrates (e.g., lactate) or grow with H^2 as electron donor and acetate as carbon source. For example, the species *Desulfovibrio profundus* was isolated from cold Japan Sea subsur-face sediments and is represented by two strains from

80 m and 500 mbsf (Bale et al., 1997). These strains show elevated sulfide production at near in situ hydrostatic pressures (100–150 bar) and produce sulfide over a wide range of pressures (up to 300–350 bar). *D. profundus* has an unusually broad growth temperature range (15–65 °C) bridging mesophilic and thermophilic conditions (Bale et al., 1997). The related species *Desulfovibrio aespoeensis* was isolated from groundwater in deep granitic rock of the Fennoscandian Shield (Motamedi and Pedersen, 1998). Strains related to *Desulfomicrobium baculatum* and *Desulfovibrio longreachii* were also obtained from the same site (Pedersen et al., 1996). Compared with other deep continental sites, the sulfate concen-trations in these granite-hosted aquifers are relatively high (>1 mM; Pedersen et al., 1996). However, sulfate availability does not necessarily indicate the occur-rence of SRPs. Two deep, anaerobic aquifers in the Columbia River flood basalts–one with high and one with near-zero sulfate concentrations–harbored similar densities of SRPs of the genus *Desulfovibrio* (Fry et al., 1997). Apparently, these sulfate reducers can grow fermentatively or by metal reduction, as alter-natives to sulfate reduction. It should also be noted that metal- and humic acid-reducing microorganisms in the subsurface are at least as diverse as sulfate reducers, and are important components of some deep subsurface microbial communties (Lovley, 2001).

SRPs in the deep subsurface frequently oxidize ancient, buried organic matter. Deeply buried shale layers can provide organic substrates for SRPs, as shown for cretaceous marine black shales and sand-stones in New Mexico. There, the organic-rich but microbially impermeable shales provided substrates for sulfate-reducing populations that live in the adjacent, organic-poor, relatively porous sandstone layers (Krumholz et al., 1997). Subsequently, the species *Desulfomicrobium hypogeium* was isolated from these deep cretaceous sandstones (Krumholz et al., 1999). In competition experiments with a subsur-face acetogen from the same habitat, *D. hypogeium* showed an extremely high affinity for hydrogen, indicating that sulfate-reduction can be the dominant electron sink for subsurface hydrogen (Krumholz et al., 1999).

Deep, geothermally heated oil fields with abundant organic matter and sulfur compounds are a promising hunting ground for unusual thermophilic SRPs. The novel genera and species *Thermodesulforhabdus norvegicus* (Beeder et al., 1995) and *Desulfacinum infernum* (Rees et al., 1995) have been isolated from deep, hot oil fields, as have hyperthermophilic archaeal sulfate reducers of the genus *Archaeoglobus* (Beeder et al., 1994; L'Haridon et al., 1995; Stetter et al., 1993). Oil fields and geothermal springs have commonly yielded moderately thermophilic members of the gram-positive, spore-forming genus *Desulfotomaculum*, including *Desulfotomaculum thermocisternum* (Nilsen and Torsvik, 1996; Rosnes et al., 1991), *Desulfotomaculum geothermicum* (Daumas et al., 1988), *Desulfotomaculum australicum* (Love et al., 1993), and *Desulfotomaculum kuznetsovii* (Nazina et al., 1987). These gram-positive sulfate reducers can utilize a wide substrate spectrum of low-molecular-weight alcohols and organic acids. They are able to grow in co-culture with thermophilic fermenting bacteria that convert complex petroleum compounds into easily degradable low-molecular-weight compounds (Rosnes et al., 1991), or in syntrophic association with hydrogenotrophic methanogens from the same habitat (Nilsen and Torsvik, 1996). Thermophilic *Desulfotomaculum* species are not limited to oil fields, however. The thermophilic species *Desulfotomaculum putei*, for example, was isolated from 2.7 km deep, terrestrial Triassic sediments (Liu et al., 1997).

5.1. Analysis of SRPs using the dsr gene

A relatively new molecular approach to analyze the diversity of SRPs is based on key genes of their distinctive physiological pathway—dissimilatory sulfate reduction. Particular targets in this approach are the α- and β-subunits of the dissimilatory sulfite reductase gene (dsrAB), which catalyzes the reduction of sulfite to sulfide. PCR amplification and sequence analysis of this highly conserved, phylogenetically informative gene allows simultaneous functional and phylogenetic identification of SRPs (Klein et al., 2001; Wagner et al., 1998). With amplification and sequencing of dsrAB genes from environmental samples, it has become possible to analyze the community composition of sulfate-reducing bacteria and archaea independent of cultivation. This functional gene approach circumvents the limitations of 16S rRNA gene analysis, which identifies the phylotype but not the phenotype, and cannot identify novel SRPs without precedent (Teske et al., 2003).

Analysis of dsrAB sequences from deep subsurface samples (e.g., groundwater from a Uranium mill tailings site or biofilms growing in deep South African gold mine boreholes) revealed several phylogenetic lineages of uncultured SRPs (Baker et al., 2003; Chang et al., 2001). Numerous *Desulfotomaculum*-related dsrAB phylotypes occurred in the clone libraries from both habitats, indicating a wide range of environments for these spore-forming, frequently thermophilic sulfate reducers. Several dsrAB phylotypes are not related to cultured sulfate reducers, and represent novel sulfate-reducing lineages with essentially unknown physiology (Baker et al., 2003; Chang et al., 2001). Continued molecular surveys will lead to a general outline of the environmental distribution of these novel deep subsurface sulfate reducers and may potentially identify geochemical controls that define their habitat range. New cultivation approaches to bring these novel SRPs into pure culture should use as much as possible the in situ geochemical and physical parameters as guidelines for designing media and incubation conditions.

6. Contamination controls

Contamination controls are essential for the stringent interpretation of microbiological deep subsurface data, in particular in experiments where a weak microbial signal is amplified (PCR amplification of subsurface DNA; microbial enrichments and cultivations); yet these controls are rarely applied consistently. Two general contamination monitoring strategies for deep subsurface drilling and sampling are bacterial mimics (microspheres) or water-soluble chemical tracers that are added to the drilling fluid. Fluorescent microspheres of a similar size to indigenous microorganisms (0.2–1.3 μm) have been used previously in land-based drilling operations to assess subsurface dispersal and transport of these bacterial mimics (Harvey et al., 1989). Perfluorocarbon tracers (PFT) have been widely used in land-based drilling applications (Colwell et al., 1992; McKinley and Colwell, 1996; Russell et al., 1992) because they are chemically inert and can be detected with high sensitivity. Detailed contamination control protocols for marine deep subsurface sediment sampling were developed during ODP leg 185 (Smith et al., 2000a; Smith et al., 2000b), and were applied for all drill sites and sediment cores on ODP leg 201 (House et al., 2003). PFT tracer results showed that the center of the sediment cores contained less PFT than the core margins and therefore had a lower potential seawater contamination; concentration differences from the outer margin to the center of 2.5 inch diameter cores ranged from factor 3 to 300. Drilling technique had a major impact; samples obtained by advanced piston coring (APC) were generally less contaminated (factor 3–9) than sediments obtained by extended core barrel (XCB) drilling, which is more disruptive and introduces more seawater into the sediment core. PFT tracer analysis showed that sediment samples from the center of APC cores contained generally less than 0.1 μL of seawater per gram of sediment. Assuming an average density of 5×10^8 microbial cells per liter of seawater, this seawater contamination corresponds to entrainment of <50 cells per gram of sediment. This is a maximum estimate, based on the assumption that bacteria-sized particles permeate sediment as easily as a PFT tracer molecule. Direct comparisons between PFT tracer results and fluorescent bacterial mimics showed that bacteria-sized particles do not permeate a sediment core as much as seawater (House et al., 2003).

After drilling fluid and seawater contamination, the second and perhaps more persistent source of contamination is the home laboratory where deep subsurface samples are analyzed in apparent safety. In particular, DNA-based studies give excellent cautionary tales where "subsurface DNA" is introduced into a sample during DNA preparation and purification (Webster et al., 2003). Sediment samples with low natural DNA content are particularly sensitive to amplification of contaminant DNA, as seen by PCR amplification and sequencing of trace DNA from negative controls (Tanner et al., 1998). Contamination sources include E.coli, the host bacterium that is used for cloning subsurface DNA amplificates (Kormas et al., 2003), or diverse freshwater bacteria that live in water mains or even in water purification columns that are used to produce "molecular biology-grade" water (Matsuda et al., 1996; Percival et al., 1998). The safest way to identify contaminants is to compare clone libraries of subsurface samples to clone libraries of negative controls and published contaminant sequen-

ces; suspect contaminants should be excluded from the analysis, or clearly identified (Kormas et al., 2003).

7. Concluding remarks

Life is not strictly a surface phenomenon. Deep-sea volcanic systems, where rocks and fluids bring microorganisms serendipitously to the surface, are windows into the subsurface. Continental flood basalts, granites, sediments, and mine shafts are further hunting grounds for deep life. Our knowledge base of the subsurface biosphere is expanding, but many questions remain unanswered. How deep does the biosphere extend? How are subsurface organisms and their nutrients transported? What are the energy sources in deep rocks where neither sunlight nor the products of photosynthesis can penetrate? How different, both genetically and physiologically, are subsurface microbes from surface ones? How metabolically active are the organisms that inhabit subsurface rocks and aquifers? Deep drilling will provide additional samples to help address these questions, but geobiologists, in collaboration with geophysicists (rock properties, fluid flow), geochemists (composition of rocks and fluids), and engineers (drilling and sample acquisition) are needed to understand the complexities of the subsurface and its microbial inhabitants.

The terrestrial (marine and continental) subsurface biosphere also serves as an analog to extraterrestrial environments that may harbor unicellular life forms of completely unknown physiology and phylogeny. A recent review (McKay, 2001) sketched out the general constraints and characteristics of subsurface life on Mars and Europa, including potential sources of chemical energy, water, carbon, and other nutrients. Several studies investigated in more detail the plausibility of autotrophic methanogenesis and other redox reactions as sources of geochemical energy for life on Mars and Europa (e.g., Boston et al., 1992; Fisk and Giovannoni, 1999; Gaidos et al., 1999; Jakosky and Shock, 1998; McCollom, 1999). A consensus conclusion is that the total biomass (if any exists) on these planetary bodies is substantially less than that supported by photosynthesis on Earth. Nevertheless, on Mars and Europa, the reduction with H_2 of oxidized compounds may suffice to support the emergence of life and its proliferation, at least in localized niches. As an example, McCollom (1999) used thermodynamic models to conclude that several plausible scenarios for a Europan geochemistry, including both a reduced and a relatively oxidized ocean, lead to conditions under which methanogenesis is exergonic. Sulfate-reduction coupled to H_2-oxidation may also supply energy for microorganisms in Europan hydrothermal systems. Gaidos et al. (1999) disagree with these assessments, but blame the apparent lack of geochemical energy on the dearth of TEAs, not on an insufficient amount of H_2. However, in comparable extraterrestrial subsurface environments and in deep continental aquifers on Earth discussed above, this may be a moot point since CO_2 (or HCO_3^-) is the predominant form of stable carbon, which can serve as both an oxidant and carbon source.

As we write this, no fewer than eight NASA Mars exploration missions are underway or firmly scheduled for launch: Mars Global Surveyor, Mars Odyssey, and Mars Express are presently orbiting the Red Planet, collecting data; the Mars Explorer Rovers– 'Spirit' and 'Opportunity'–have been on the surface of Mars since January 2004 where they have completed their primary missions and are now into their extended missions; and the launches of Mars Reconnaissance Orbiter, Phoenix (the first mission in NASA's Scout program), and Mars Science Laboratory are scheduled for August 2005, August 2007, and late 2009, respectively. In addition to these eight missions, NASA plans other scientific orbiters, rovers, and landers in the second decade of this century, including the first sample return from Mars. Data collected by 'Opportunity' suggest that an expansive, shallow body of water was once present at Terra Meridiani, giving substantial momentum to the 'life on Mars' hypothesis̄whether now or in the past (Arvidson, 2004; Morris et al., 2004; Squyres, 2004). In addition, it was recently reported that spectral data acquired by Mars Explorer indicate methane in the martian atmosphere (Kerr, 2004). The two most likely sources are degassing of trapped methane through unknown volcanic systems or methane-generating life in the martian subsurface.

Though the current and planned missions to Mars have a multitude of objectives, much of the scientific (and general public) interest is driven by the search for life. Owing to the less than hospitable surface characteristics of Mars, the most promising search for habitable zones may require looking below the Martian surface, and our ability to accurately interpret signs of past or present life there will be vastly improved by a better understanding of the richness and diversity of the subsurface biota and environments on Earth.

Acknowledgements

Previous work and new perspectives that contributed to this review were supported by NSF Life in Extreme environments (LExEn) grants OCE-9714288 (to JPA) and OCE-9714195 and OCE-9978310 (to AT); by NSF OCE grants 0221417 and 0234460 (to JPA); and by the NASA Astrobiology Institute Centers at the University of Rhode Island (Subsurface Biospheres) and The Marine Biological Laboratory (Environmental Genomes).

References

Alt, J.C., 1995. Subseafloor processes in mid-ocean ridge hydrothermal systems. In: Humphris, S.E., Zierenberg, R.A., Mullineaux, L.S., Thomson, R.E. (Eds.), Seafloor Hydrothermal Systems: Physical, Chemical, Biological and Geological Interactions. American Geophysical Union, Washington, DC, pp. 85–114.

Amann, R.I., 1995. In situ identification of micro-organisms by whole cell hybridization with rRNA-targeted nucleic acids probes. Mol. Microb. Ecol. Man. 3, 1–15.

Amend, J.P., Shock, E.L., 2001. Energetics of overall metabolic reactions of thermophilic and hyperthermophilic archaea and bacteria. FEMS Microbiol. Rev. 25, 175–243.

Amend, J.P., Rogers, K.L., Shock, E.L., Gurrieri, S., Inguaggiato, S., 2003. Energetics of chemolithoautotrophy in the hydrothermal system of Vulcano Island, southern Italy. Geobiology 1, 37–58.

Anderson, R.T., Chapelle, F.H., Lovley, D.R., 1998. Evidence against hydrogen-based microbial ecosystems in basalt aquifers. Science 281, 976–977.

Arvidson, R.E., 2004. Geology of Meridiani Planum as inferred from Mars Exploration Rover Observations. Lunar and Planetary Science Conference, Houston, Texas, Abstract #2165.

Bach, W., Edwards, K.J., 2003. Iron and sulfide oxidation within the basaltic ocean crust: implications for chemolithoautotrophic microbial biomass production. Geochim. Cosmochim. Acta 67, 3871–3887.

Baker, B.J., et al., 2003. Related assemblages of sulphate-reducing bacteria associated with ultradeep gold mines of South Africa and deep basalt aquifers of Washington State. Environ. Microbiol. 5, 267–277.

Bale, S.J., et al., 1997. *Desulfovibrio profundus* sp. nov., a novel barophilic sulphate-reducing bacterium from deep sediment layers in the Japan Sea. Int. J. Syst. Bacteriol. 47, 515–521.

Balkwill, D.L., et al., 1997. Phylogenetic characterization of bacteria in the subsurface microbial culture collection. FEMS Microbiol. Rev. 20, 201–216.

Banwart, S., et al., 1994. Large-scale intrusion of shallow water into a granite aquifer. Water Resour. Res. 30, 1747–1763.

Banwart, S., et al., 1996. Organic carbon oxidation induced by largescale shallow water intrusion into a vertical fracture zone at the Äspö Hard Rock Laboratory (Sweden). J. Contam. Hydrol. 21, 115–125.

Bazylinski, D.A., Farrington, J.W., Jannasch, H.W., 1988. Hydrocarbons in surface sediments from a Guaymas Basin hydrothermal vent site. Org. Geochem. 12, 547–558.

Beeder, J., Nilsen, R.K., Rosnes, J.T., Torsvik, T., Lien, T., 1994. *Archaeoglobus fulgidus* isolated from hot North Sea oil field waters. Appl. Environ. Microbiol. 60, 1227–1231.

Beeder, J., Torsvik, T., Lien, T., 1995. *Thermodesulforhabdus norvegicus*, gen. nov. sp. nov. a novel thermophilic sulfate-reducing bacterium from oil field water. Arch. Microbiol. 164, 331–336.

Blöchl, E., et al., 1997. *Pyrolobus fumarii*, gen. and sp. nov., represents a novel group of archaea, extending the upper temperature limit for life to 113 °C. Extremophiles 1, 14–21.

Boetius, A., et al., 2000. A marine microbial consortium apparently mediating anaerobic oxidation of methane. Nature 407, 623–626.

Boston, P.J., Ivanov, M.V., McKay, C.P., 1992. On the possibility of chemosynthetic ecosystems in subsurface habitats on Mars. Icarus 95, 300–308.

Brannan, D.K., Caldwell, D.E., 1980. *Thermothrix thiopara*: growth and metabolism of a newly isolated thermophile capable of oxidizing sulfur and sulfur compounds. Appl. Environ. Microbiol. 40, 211–216.

Bridge, T.A.M., Johnson, D.B., 1998. Reduction of soluble iron and reductive dissolution of ferric iron-containing minerals by moderately thermophilic iron-oxidizing bacteria. Appl. Environ. Microbiol. 64, 2181–2186.

Brock, T.D., Gustafson, J., 1976. Ferric iron reduction by sulfur- and iron-oxidizing bacteria. Appl. Environ. Microbiol. 32, 567–571.

Brocks, J.J., Logan, G.A., Buick, R., Summons, R.E., 1999. Archean molecular fossils and the early rise of eukaryotes. Science 285, 1033–1036.

Burggraf, S., Jannasch, H.W., Nicolaus, B., Stetter, K.O., 1990. *Archaeoglobus profundus* sp. nov. represents a new species within the sulfate-reducing archaebacteria. Syst. Appl. Microbiol. 13, 24–28.

Butterfield, D.A., McDuff, R.E., Franklin, J., Wheat, C.G., 1994. Geochemistry of hydrothermal vent fluids from Middle Valley, Juan de Fuca Ridge. Proceedings of the ocean drilling program. Scientific results, vol. 139. Ocean Drilling Program, College Station, TX, pp. 395–410.

Caldwell, D.E., Caldwell, S.J., Laycock, J.P., 1976. *Thermothrix thioparus* gen. et sp. nov. a facultatively anaerobic facultative chemolithotroph living at neutral pH and high temperature. Can. J. Microbiol. 22, 1509–1517.

Calvert, S.E., 1966. Origin of diatom-rich varved sediments from the Gulf of California. J. Geol. 76, 546–565.

Canganella, F., Jones, W.J., Gambacorta, A., Antranikian, G., 1998. *Thermococcus guaymasensis* sp. nov. and *Thermococcus aggregans*, sp. nov., two novel thermophilic archaea isolated from the Guaymas Basin hydrothermal vent site. Int. J. Syst. Bacteriol. 48, 1181–1185.

Chang, J.-Y., et al., 2001. Diversity and characterization of sulfate-reducing bacteria in groundwater at an uranium mill tailings site. Appl. Environ. Microbiol. 67, 3149–3160.

Chapelle, F.H., et al., 2002. A hydrogen-based subsurface microbial community dominated by methanogens. Nature 415, 312–315.

Colwell, F.S., 2001. Constraints on the distribution of microorganisms in subsurface environments. In: Fredrickson, J.K., Fletcher, M. (Eds.), Subsurface microbiology and biogeochemistry. Wiley-Liss, New York, pp. 71–95.

Colwell, F.S., et al., 1992. Innovative techniques for collection of saturated and unsaturated subsurface basalts and sediments for microbiological characterization. J. Microbiol. Methods 15, 279–292.

Cowen, J.P., et al., 2003. Fluids from aging ocean crust that support microbial life. Science 299, 120–123.

Cragg, B.A., Parkes, R.J., 1994. Bacterial profiles in hydrothermally active deep sediment layers from Middle Valley (N.E. Pacific) sites 857 and 858. Proceedings of the ocean drilling program. Scientific results, vol. 139. Ocean Drilling Program, College Station, TX, pp. 509–516.

Cragg, B.A., Summit, M., Parkes, R.J., 2000. Bacterial profiles in a sulfide mount (site 1035) and an area of active fluid venting (site 1036) in hot hydrothermal sediments from Middle Valley (Northeast Pacific). Proceedings of the ocean drilling program. Scientific results, vol. 169. Ocean Drilling Program, College Station, TX, pp. 1–18.

Curray, J.R., et al., 1982. Initial reports of the deep sea drilling project. U.S. Government Printing Office, Washington, D.C.

Daumas, S., Cord-Ruwisch, R., Garcia, J.L., 1988. *Desulfobacterium geothermicum* sp. nov., a thermophilic, fatty-acid degrading, sulfate-reducing bacterium isolated with H_2 from geothermal ground water. Antonie van Leeuwenhoek 54, 165–178.

Davis, E.E., et al., 1992. Proc. Ocean Drill. Program, A, Initial Rep. 139.

Davis, E.E., et al., 1997. Seafloor heat flow on the eastern flank of the Juan de Fuca Ridge: data from "Flankflux" studies through 1995. Proceedings of the ocean drilling program. Initial reports, vol. 168, pp. 23–33 (College Station, TX).

Deming, J.W., Baross, J.A., 1993. Deep-sea smokers: windows to a subsurface biosphere. Geochim. Cosmochim. Acta 57, 3219–3230.

Dhillon, A., Teske, A., Dillon, J., Stahl, D.A., Sogin, M.L., 2003. Molecular characterization of sulfate-reducing bacteria in the Guaymas Basin. Appl. Environ. Microbiol. 69, 2765–2772.

D'Hondt, S., Rutherford, S., Spivack, A., 2002. Metabolic activity of subsurface life in deep-sea sediments. Science 295, 2067–2070.

D'Hondt, S., et al., 2003. Controls on microbial communities in deeply buried sediments, Eastern Equatorial Pacific and Peru Margin. Proceedings of the ocean drilling program. Initial report, vol. 201. Ocean Drilling Program, College Station, TX.

Edgcomb, V.P., Kysela, D.T., Teske, A., Gomez, A.d.V., Sogin, M.L., 2002. Benthic eukaryotic diversity in the Guaymas Basin hydrothermal vent environment. Proc. Natl. Acad. Sci. U. S. A. 99, 7658–7662.

Elderfield, H., Wheat, C.G., Mottl, M.J., Monnin, C., Spiro, B., 1999. Fluid and geochemical transport through oceanic crust: a transect across the eastern flank of the Juan de Fuca Ridge. Earth Planet. Sci. Lett. 172, 151–165.

Elsgaard, L., Isaksen, M.F., Jørgensen, B.B., Alayse, A.-M., Jannasch, H.W., 1994. Microbial sulfate reduction in deep-sea sediments at the Guaymas basin hydrothermal vent area: influence of temperature and substrates. Geochim. Cosmochim. Acta 58, 3335–3343.

Finster, K., Liesack, W., Thamdrup, B., 1998. Elemental sulfur and thiosulfate disproportionation by *Desulfocapsa sulfoexigens* sp. nov., a new anaerobic bacterium isolated from marine surface sediments. Appl. Environ. Microbiol. 64, 119–125.

Fisher, A.T., et al., 2003. Hydrothermal recharge and discharge across 50 km guided by seamounts on a young ridge flank. Nature 421, 618–621.

Fisk, M.R., Giovannoni, S.J., 1999. Sources of nutrients and energy for a deep biosphere on Mars. J. Geophys. Res., (Planets) 104, 11805–11815.

Fossing, H., et al., 1995. Concentration and transport of nitrate by the mat-forming sulfur bacterium *Thioploca*. Nature 374, 713–715.

Fouquet, Y., et al., 1998. Proc. Ocean Drill. Program, Initial Rep. 169.

Fredrickson, J.K., Onstott, T.C., 2001. Biogeochemical and geological significance of subsurface microbiology. In: Fredrickson, J.K., Fletcher, M. (Eds.), Subsurface Microbiology and Biogeochemistry. Wiley-Liss, New York, pp. 3–37.

Fry, N.K., Fredrickson, J.K., Fishbain, S., Wagner, M., Stahl, D.A., 1997. Population structure of microbial communities associated with two deep, anaerobic, alkaline aquifers. Appl. Environ. Microbiol. 63, 1498–1504.

Gaidos, E.J., Nealson, K.H., Kirschvink, J.L., 1999. Life in ice-covered oceans. Science 284, 1631–1633.

Goetz, F.E., Jannasch, H.W., 1993. Aromatic hydrocarbon-degrading bacteria in the petroleum-rich sediments of the Guaymas Basin hydrothermal vent site: preference for aromatic carboxylic acids. Geomicrobiol. J. 11, 1–18.

Gold, T., 1992. The deep, hot biosphere. Proc. Natl. Acad. Sci. U. S. A. 89, 6045–6049.

Gold, T., 1999. The deep hot biosphere. Springer-Verlag, New York (235 pp.).

Gold, T., Soter, S., 1982. Abiogenic methane and the origin of petroleum. Energy Explor. Exploit. 1, 89–104.

Götz, D., et al., 2002. *Persephonella marina* gen. nov., sp. nov. and *Persephonella guaymasensis* sp. nov., two novel, thermophilic, hydrogen-oxidizing microaerophiles from deep-sea hydrothermal vents. Int. J. Syst. Evol. Microbiol. 52, 1349–1359.

Guezennec, J.G., et al., 1996. Bacterial community structure in sediments from Guaymas Basin, Gulf of California, as determined by analysis of phospholipid ester-linked fatty acids. J. Mar. Biotechnol. 4, 165–175.

Habicht, K.S., Gade, M., Thamdrup, B., Berg, P., Canfield, D.E., 2002. Calibration of sulfate levels in the Archean ocean. Science 298, 2372–2374.

Hafenbradl, D., et al., 1996. *Ferroglobus placidus* gen. nov., sp. nov., a novel hyperthermophilic archaeum that oxidizes Fe^{2+} at neutral pH under anoxic conditions. Arch. Microbiol. 166, 308–314.

Harmsen, H.J.M., Prieur, D., Jeanthon, C., 1997. Group-specific 16S rRNA-targeted oligonucleotide probes to identify thermophilic bacteria in marine hydrothermal vents. Appl. Environ. Microbiol. 63, 4061–4068.

Harvey, R.W., George, L.H., Smith, R.L., LeBlanc, D.R., 1989. Transport of microspheres and indigenous bacteria through a sandy aquifer. Results of natural- and forced-gradient tracer experiments. Environ. Sci. Technol. 23, 51–56.

Hinrichs, K.-U., Hayes, J.M., Sylva, S.P., Brewer, P.G., DeLong, E.F., 1999. Methane-consuming archaebacteria in marine sediments. Nature 398, 802–805.

Hoek, J., Banta, A., Hubler, F., Reysenbach, A.-L., 2003. Microbial diversity of a sulphide spire located in the Edmond deep-sea hydrothermal vent field on the Central Indian Ridge. Geobiology 1, 119–127.

Holden, J.F., Summit, M., Baross, J.A., 1998. Thermophilic and hyperthermophilic microorganisms in 3–30 °C hydrothermal fluids following a deep-sea volcanic eruption. FEMS Microbiol. Immunol. 25, 33–41.

Holm, N.G., Charlou, J.L., 2001. Initial indications of abiotic formation of hydrocarbons in the rainbow ultramafic hydrothermal system. Earth Planet. Sci. Lett. 191, 1–8.

House, C., Cragg, B.A., Teske, A., Party, L.S., 2003. Drilling contamination tests during ODP Leg 201 using chemical and particulate tracers. Proceedings of the ocean drilling program. Initial reports, vol. 201. Ocean Drilling Program, College Station, TX, pp. 1–19.

Huber, R., Kurr, M., Jannasch, H.W., Stetter, K.O., 1989. A novel group of abyssal methanogenic archaebacteria (*Methanopyrus*) growing at 110 °C. Nature 342, 833–834.

Huber, R., Stoffers, P., Cheminee, J.L., Richnow, H.H., Stetter, K.O., 1990. Hyperthermophilic archaebacteria within the crater and open-sea plume of erupting MacDonald Seamount. Nature 345, 179–181.

Huber, R., et al., 1992. *Aquifex pyrophilus* gen. nov. sp. nov., represents a novel group of marine hyperthermophilic hydrogen-oxidizing bacteria. Syst. Appl. Microbiol. 15, 340–351.

Huber, R., et al., 1995. *Thermococcus chitonophagus* sp. nov., a novel, chitin-degrading, hyperthermophilic archaeum from a deep-sea hydrothermal environment. Arch. Microbiol. 164, 255–264.

Huber, J.A., Butterfield, D.A., Baross, J.A., 2003. Bacterial diversity in a subseafloor habitat following a deep-sea volcanic eruption. FEMS Microbiol. Immunol. 43, 393–409.

Jakosky, B.M., Shock, E.L., 1998. The biological potential of Mars, the early Earth, and Europa. J. Geophys. Res. 103, 19359–19364.

Jannasch, H.W., Nelson, D.C., Wirsen, C.O., 1989. Massive natural occurrence of unusually large bacteria (*Beggiatoa spp.*) at a hydrothermal deep-sea vent site. Nature 342, 834–836.

Jannasch, H.W., Wirsen, C.O., Molyneaux, S.J., Langworthy, T.A., 1992. Comparative physiological studies on hyperthermophilic archaea isolated from deep-sea hot vents with special emphasis on *Pyrococcus* strain GB-D. Appl. Environ. Microbiol. 58, 3472–3481.

Jannsen, P.H., Schuhmann, A., Bak, F., Liesack, W., 1996. Disproportionation of inorganic sulfur compounds by sulfate-reducing bacterium *Desulfocapsa thiozymogenes* gen. nov., sp. nov. Arch. Microbiol. 166, 184–192.

Jeanthon, C., L'Haridon, S., Pradel, N., Prieur, D., 1999. Rapid identification of hyperthermophilic methanococci isolated from deep-sea hydrothermal vents. Int. J. Syst. Bacteriol. 49, 591–594.

Jeanthon, C., et al., 2002. *Thermodesulfobacterium hydrogenophilum* sp. nov., a thermophilic, chemolithoautotrophic, sulfate-reducing bacterium isolated from a deep-sea hydrothermal vent at Guaymas Basin, and emendation of the genus *Thermodesulfobacterium*. Int. J. Syst. Evol. Microbiol. 52, 765–772.

Jones, W., Leigh, A., Mayer, F., Woese, C., Wolfe, R., 1983. *Methanococcus jannaschii* sp. nov., an extremely thermophilic methanogen from a submarine hydrothermal vent. Arch. Microbiol. 136, 254–261.

Jones, W.J., Stugard, C.E., Jannasch, H.W., 1989. Comparison of thermophilic methanogens from submarine hydrothermal vents. Arch. Microbiol. 151, 314–319.

Jørgensen, B.B., 1982. Mineralization of organic matter in the sea bed—the role of sulfate reduction. Nature 296, 643–645.

Jørgensen, B.B., Zawacki, L.X., Jannasch, H.W., 1990. Thermophilic bacterial sulfate reduction in deep-sea sediments at the Guaymas Basin hydrothermal vents (Gulf of California). Deep-Sea Res. I 37, 695–710.

Karl, D.M., 1995. Ecology of free-living, hydrothermal vent microbial communities. In: Karl, D.M. (Ed.), The microbiology of deep-sea hydrothermal vents. CRC Press, Boca Raton, pp. 35–124.

Kashefi, K., Lovley, D.R., 2003. Extending the upper temperature limit for life. Science 301, 934.

Kashefiet, K., et al., 2002. *Geoglobus ahangari* gen. nov., sp. nov., a novel hyperthermophilic archaeon capable of oxidizing organic acids and growing autotrophically on hydrogen with Fe(III) serving as the sole electron acceptor. Int. J. Syst. Evol. Microbiol. 52, 719–728.

Kawka, O.E., Simoneit, B.R.T., 1987. Survey of hydrothermally-generated petroleums from the Guaymas Basin spreading center. Org. Geochem. 11, 311–328.

Kerr, R.A., 2004. Life or volcanic belching on Mars? Science 303, 1953.

Kieft, T.L., Phelps, T.J., 1997. Life in the slow lane. In: Amy, P.S., Haldeman, D.L. (Eds.), The microbiology of the terrestrial subsurface. CRC Press, Boca Raton, FL, pp. 137–164.

Klein, M., et al., 2001. Multiple lateral transfer events of dissimilatory sulfite-reductase genes between major lineages of bacteria. J. Bacteriol. 183, 6028–6034.

Kormas, K.A., Smith, D.C., Edgcomb, V., Teske, A., 2003. Molecular analysis of deep subsurface microbial communities in Nankai Trough sediments (ODP Leg 190, Site 1176). FEMS Microbiol. Immunol. 45, 115–125.

Kotelnikova, S., Pedersen, K., 1997. Evidence for methanogenic archaea and homoacetogenic bacteria in deep granitic rock aquifers. FEMS Microbiol. Rev. 20, 327–339.

Kotelnikova, S., Pedersen, K., 1998. Distribution and activity of methanogens and homoacetogens in deep granitic aquifers at Äspö Hard Rock Laboratory, Sweden. FEMS Microbiol. Immunol. 26, 121–134.

Kotelnikova, S., Macario, A.J.L., Pedersen, K., 1998. *Methanobacterium subterraneum* sp. nov., a new alkaliphilic, eurythermic and halotolerant methanogen isolated from deep granitic groundwater. Int. J. Syst. Bacteriol. 48, 357–367.

Krumholz, L., 2000. Microbial communities in the deep subsurface. Hydrogeol. J. 8, 4–10.

Krumholz, L.R., McKinley, J.P., Ulrich, G.A., Suflita, J.M., 1997. Confined subsurface microbial communities in Cretaceous rock. Nature 386, 64–66.

Krumholz, L.R., Harris, S.H., Tay, S.T., Suflita, J.M., 1999. Characterization of two subsurface H_2-utilizing bacteria, *Desulfomicrobium hypogeium* sp. nov. and *Acetobacterium psammolithicum* sp. nov., and their ecological roles. Appl. Environ. Microbiol. 65, 2300–2306.

Kurr, M., et al., 1991. *Methanopyrus kandleri*, gen. and sp. nov. represents a novel group of hyperthermophilic methanogens, growing to 110 °C. Arch. Microbiol. 156, 239–247.

Laaksoharju, M., et al., 1995. Sulphate reduction in the Äspö HRL tunnel. Swedish Nuclear Fuel and Waste Management, Stockholm.

L'Haridon, S., Reysenbach, A.-L., Glenat, P., Prieur, D., Jeanthon, C., 1995. Hot subterranean biosphere in a continental oil reservoir. Nature 377, 223–224.

Lin, L.-H., Slater, G.F., Sherwood Lollar, B., Lacrampe-Couloume, G., Onstott, T.C., 2004. The yield and isotopic composition of radiolytic H_2, a potential energy source for the deep subsurface biosphere. Geochim. Cosmochim. Acta (in press).

Liu, Y., et al., 1997. Description of two new thermophilic *Desulfotomaculum* spp., *Desulfotomaculum putei* sp. nov., from a deep terrestrial subsurface , and *Desulfotomaculum luciae* sp. nov., from a hot spring. Int. J. Syst. Bacteriol. 47, 615–621.

Love, C.A., Patel, B.K.C., Nichols, P.D., Stackebrandt, E., 1993. *Desulfotomaculum australicum* sp. nov., a thermophilic sulfate-reducing bacterium isolated from the great artesian basin of Australia. Syst. Appl. Microbiol. 16, 244–251.

Lovley, D.R., 2001. Reduction of iron and humics in subsurface environments. In: Fredrickson, J.K., Fletcher, M. (Eds.), Subsurface microbiology and biogeochemistry. Wiley-Liss, New York, pp. 193–217.

Lyons, T.W., Kah, L.C., Gellatly, A.M., 2004. The Proterozoic sulfur isotope record of evolving atmospheric oxygen. In: Eriksson, P.G.E.A. (Ed.), The Precambrian Earth: Tempos and Events: Developments in Precambrian Geology. Elsevier, pp. 421–439.

Martens, C.S., 1990. Generation of short chain organic acid anions in hydrothermally altered sediments of the Guaymas Basin, Gulf of California. Appl. Geochem. 5, 71–76.

Matsuda, N., et al., 1996. Gram-negative bacteria viable in ultra pure water and effect of temperature on their behaviour. Colloids Surf., B Biointerfaces 5, 279–289.

McCollom, T.M., 1999. Methanogenesis as a potential source of chemical energy for primary biomass production by autotrophic organisms in hydrothermal systems on Europa. J. Geophys. Res. 104, 30729–30742.

McCollom, T.M., Shock, E.L., 1997. Geochemical constraints on chemolithoautotrophic metabolism by microorganisms in seafloor hydrothermal systems. Geochim. Cosmochim. Acta 61, 4375–4391.

McCollom, T.M., Ritter, G., Simoneit, B.R.T., 1999. Lipid synthesis under hydrothermal conditions by Fischer–Tropsch-type reactions. Orig. Life Evol. Biosph. 29, 153–166.

McHatten, S.C., Barry, J.P., Jannasch, H.W., Nelson, D.C., 1986. High nitrate concentrations in vacuolate, autotrophic marine *Beggiatoa*. Appl. Environ. Microbiol. 62, 954–958.

McKay, C.P., 2001. The deep biosphere: lessons from planetary exploration. In: Fredrickson, J.K., Fletcher, M. (Eds.), Subsurface Microbiology and Biogeochemistry. Wiley-Liss, New York, pp. 315–327.

McKinley, J.P., Colwell, F.S., 1996. Application of perfluorocarbon tracers to microbial sampling in subsurface environments using mud-rotary and air-rotary drilling techniques. J. Microbiol. Methods 26, 1–9.

McKinley, J.P., Stevens, T.O., Westall, F., 2000. Microfossils and paleoenvironments in deep subsurface basalt samples. Geomicrobiol. J. 17, 43–54.

Morris, R.V., et al., 2004. A first look at the mineralogy and geochemistry of the MER-B landing site in Meridiani Planum. Lunar and Planetary Science Conference, Houston, Texas, Abstract # 2179.

Motamedi, M., Pedersen, K., 1998. *Desulfovibrio aespoeensis* sp. nov., a mesophilic sulfate-reducing bacterium from deep groundwater at Äspö hard rock laboratory, Sweden. Int. J. Syst. Bacteriol. 48, 311–315.

Nazina, T.N., Ivanova, A.E., Kanchaveli, L.P., Rozanova, E.P., 1987. A new spreforming thermophilic methylogrophic bacterium, *Desulfotomaculum kuznetsovii* sp. nov. Microbiology (engl. transl. Mikrobiologiya) 47, 773–778.

Nelson, D.C., Wirsen, C.O., Jannasch, H.W., 1989. Characterization of large, autotrophic *Beggiatoa* spp. abundant at hydrothermal vents of the Guaymas Basin. Appl. Environ. Microbiol. 55, 2909–2917.

Nilsen, R.K., Torsvik, T., 1996. *Methanococcus thermolithotrophicus* isolated from North Sea oil field reservoir water. Appl. Environ. Microbiol. 62, 728–731.

Orphan, V.J., Howes, C.H., Hinrichs, K.-U., McKeegan, K.D., DeLong, E.F., 2001. Methane-consuming archaea revealed by directly coupled isotopic and phylogenetic analysis. Science 293, 484–487.

Otte, S., et al., 1999. Nitrogen, carbon and sulfur metabolism in natural *Thioploca* samples. Appl. Environ. Microbiol. 65, 3148–3157.

Parkes, R.J., Wellsbury, P., 2004. Deep biospheres. In: Bull, A.T. (Ed.), Microbial diversity and bioprospecting. ASM Press, Washington, D.C, pp. 120–129.

Parkes, R.J., Cragg, B.A., Wellsbury, P., 2000. Recent studies on bacterial populations and processes in subseafloor sediments: a review. Hydrogeol. J. 8, 11–28.

Pedersen, K., 1993. Bacterial processes in nuclear waste disposal. Microbiol. Eur. 1, 18–23.

Pedersen, K., 1997. Microbial life in deep granitic rock. FEMS Microbiol. Rev. 20, 399–414.

Pedersen, K., 2000a. Exploration of deep intraterrestrial microbial life: current perspectives. FEMS Microbiol. Lett. 185, 9–16.

Pedersen, K., 2000b. The hydrogen driven intra-terrestrial biosphere and its influence on the hydrochemical conditions in crystalline bedrock aquifers. In: Stober, I., Bucher, K. (Eds.), Hydrogeology of crystalline rocks. Kluwer Academic Publishers, pp. 249–259.

Pedersen, K., 2001. Diversity and activity of microorganisms in deep igneous rock aquifers of the fennoscandian shield. In: Fredrickson, J.K., Fletcher, M. (Eds.), Subsurface Microgeobiology and Biogeochemistry. Wiley-Liss, New York, pp. 97–139.

Pedersen, K., Ekendahl, S., 1990. Distribution and activity of bacteria in deep granitic groundwaters of southeastern Sweden. Microb. Ecol. 20, 37–52.

Pedersen, K., Arlinger, J., Ekendahl, S., Hallbeck, L., 1996. 16S rRNA gene diversity of attached and unattached groundwater bacteria along the Access tunnel to the Äspö Hard Rock Laboratory, Sweden. FEMS Microbiol. Ecol. 19, 249–262.

Percival, S.L., Knapp, J.S., Edyvean, R., Wales, D.S., 1998. Biofilm development on stainless steel in mains water. Water Res. 32, 243–253.

Rees, G.N., Grassia, G.S., Sheeny, A.J., Dwivedi, P.P., Patel, B.K.C., 1995. *Desulfacinum infernum* gen. nov., sp. nov., a thermophilic sulfate-reducing bacterium from a petroleum reservoir. Int. J. Syst. Bacteriol. 45, 85–89.

Reysenbach, A.-L., Holm, N.G., Hershberger, K., Prieur, D., Jeanthon, C., 1998. In search of a subsurface biosphere at a slow-spreading ridge. Proceedings of the ocean drilling program. Scientific results, vol. 158. Ocean Drilling Program, College Station, TX, pp. 355–360.

Rosnes, J.T., Torsvik, T., Lien, T., 1991. Spore-forming thermophilic sulfate-reducing bacteria isolated from North Sea oil field waters. Appl. Environ. Microbiol. 57, 2302–2307.

Rueter, P., et al., 1994. Anaerobic oxidation of hydrocarbons in crude oil by new types of sulphate-reducing bacteria. Nature 372, 455–458.

Rushdi, A.I., Simoneit, B.R.T., 2002. Hydrothermal alteration of organic matter in sediments of the Northeastern Pacific Ocean: part 2. Escanaba Trough, Gorda Ridge. Appl. Geochem. 17, 1467–1494.

Russell, B.F., Phelps, T.J., Griffin, W.T., Sargent, K.A., 1992. Procedures for sampling deep subsurface microbial communities in unconsolidated sediments. Ground Water Monit. Rev. 12, 96–104.

Schouten, S., Wakeham, S.G., Hopmans, E.C., Sinninghe Damsté, J.S., 2003. Biogeochemical evidence that thermophilic archaca mediate the anaerobic oxidation of methane. Appl. Environ. Microbiol. 69, 1680–1686.

Schrader, H., 1982. Diatom biostratigraphy and laminated diatomaceous sediments from the Gulf of California, Deep Sea Drilling

Project leg 64. Initial reports of the deep sea drilling project, vol. 64. U.S. Government Printing Office, Washington, D.C, pp. 973–981.

Schrenk, M.O., Kelley, D.S., Delaney, J.R., Baross, J.A., 2003. Incidence and diversity of microorganisms within walls of an active deep-sea sulfide chimney. Appl. Environ. Microbiol. 69, 3580–3592.

Shipboard Scientific Party, 1998. Dead Dog Area (Site 1036). Proceedings of the ocean drilling program. Initial reports, vol. 169. Ocean Drilling Program, College Station, TX.

Shock, E.L., 1990. Geochemical constraints on the origin of organic compounds in hydrothermal systems. Orig. Life Evol. Biosph. 20, 331–367.

Shock, E.L., Schulte, M.D., 1998. Organic synthesis during fluid mixing in hydrothermal systems. J. Geophys. Res. 103, 28513–28527.

Shock, E.L., Holland, M., Meyer-Dombard, D.R., Amend, J.P., 2005. Geochemical sources of energy for microbial metabolism in hydrothermal ecosystems: Obsidian Pool, Yellowstone National Park, USA. Proceedings of the geothermal biology and geochemistry in Yellowstone National Park conference. Thermal Biology Institute (in press).

Simoneit, B.R.T., 2002. Carbon isotope systematics of individual hydrocarbons in hydrothermal petroleum from Middle Valley, Northeastern Pacific Ocean. Appl. Geochem. 17, 1429–1433.

Simoneit, B.R.T., 2002. Carbon isotope systematics of individual hydrocarbons in hydrothermal petroleum from Middle Valley, Northeastern Pacific Ocean. Appl. Geochem. 17, 1429–1433.

Simoneit, B.R.T., Fetzer, J.C., 1996. High molecular weight polycyclic aromatic hydrocarbons in hydrothermal petroleums from the Gulf of California and Northeast Pacific Ocean. Org. Geochem. 24, 1065–1077.

Simoneit, B.R.T., Sparrow, M.A., 2002. Dissolved organic carbon in interstitial waters from sediments of Middle Valley and Escanaba Trough, Northeast Pacific, ODP Legs 139 and 169. Appl. Geochem. 17, 1495–1502.

Simoneit, B.R.T., Schoell, M., Kvenvolden, K.A., 1997. Carbon isotope systematics of individual hydrocarbons in hydrothermal petroleum from Escanaba Trough, Northeastern Pacific Ocean. Org. Geochem. 26, 511–515.

Smith, D.A., et al., 2000a. Methods for quantifying potential microbial contamination during deep ocean coring. Ocean Drill. Program Tech. Note 28.

Smith, D.A., et al., 2000b. Tracer-based estimates of drilling-induced microbial contamination of deep sea crust. Geomicrobiol. J. 17, 207–219.

Squyres, S.W., 2004. Initial results from the MER Athena Science Investigation at Gusev Crater and Meridiani Planum. Lunar and planetary science conference, Houston, Texas, Abstract # 2187.

Stein, J.S., et al., 1998. Fine scale heat flow, shallow heat sources, and decoupled circulation systems at two sea-floor hydrothermal sites, Middle Valley, northern Juan de Fuca Ridge. Geology 26, 1115–1118.

Stetter, K.O., 1999. Extremophiles and their adaptation to extreme environments. FEBS Lett. 452, 22–25.

Stetter, K.O., et al., 1993. Hyperthermophilic archaea are thriving in deep North Sea and Alaskan oil reservoirs. Nature 365, 743–745.

Stevens, T., 1997. Lithoautotrophy in the subsurface. FEMS Microbiol. Rev. 20, 327–337.

Stevens, T.O., McKinley, J.P., 1995. Lithoautotrophic microbial ecosystems in deep basalt aquifers. Science 270, 450–454.

Stumm, W., Morgan, J.J., 1996. Aquatic chemistry: chemical equilibria and rates in natural waters. John Wiley & Sons, New York (1022 pp.).

Summit, M., Baross, J.A., 1998. Thermophilic subseafloor micro-organisms from the 1996 North Gorda Ridge Eruption. Deep-Sea Res. II 45, 2751–2766.

Summit, M., Baross, J.A., 2001. A novel microbial habitat in the mid-ocean ridge subseafloor. Proc. Natl. Acad. Sci. U. S. A. 98, 2158–2163.

Summit, M., Peacock, A., Ringelberg, D., White, D.C., Baross, J.A., 2000. Phospholipid fatty-acid derived microbial biomass and community dynamics in hot, hydrothermally influenced sediments from Middle Valley, Juan de Fuca Ridge. Proceedings of the ocean drilling program. Scientific results, vol. 169. Ocean Drilling Program, College Station, TX.

Summons, R.E., Jahnke, L.L., Hope, J.M., Logan, G.A., 1999. 2-Methylhopanoids as biomarkers for cyanobacterial oxygenic photosynthesis. Nature 400, 554–556.

Takai, K., Horikoshi, K., 1999. Genetic diversity of archaea in deep-sea hydrothermal vent environments. Genetics 152, 1285–1297.

Takai, K., Komatsu, T., Inagaki, F., Horikoshi, K., 2001. Distribution of archaea in a black smoker chimney structure. Appl. Environ. Microbiol. 67, 3618–3629.

Tanner, M.A., Goebel, B.M., Dojka, M.A., Pace, N.R., 1998. Specific ribosomal DNA sequences from diverse environmental settings correlate with experimental contaminants. Appl. Environ. Microbiol. 64, 3110–3113.

Teske, A., et al., 2002. Microbial diversity of hydrothermal sediments in the Guaymas Basin: evidence for anaerobic methanotrophic communities. Appl. Environ. Microbiol. 68, 1994–2007.

Teske, A., Dhillon, A., Sogin, M.L., 2003. Genomic markers of ancient anaerobic microbial pathways: sulfate reduction, methanogenesis, and methane oxidation. Biol. Bull. 204, 186–191.

Tor, J.M., Amend, J.P., Lovley, D.R., 2003. Metabolism of organic compounds in anaerobic, hydrothermal sulfate-reducing marine sediments. Environ. Microbiol. 5, 583–591.

Tullborg, E.-L., 2000. Ancient microbial activity in crystalline bedrock—results from stable isotope analyses of fracture calcites. In: Stober, I., Bucher, K. (Eds.), Hydrogeology of Crystalline Rocks. Kluwer, Dordrecht, pp. 261–275.

Völkl, P., et al., 1993. *Pyrobaculum aerophilum* sp. nov., a novel nitrate-reducing hyperthermophilic archaeum. Appl. Environ. Microbiol. 59, 2918–2926.

Von Damm, K.L., et al., 1990. Seafloor hydrothermal activity: black smoker chemistry and chimneys. Annu. Rev. Earth Planet. Sci. 18, 173–204.

Von Damm, K., 1995. Controls on the chemistry and temporal variability of seafloor hydrothermal fluids. In: Humphris, S.E., Zierenberg, R.A., Mullineaux, L.S., Thomson, R.E. (Eds.), Seafloor Hydrothermal Systems: Physical, Chemical, Biological and Geological Interactions. American Geophysical Union, Washington, D.C, pp. 222–247.

Von Damm, K.L., et al., 1985. Chemistry of submarine hydro-thermal solutions at 21°N, East Pacific Rise. Geochim. Cosmochim. Acta 49, 2197–2220.

Wagner, M., Roger, A.J., Flax, J.L., Brusseau, G.A., Stahl, D.A., 1998. Phylogeny of dissimilatory sulfite reductases supports an early origin of sulfate reduction. J. Bacteriol. 180, 2975–2982.

Weber, A., Jørgensen, B.B., 2002. Bacterial sulfate reduction in hydrothermal sediments of the Guaymas Basin, Gulf of California, Mexico. Deep-Sea Res. I 49, 827–841.

Webster, G., Newberry, C.J., Fry, J.C., Weightman, A.J., 2003. Assessment of bacterial community structure in the deep sub-seafloor biosphere by 16S rDNA-based techniques: a cautionary tale. J. Microbiol. Methods 55, 155–164.

Welhan, J.A., 1988. Origins of methane in hydrothermal systems. Chem. Geol. 71, 183–198.

Whelan, J.K., Simoneit, B.R.T., Tarafa, M.E., 1988. C1–C8 hydrocarbons in sediments from Guaymas Basin, Gulf of California—comparison to Peru Margin, Japan Trench and California Borderlands. Org. Geochem. 12, 171–194.

Whitman, W.B., Coleman, D.C., Wiebe, W.J., 1998. Prokaryotes: the unseen majority. Proc. Natl. Acad. Sci. U. S. A. 95, 6578–6583.

You, C.-F., et al., 1994. Boron and halide systematics in submarine hydrothermal systems: effects of phase separation and sedi-mentary contributions. Earth Planet. Sci. Lett. 123, 227–238.

Available online at www.sciencedirect.com

Palaeogeography, Palaeoclimatology, Palaeoecology 219 (2005) 157–170

www.elsevier.com/locate/palaeo

Microbial transformations of organic matter in black shales and implications for global biogeochemical cycles

S.T. Petsch[a,*], K.J. Edwards[b], T.I. Eglinton[b]

[a]*Department of Geosciences, University of Massachusetts-Amherst, Amherst, MA 01003, United States*
[b]*Department of Marine Chemistry and Geochemistry, Woods Hole Oceanographic Institution, Woods Hole, MA 02543, United States*

Received 3 June 2003; accepted 29 October 2004

Abstract

The various roles that microorganisms play in transformation of organic matter in geologic environments are yet to be fully revealed. Many of these roles influence and perhaps control the composition of earth's atmosphere over geologic time by directly impacting global-scale cycling of carbon dioxide, oxygen, and methane. One example is the weathering of black shales. Exposure of organic matter-rich sedimentary rocks to oxygenated earth surface conditions results in significant changes in outcrop-scale rock geochemistry with implications for element cycling on a global scale. This article reviews the progress of ongoing research of a black shale weathering profile exposed near Clay City, Kentucky, USA. Using tools of molecular biology, microbial ecology, isotope geochemistry, and organic geochemistry, this research explores the role of microorganisms in utilization and oxidation of sedimentary organic matter and sulfide mineral oxidation, and examines the communities of microorganisms that may live in this environment. From this and related studies, we are developing greater awareness of the importance of microorganisms in transfer of organic materials among various reservoirs of the geosphere, biosphere, hydrosphere, and atmosphere.
© 2004 Elsevier B.V. All rights reserved.

Keywords: Organic matter; Kerogen; Shale; Assimilation; Microorganisms; Lipids

1. Introduction

To a large degree, the composition of Earth's atmosphere is controlled by biological activity operating on a global scale. The concentration and even the existence of several atmospheric gases can be linked directly to the interaction of biological and chemical process acting over geologic temporal and spatial scales (i.e. Berner, 1987; Berner and Canfield, 1989; Berner et al., 2000; Van Cappellen and Ingall, 1996; Petsch and Berner, 1998). Foremost among these gases are oxygen [O_2], carbon dioxide [CO_2], and methane [CH_4]. Study of controls on atmospheric O_2 concentration is relevant because of the central role that O_2 plays in aerobic heterotrophy (O_2 is a necessary ingredient for life as we humans experience it) and because O_2 is the second most abundant component of

---

* Corresponding author.
E-mail address: spetsch@geo.umass.edu (S.T. Petsch).

S.T. Petsch et al. / Palaeogeography, Palaeoclimatology, Palaeoecology 219 (2005) 157–170

our atmosphere today (slightly less than 21%, compared with 78% N_2). CO_2 and CH_4 are only trace components of the modern atmosphere, at ~380 ppmv and ~1.7 ppmv, respectively. However, both of these gases contribute significantly to Earth's greenhouse effect, and variability in the concentration of these gases in the atmosphere through geologic time may have played a central role in controlling global climate. Reconstruction of atmospheric CO_2 and CH_4 concentrations in Earth's atmosphere through time has been the focus of many lines of research, either through direct sampling of Pleistocene/Holocene ice cores (i.e. Petit et al., 1999; Thompson et al., 2002), proxy CO_2 (Cerling, 1992; Mora et al., 1996; Pagani et al., 1999; Yapp and Poths, 1992; Royer et al., 2001a,b) and CH_4 (Dickens, 2001; Jahren et al., 2001; Katz et al., 1999) records obtained through isotopic and/or molecular signals preserved in sediments and paleosols, or numerical models of global-scale carbon cycling that yield estimates of CO_2 through time (Berner et al., 1983; Berner, 1994; Berner and Kothavala, 2001). Studies on the rise and subsequent evolution of atmospheric O_2 are central to evolving understanding of geobiology and global biogeochemistry (i.e. Graham et al., 1995; Canfield and Teske, 1996; Canfield, 1998; Dudley, 1998; Canfield et al., 2000; Dudley, 2000; Habicht et al., 2002; Anbar and Knoll, 2002; Berner et al., 2002; Holland, 2002).

O_2 and CO_2 are also linked in the mirrored role each plays in the geochemical carbon cycle. During photosynthesis, organisms fix CO_2, chemically reducing it to carbohydrates and other biomass. Electrons donated to CO_2 during photosynthesis derive from the splitting of water, which releases free O_2. Thus during photosynthesis, CO_2 is consumed and O_2 is produced. Aerobic respiration is the gain of energy obtained through oxidation of biomass with O_2 to produce CO_2. Global-scale rates of photosynthesis and respiration very nearly balance each other; however, there is a slight excess of carbon fixation relative to respiration. This is evidenced by the sinking of organic matter [OM] through the oceans of the world and accumulation in marine sediments. The small fraction of organic carbon buried in sediments represents the photosynthetic carbon fixation flux that escapes respiration. Thus burial of organic matter in sediments can be equated to net O_2 production and CO_2 consumption over geologic time, outside of the very rapid O_2–CO_2 cycling associated with photosynthesis and respiration in the active surface biosphere.

If burial of organic matter in sediments represents addition of O_2 to the atmosphere, then removal of O_2 is accomplished by the oxidative weathering of organic matter stored in ancient sedimentary rocks. Many lacustrine, deltaic, and shelf sediments contain measurable concentrations of organic carbon. After lithification, these become the organic carbon-rich shales, mudrocks, and marls that can be important fossil fuel source rocks. Among these are black shales: fine-grained laminated sedimentary rocks rich in organic carbon (~3% by weight or more) that also usually contain chemically distinct distributions of redox sensitive metals and sulfide minerals. When ancient sedimentary rocks rich in organic carbon are exposed on Earth's continents through uplift and erosion, these rocks become exposed to a very different chemical environment than conditions of the Earth's subsurface. During weathering of black shales, a highly reducing environment rich in organic carbon and sulfide minerals is brought in close contact with the strongly oxygenated conditions of Earth's surface. Organic matter and sulfide minerals in the rock become oxidized, consuming O_2, liberating inorganic carbon and sulfate, and generating acidity. Thus, as with photosynthesis and respiration on the short term, burial and weathering of sedimentary organic matter provide two strong controls on O_2 and CO_2 in the Earth's atmosphere (Berner, 1987, 1999; Petsch and Berner, 1998; Berner et al., 2000, 2002).

In spite of this logical argument for carbon cycling through sedimentary organic matter providing a strong control on atmospheric composition through geologic time, direct knowledge of the detailed processes and controls of black shale weathering remains limited. Specifically, the rates, mechanisms, and pathways of organic carbon loss and transformation during weathering are virtually unknown. Building on several previous studies that examined organic carbon loss within shale weathering profiles (Leythäuser, 1973; Clayton and Swetland, 1978; Lewan, 1980; Littke et al., 1991), in recent years we have sought to evaluate the specific chemical and microbiological transformations of sedimentary organic matter during weathering of black shales, with a goal of applying knowledge gained towards improved understanding of the geochemical carbon cycle and the role of rock weathering in atmospheric evolution through geologic time (Petsch et al., 2000, 2001a,b, 2003).

This paper summarizes several lines of investigation examining the degradation and transformation of organic matter during rock weathering at a site near Clay City, Kentucky (USA). Chemical analyses, including elemental and isotopic analysis, analytical pyrolysis, and nuclear magnetic resonance, have been used to establish the changes in organic matter abundance and composition within this weathering profile. The abundance and $\delta^{13}C$ and $\Delta^{14}C$ ratios of cell membrane lipids have been used to trace carbon pathways within this weathering profile and enrichment cultures generated from this site. Membrane lipids and DNA harvested from enrichment cultures have been used to describe the microbial communities found in both environmental samples and enrichment cultures.

2. Clay City, Kentucky: the field site

Late Devonian marine sedimentary rocks outcrop as a 5- to 10-km-wide band extending ~300 km around the edge of the Jessamine Dome in central Kentucky. Among these rocks are successions of finely laminated pyritic black shales upwards of 30 m thick, termed the New Albany, Chattanooga and Ohio Shales (in the west, south, and northeast portions of Kentucky, respectively). In Powell County, approximately 50 km ESE of Lexington (KY), a suite of road cuts through these Late Devonian black shales is located near the town of Clay City. One particular road cut has been the focus of much attention due to the obvious and well-developed weathering profile exposed (Fig. 1). A well-developed, vertical weathering front into this black shale was exposed ~40 years ago (private landowner, personal communication) when the hillside was excavated to enlarge the roadway and provide fill for construction. The exposed weathering profile exhibits a distinctly lighter brown color and a friable unconsolidated physical texture compared with unweathered rock in the center of the roadcut. This weathered material is not the result of deposition of soil from upslope because individual rock strata are discernibly continuous in horizontal layers extending from unweathered rock through the weathering profile. Our efforts have

Fig. 1. Exposure of ~4 m thick black shale weathering profile in roadcut through L. Devonian black shale near Clay City, Kentucky, USA.

focused on evaluating the chemical and microbiological variations associated with color and textural changes within this and other black shale weathering profiles.

3. Black shale weathering: organic geochemistry

The first suite of samples examined for this study were removed from the shale, along a horizontal profile extending ~5 m inward from the northern side of the road cut. Samples were taken from a single horizon so that subsequent chemical analyses would not be confounded by bed-to-bed chemical heterogeneity. Detailed methods and results can be found in Petsch et al. (2000) and Petsch et al. (2001a). In brief, five key characteristics from these samples are described below:

1. Weight percent total organic carbon [TOC] falls from nearly 10% in unweathered samples to ~1% in weathered samples (Fig. 2). Low TOC is focused mainly between 1 and 2 m in from the edge of the profile, reflecting a balance between weathering (removal of OM) and erosion (delivery of higher TOC-bearing rock towards the surface of the weathering profile).

2. Pyrite (FeS_2) content falls from 3% in unweathered samples to zero in weathered samples (Fig. 2). Pyrite loss is focused between 4 and 5 m,

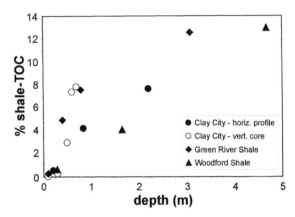

Fig. 3. Weight percent TOC derived from kerogen in black shale weathering profile, calculated as total OM minus contribution from modern soil OM. Relative amount of ancient and modern OM was estimated using a ^{14}C mixing model, assuming soil OM with a $\Delta^{14}C = -150‰$ (most "modern" value for OM in Clay City core, and a conservative estimate for bulk soil OM ^{14}C content) and ancient kerogen $\Delta^{14}C = -1000‰$. Green River Shale and Woodford Shale sampling localities are described in Petsch et al. (2000); radiocarbon analyses are presented in Petsch et al. (2001a).

substantially below the zone of TOC loss. The sequence of pyrite loss prior to organic matter loss strongly suggests a decoupling in the oxidation of these two materials, and specifically that the kinetics of pyrite oxidation exceed organic matter loss at this site.

3. Investigations at Clay City and several other similar sites reveal that as a general rule, TOC content does not fall to zero within black shale weathering profiles. At most sites, samples from the surfaces of shale weathering profiles contain 1–4% TOC (Petsch et al., 2000). [14]C analysis of bulk weathering profile TOC have been used to construct simple isotope mass balance models. These models reveal that while some carbon in the profile derives from humic materials, that originate from modern terrestrial biomass, especially in near-surface layers, much of the organic carbon must derive from the rock (Fig. 3).

4. Kerogen (defined as solvent-insoluble sedimentary organic matter) isolated from the weathering profile reveals a largely constant elemental composition (N/C and S/C) across the weathering profile, in spite of substantial loss of total organic matter (Fig. 4). This constancy in composition was unexpected, because heteroatomic N- and S-bearing component within kerogen were thought likely

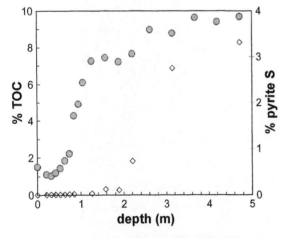

Fig. 2. Weight percent total organic carbon [TOC ●] and pyrite sulfur [FeS_2 ◇] along horizontal transect through black shale weathering profile. Data from Petsch et al. (2000, 2001a,b).

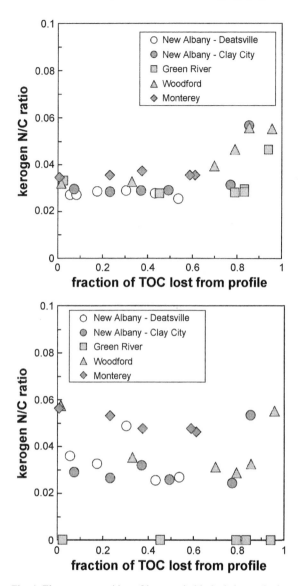

Fig. 4. Element composition of kerogen in black shale weathering profiles, expressed as mole ratio S/C (upper panel) and N/C (lower panel) relative to fraction of total organic carbon lost from the profile. Data from Petsch et al. (2000).

to be more reactive and susceptible to oxidation than the highly aliphatic (saturated hydrocarbon) kerogen framework.

5. More detailed structural investigations of kerogen composition across this and other weathering profiles, using infrared spectroscopy, analytical pyrolysis (Petsch et al., 2000), and ^{13}C nuclear magnetic resonance (Petsch et al., 2001a) have

confirmed a remarkable degree of consistency in kerogen composition within black shale weathering profiles (Fig. 5). This contrasts strongly with the changes observed in organic matter composition during sediment burial (Goth et al., 1988; Derenne et al., 1991, 1992; Flaviano et al., 1994; Lichtfouse et al., 1998; Hatcher et al., 1983) during which the overall trend in OM composition is selective loss of heteroatomic components (carbohydrates, proteinaceous material) and preservation of highly aliphatic components.

What can these features reveal about black shale weathering in general and its role in the biogeochemical cycles of O_2 and CO_2? [14]C isotope mass balance reveals that under contemporary pO_2, ~10% of shale OM at Clay City is transported out of a given horizon as particulate, ^{14}C-free relic OM (Petsch et al., 2000, 2001a). ~90% is lost from the shale through oxidation and/or dissolution. These proportions are controlled by a balance between chemical reactions (oxidation, dissolution) and physical transport (denudation, erosion). For a given column of shale exposed to weathering, the denudation rate controls the length of time any single horizon spends within the weathered zone of the column; slow denudation rates lead to long exposure times. Chemical reactions are limited by time and other terms in the reaction rate law (these may include surface area, pH, O_2 availability). Thus to first order, both the depth of a weathering profile (how far weathering has propagated into the shale) and the intensity of weathering (what proportion of shale OM is not dissolved or oxidized during weathering) should relate in some way to denudation rate. In part, this holds true. Qualitative estimates of denudation rate at several examined shale weathering profiles based on local relief and climate relate to both depth and intensity of weathering. The high denudation rate cliff exposure of the Monterey Shale exhibits a compressed weathering profile (TOC falls from 12% to 5%) compared with the exposures of both Green River and Woodford Shale (TOC falls from ~20% to <1%) where slopes are modest, climates are semiarid, and denudation rates are probably low. At even low denudation rates, however, a small proportion of shale OM escapes chemical reaction and is lost from the profile surface as relic particulate OM. This reveals that rates of OM reaction during weathering are not so

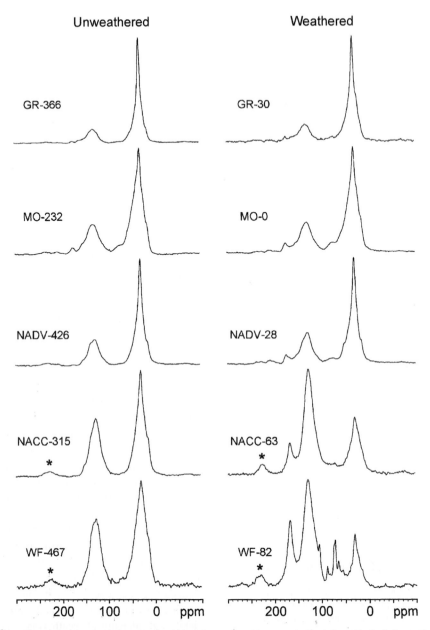

Fig. 5. CP/MAS ^{13}C NMR spectra for unweathered and weathered samples of kerogen isolated from black shale weathering profiles, from Petsch et al. (2001a). GR—Green River Shale; MO—Monterey Shale; NACC—New Albany Shale (Clay City site); NADV—New Albany Shale (Deatsville site); WF—Woodford Shale. Sample depths within profile are given in centimeters.

fast as to be completely limited by denudation rate and thus transport controlled.

There is never enough time for either oxidation or dissolution to completely reaction all shale OM, even in low denudation rate profiles. This opens the possibility for a response of black shale weathering rates to changes in atmospheric O_2. Might there be more chemical reaction (oxidation and dissolution) of shale OM during weathering under elevated pO_2? Previous study (Chang and Berner, 1999) has suggested that rates of abiotic oxidation of bituminous coal depend on dissolved O_2 concentration, with a reaction order of ~0.5. If applicable to shale OM oxidation and dissolution in environmental settings, this reaction rate law would suggest that weathering would be more intense under elevated pO_2. As long as denudation rates are rapid enough to prevent complete oxidation of shale OM within the weathering profile (which they are for all profiles examined in Petsch et al., 2000, 2001a), a greater oxidation/dissolution flux during weathering under elevated pO_2 is possible. Alternatively, shale OM oxidation rates may be insensitive to increasing pO_2 above some threshold concentration, if microbial processes that follow Michaelis–Menten type kinetics dominate kerogen degradation. Furthermore, oxidation of shale-derived OM transported out of weathering profiles as particulate or dissolved phases is also a significant component of the coupled C–O biogeochemical cycle. Several recent studies have shown that weathering of ancient, OM-rich sedimentary rocks may provide a significant source of organic carbon to river and estuary OM pools and coastal marine sediments (Leithold and Blair, 2001; Masiello and Druffel, 2001; Raymond and Bauer, 2001b; Blair et al., 2003). In addition to a distinct ^{14}C signature, ancient shale-derived pool in modern systems likely exhibits substantially different reactivity and bioavailability compared with other dissolved and particulate OM pools. Other research has indicated that ancient OM released from shales to rivers during weathering is to some extent labile and is assimilated or remineralized during river and estuarine transport (Raymond and Bauer, 2000, 2001a,c). The response of OM oxidation rates to elevated pO_2 is likely to be similar in both shale weathering profiles and during downstream transport.

4. Black shale weathering: microbial utilization of ancient organic matter

Of course, in reality, a weathering profile cannot be perfectly represented by abiotic laboratory simulations of kerogen degradation. If microorganisms are present within black shale weathering profile, their activity may hold significant implications for understanding controls on the chemical reactions of OM occurring during shale weathering. The kinetic rate constants, rate laws and reaction order with respect to dissolved

O_2 may be very different for sterile versus microbially mediated reactions. Rates of microbially mediated reactions may depend on the concentrations of other reagents such as SO_4^{2-}, H^+, or N, P or other nutrients. Reaction of shale OM may not necessarily be an aerobic process, if anaerobic microorganisms can access shale OM using alternative electron acceptors such as sulfate or ferric iron. In such cases, OM oxidation would not follow a simple 0.5 reaction order with respect to O_2. Thus evaluating the activity and limitations of microorganisms within black shale weathering profiles is necessary to fully understand transport and transformations of OM during shale weathering.

There are several reasons why shale OM within a weathering profile may be accessible for microorganisms.

- Black shales contain significant concentrations of organic carbon, oxidation of which may supply energy for growth of heterotrophs.
- Weathering profiles in black shales represent sharp chemical gradients, whereas microbial activity may be concentrated to harness transfer of chemical energy.
- Weathered shales are fractured rocks within meters of the earth's surface, and as such cannot remain sterile or isolated from surficial soil communities.

The goal of this research is to show that there are organisms capable of using shale OM as a carbon source. Once established, efforts can expand to examine how active or widespread these organisms may be, what controls their growth, and what reactions using shale OM are performed by their activity.

The possibility of microbial activity within the Clay City shale weathering profile has been investigated from several different perspectives. First is direct observation through microscopy. All environmental samples collected from a 70-cm vertical core into the Clay City weathering profile reveal the presence of microorganisms (Petsch et al., 2001b), including sections where geochemical evidence indicates that the organic carbon is ^{14}C-free and thus inherited from the shale and not added more recently as soil carbon (Table 1, from Petsch et al., 2001b, 2003). Kerogen-utilizing microorganisms were enriched from environmental samples using an

Table 1

Weight percent, ^{13}C ratio, and ^{14}C ratios in organic matter from 70-cm core into black shale at Clay City, Kentucky

Depth (cm)	Percent (%) TOC	δ^{13}C	Δ^{14}C[a]	Percent (%) shale OM[b]	Total PLFA (nmol/g)
Surface litter	1.6				226.8
0–10	1.4	−25.8			84.3
10–20	1.3	−25.7	−152	0	53.2
20–30	0.7	−26.6	−473	38	29.9
40–50	4.3	−26.7	−727	68	30.9
50–60	7.7	−29.2	−996	100	12.2
60–70	7.4	−27.3			16.8

[a] Radiocarbon analyses described in Petsch et al. (2001b).

[b] Calculated using a simple isotope mixing model [% shale OM=100*(Δ^{14}C$_{sample}$−Δ^{14}C$_{10-20\ cm}$)/(−1000‰−Δ^{14}C$_{10-20\ cm}$).

organic carbon-free mineral salts medium and sterilized kerogen as carbon substrate. These enrichments sustained growth up to 1 year after initiation, following several transfers and dilutions. The presence of microorganisms in environmental samples is necessary but not sufficient to indicate microbial utilization of ancient shale organic matter. Further support is provided by detection of organisms at core depths where ^{14}C analysis indicates entirely ancient (rock-derived) organic matter and growth of organisms in enrichment cultures in which the only available carbon source was kerogen.

Utilization of kerogen by enrichment culture microorganisms was evaluated by analysis of cellular phospholipids. Phospholipids are integral components of cell membranes that are rapidly degraded after cell

death and serve as indicators of living biomass (Baird and White, 1985; Baird et al., 1985) and are not inherited from shale organic matter (Petsch et al., 2001b, 2003). The abundance of cell membrane phospholipids provides an index of the microbial biomass and thus cell density within the core, and the distribution of phospholipids in both enrichment culture and core samples provides information about the taxonomic composition of the microbial communities present and actively metabolizing in these environments (Table 2). Although limited in utility compared with DNA-based communities analyses, phospholipid profiling provides a first-order approximation of the community structure of microorganisms from a suite of related environmental samples. Distinct zonations of phospholipid distributions indicate a succession of microbial communities within this short 70-cm core, corresponding to zones of ample soil organic matter, well-ventilated organic matter-lean sections, and the onset of mixed aerobic/anaerobic utilization of purely shale organic matter (Petsch et al., 2003). What metabolic roles these organisms play, and their phylogenetic affinities, remain unclear.

Targeted ^{14}C and ^{13}C analysis of phospholipid-derived fatty acid methyl esters [FAMEs] was used to confirm microbial degradation of shale OM (Petsch et al., 2001b). Many studies of biochemical compounds and compound classes have demonstrated the utility of carbon isotopes in tracking the pathways of carbon through organisms and communities. In this study, the magnitude of the ^{14}C isotopic offset between Devon-

Table 2

Concentration of phospholipid fatty acids (PLFA) and relative abundance of PLFA compound classes in 70-cm core into black shale at Clay City, Kentucky

Depth (cm)	Total PLFA (μmol/g)	μmol_{PLFA}/ mol_{TOC}[a]	Relative abundance of PLFA compound classes (%)[b]						
			C16:0+ C18:0	C16:1+ C18:1	C18:2	n-alkyl> C18:0	n-alkyl< C16:0	Methyl-branched	Cyclopropyl
0	460.9	2.13	17.9	14.3	9.3	22.8	16.6	15.1	1.9
0–10	198.2	1.18	14.7	14.2	8.9	21.7	7.2	22.6	4.7
10–20	122.3	0.783	11.7	38.5	22.7	10.5	3.8	9.6	0.3
20–30	88.4	1.05	14.2	38.2	26.9	6.9	2.6	7.4	0.1
40–50	59.9	0.116	13.2	7.3	4.2	19.3	11.9	19.3	1.4
50–60	31.5	0.034	18.9	30.2	7.6	9.1	4.8	17.9	0.4
60–70	38.9	0.041	16.8	34.1	13.7	10.7	5.9	9.3	0.5

[a] PLFA concentration normalized to total organic carbon content.

[b] Calculated as percentage abundance of PLFA compound classes normalized to total PLFA mass.

ian age shale organic matter (~370 Ma) used as substrate and modern atmospheric CO_2 makes ^{14}C analysis of biomass in enrichment cultures a clear signal of microbial uptake of ancient shale carbon (Fig. 6). Analysis of four phospholipid-derived FAME compound classes isolated from kerogen enrichment cultures revealed $\Delta^{14}C$ values between −771 and −921‰. This equates to 74–94% of cellular carbon deriving from an ancient, ^{14}C-free carbon source. The only carbon sources available to the enrichments were

kerogen, modern atmospheric CO_2 and any inorganic carbon produced by oxidation of shale OM; the shale substrate employed in this experiment was free of carbonate minerals. The ^{14}C-depleted ratios in cellular phospholipids show that modern CO_2 could not be a significant source of cellular carbon, leaving only kerogen or inorganic carbon produced through kerogen oxidation as possible carbon sources. These two sources were subsequently distinguished via ^{13}C isotopic analysis of FAMEs. Direct assimilation of

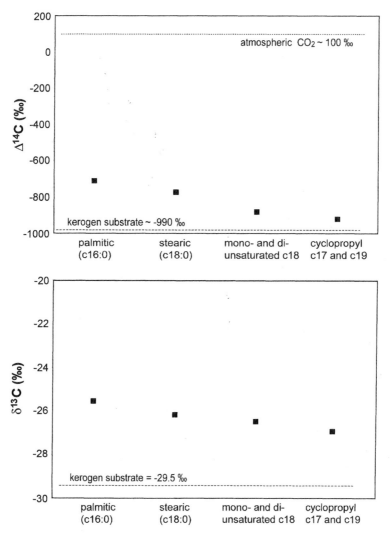

Fig. 6. Isotopic composition of phospholipid-derived fatty acid methyl esters isolated from culture of microorganisms grown using kerogen as sole organic carbon source. Upper panel: ^{14}C ratios contrasted against kerogen substrate and atmospheric CO_2, indicating that most cellular carbon in this culture derives from a ^{14}C-free source. Lower panel: ^{13}C ratios, indicating that heterotrophic assimilation (and not fixation of inorganic carbon) is the source of carbon to these organisms. Data from Petsch et al. (2001b).

kerogen into cellular biomass would yield FAME ^{13}C ratios similar to the kerogen, while autotrophic fixation of inorganic carbon would be expressed through significantly lower $\delta^{13}C$ ratios in FAMEs compared with kerogen. Analysis showed that FAMEs and kerogen exhibited similar $\delta^{13}C$ ratios, and thus kerogen assimilation, and not autotrophic fixation of inorganic carbon derived from kerogen oxidation, was the source of carbon to these organisms (Fig. 6).

The identities of microorganisms present in enrichment culture have been obtained through DNA sequence analysis. DNA extracted from enrichment cultures was extracted and purified using a Qiagen Miniprep DNA extraction kit (Qiagen, Chatsworth, CA). 16*S* rRNA genes were amplified from chromosomal DNA by PCR with universal primers 27 F and 1492 R. PCR products were ligated into the pCR-TOPO vector using the TOPO TA Cloning kit (Invitrogen, Carlsbad, CA), and transformed into competent *Escherichia coli* strain DH5α as per manufacturer instructions. White ampicillin-resistant colonies with plasmids containing cloned inserts were selected following overnight growth on Luria-Bertani plates amended with 100 μg/ml ampicillin. Selected clones were grown overnight in ampicillin-amended LB broth. Plasmid DNA was isolated using a Qiagen plasmid extraction kit (Qiagen, Chatsworth, CA), and inserts were amplified by PCR using M13F and M13R primers. Amplified inserts were digested using the tetrameric restriction endonucleases pairs (1 U ea.) HinP1 I (5′-G^CGC) and MspI (5′-C^CGG) in 1×NEB2 buffer, 0.01% final concentration Triton X-100 (New England BioLabs, Beverly, MA) for 3 h

at 37 °C, followed by heat shock inactivation at 60 °C. The digested rDNA fragments were then separated by agarose (4% Metaphor, FMC Bioproducts) gel electrophoresis with a 100-bp DNA marker (New England BioLabs), which revealed several non-duplicate clones. Non-duplicate inserts were sequenced at the Univ. Maine Sequencing Facility. The resulting single-stranded 16*S* rRNA gene sequences were analyzed for similarity with known sequences using a BLAST search of the National Center for Biotechnology Information database (http://www.ncbi.nlm.nih.gov). These results point to close similarity of enrichment culture microorganisms with representatives of common soil genera, widespread in many environments, and capable of a range of respiratory and fermentative modes of metabolism (Table 3). The presence of organisms in enrichment cultures incubated for several months with only kerogen as a carbon source, and in which ^{14}C analysis of cellular phospholipids demonstrates microbial assimilation of kerogen, shows that several common microorganisms capable of utilizing diverse carbon sources in soil environments are also capable of using refractory macromolecular kerogen when conditions require.

5. Microbial communities in black shales: diversity of metabolic roles

In addition to a rich source of organic matter, black shales often containing significant concentrations of sulfide minerals, notably pyrite (FeS_2). Oxidation of

Table 3

Similarity of 16*S* rRNA sequences of organisms in shale OM enrichment culture with known organisms, their phylogenetic relationships, and metabolic roles

Sample	Similarity (%)	Genus	Phylogenetic position	Metabolic roles
ccsp-6	98	*Acinetobacter*	γ proteobacteria	aerobic, organotrophic
ccsp-2	96	*Clostridium*	Gram+ eubacteria	anaerobic, fermentative
ccsp-1	98	*Comamonas*	β proteobacteria	aerobic (respiratory), organotrophic
				some partially chemoautotrophic, using H_2 or CO
				may be anaerobic, using NO_3 as e^- acceptor.
ccsp-9	97	*Dechloromonas*	β proteobacteria	anaerobic, organotrophic, may use NO_3 as e^- acceptor
ccsp-17	94			can access aromatic hydrocarbons
ccsp-8	99	*Pseudomonas*	γ proteobacteria	mainly aerobic (respiratory), organotrophic
ccsp-11	98			some partially chemoautotrophic, using H_2 or CO
ccsp-12	97			some anaerobic, using NO_3 as e^- acceptor

pyrite in black shale weathering profiles generates a source of sulfate, ferric iron, and acidity. Near the field site in Clay City, extensive mats of white, microscopic biopolymeric sheaths similar to those produced by S- and Fe-oxidizing bacteria are observed where anoxic groundwaters impinge on the surface along stream channels. Thick coatings of amorphous ferric oxyhydroxides are also observed in stream channels draining the pyritic shale. Highly acidic waters (pH 2–3) have been pulled from shallow (3–60 cm) lysimeters installed in the weathering profile and collected from water draining out of the face of the roadcut. These are strong suggestions that microbial iron and sulfur oxidation may play active roles in oxidation of sulfide minerals during weathering of black shales. Comparing the concentration of pyrite and TOC within the Clay City weathering profile reveals that pyrite loss occurs much deeper than TOC loss, suggesting that pyrite oxidation is occurring in deeper, oxygen-deficient zones of the weathering profile. Pyrite oxidation in shale weathering profiles may not in fact require free molecular oxygen. It has been shown that significant pyrite oxidation can be maintained using Fe^{3+} as an electron acceptor in both the presence and absence of O_2 (Williamson and Rimstidt, 1994). Iron- and sulfide-oxidizing bacteria may live and be active at very low $[O_2]$, much lower than required for shale OM oxidation.

The distribution of phospholipids within the weathering profile suggest the presence of several classes of anaerobic microorganisms, including both fermentative and sulfate-reducing bacteria. Organisms from the enrichment culture contain $16S$ rRNA sequences similar to both aerobic and anaerobic (denitrifying) organotrophic soil microorganisms. Efforts at DNA-based community analysis of shale weathering profile organisms are currently underway. Certainly, the episodically wet–dry conditions of the profile may provide nitrate, sulfate, and ferric iron for anaerobic heterotrophs, and sulfide and ferrous iron for micro-aerophilic autotrophs within the profile. Other metabolic roles (fermentation, methanogenesis) are also possible in this environment. Biogenic methane has been detected in deep subsurface brines within black shales in the Michigan Basin (Martini et al., 1996, 1998), and both shallow and deep methanogenesis may be sources of methane from black shales

throughout the Appalachian Basin as well. Many organisms have been identified in anoxic marine and freshwater sediments that are capable of degradation of hydrocarbons (e.g. Coates et al., 2001; So and Young, 1999, 2001; Perez-Jimenez et al., 2001) using SO_4, NO_3, and possibly Fe(III) as electron acceptors. Possibly, these or other organisms may be active within shale weathering profiles, facilitating the degradation of OM during weathering. If so, then the thickness and intensity of the weathered zone on black shales may not directly correlate with O_2 availability, but instead with delivery of other nutrients or substrates.

6. Conclusions

Recent efforts have revealed that, perhaps surprisingly given the low porosity and permeability of these rocks, black shales may host intriguing examples of geobiological activity. Organisms in black shale weathering profiles can utilize shale organic matter as a carbon source, and as such play a role in controlling oxidation of sedimentary organic matter during rock weathering and ultimately, the composition of earth's atmosphere through geologic time. The rate of organic matter oxidation during shale weathering may be controlled by oxygen availability, local erosion rate, and availability of substrates necessary for microbial degradation of OM. In addition to relevance to global cycles of carbon and oxygen, weathering of black shale plays a more immediate role in groundwater and surface water quality. Oxidation of sulfide minerals in black shales may be linked to supply of metals including Mo and As to groundwater supplies in the Appalachian Basin and Midwest (Warner, 2001; Tuttle et al., 2001, 2002; Goldhaber et al., 2002). Many groundwaters in black shales are anoxic, sulfidic, and acidic. Thus areas underlain by black shale may suffer from poor environmental water quality.

The breadth of metabolic and physiologic capacities that can be supported within black shales extends beyond aerobic heterotrophs to include anaerobic heterotrophs such as sulfate-reducing and -fermentative bacteria, chemoautotrophs such as sulfur- and iron-oxidizing bacteria, and methanogenic archaea. Recognition of the roles microorganisms may play in transforming organic matter within black shale is thus important for developing understanding of potential fossil fuel resources, as well as the biogeochemical cycles of elements C, O, Fe, S, and others.

Acknowledgements

The authors acknowledge R. A. Berner, J. M. Hayes, M. Oades, B. Peucker-Ehrenbrink, and R. Smernik for their helpful participation in this research. We also thank Nora Noffke, David Burdige and an anonymous reviewer for their comments to improve this manuscript. We also acknowledge support of the National Science Foundation through award EAR-0106707 to Petsch, Edwards and Eglinton, and the Woods Hole Oceanographic Postdoctoral Scholar program, to Petsch.

References

Anbar, A.D., Knoll, A.H., 2002. Proterozoic ocean chemistry and evolution: a bioinorganic bridge. Science 297, 1137–1142.

Baird, B.H., White, D.C., 1985. Biomass and community structure of the abyssal microbiota determined from the ester-linked phospholipids recovered from Venezuela Basin and Puerto Rico Trench sediments. Mar. Geol. 68, 217–231.

Baird, B.H., Nivens, D.E., Parker, J.H., White, D.C., 1985. The biomass, community structure, and spatial distribution of the sedimentary microbiota from a high-energy area of the deep sea. Deep-Sea Res. 32, 1089–1099.

Berner, R.A., 1987. Models for carbon and sulfur cycles and atmospheric oxygen: application to Paleozoic geologic history. Am. J. Sci. 287, 177–196.

Berner, R.A., 1994. GEOCARB II: a revised model of atmospheric CO_2 over Phanerozoic time. Am. J. Sci. 294, 56–91.

Berner, R.A., 1999. Atmospheric oxygen over Phanerozoic time. Proc. Natl. Acad. Sci. U. S. A. 96, 10955–10957.

Berner, R.A., Canfield, D.E., 1989. A new model for atmospheric oxygen over Phanerozoic time. Am. J. Sci. 289, 333–361.

Berner, R.A., Kothavala, Z., 2001. GEOCARB III: a revised model of atmospheric CO_2 over Phanerozoic time. Am. J. Sci. 301, 182–204.

Berner, R.A., Lasaga, A.C., Garrels, R.M., 1983. The carbonate-silicate geochemical cycle and its effect on atmospheric carbon dioxide over the past 100 million years. Am. J. Sci. 283, 641–683.

Berner, R.A., Petsch, S.T., Lake, J.A., Beerling, D.J., Popp, B.N., Lane, R.S., Laws, E.A., Westley, M.B., Cassar, N., Woodward, F.I., Quick, W.P., 2000. Isotope fractionation and atmospheric oxygen: implications for Phanerozoic O_2 evolution. Science 287, 1630–1633.

Berner, R.A., Beerling, D.J., Dudley, R., Robinson, J.M., Wildman, R.A., 2002. Phanerozoic atmospheric oxygen. Annu. Rev. Earth Planet. Sci. 31, 105–134.

Blair, N.E., Leithold, E.L., Ford, S.T., Peeler, K.A., Holmes, J.C., Perkey, D.W., 2003. The persistence of memory: the fate of ancient sedimentary organic carbon in a modern sedimentary system. Geochim. Cosmochim. Acta 67, 63–73.

Canfield, D.E., 1998. A new model for Proterozoic ocean chemistry. Nature 396, 450–453.

Canfield, D.E., Teske, A., 1996. Late Proterozoic rise in atmospheric oxygen concentration inferred from phylogenetic and sulphur-isotope studies. Nature 382, 127–132.

Canfield, D.E., Habicht, K.S., Thamdrup, B., 2000. The Archean sulfur cycle and the early history of atmospheric oxygen. Science 288, 658–661.

Cerling, T.E., 1992. Use of carbon isotopes in paleosols as an indicator of the pCO_2 of the paleoatmosphere. Glob. Biogeochem. Cycles 6, 307–314.

Chang, S.B., Berner, R.A., 1999. Coal weathering and the geochemical carbon cycle. Geochim. Cosmochim. Acta 63, 3301–3310.

Clayton, J.L., Swetland, P.J., 1978. Subaerial weathering of sedimentary organic matter. Geochim. Cosmochim. Acta 42, 305–312.

Coates, J.D., Chakraborty, R., Lack, J.G., O'Connor, S.M., Cole, K.A., Bender, K.S., Achenbach, L.A., 2001. Anaerobic benzene oxidation coupled to nitrate reduction in pure culture by two strains of Dechloromonas. Nature 411, 1039–1043.

Derenne, S., Largeau, C., Casadevall, E., Berkaloff, C., Rousseau, B., 1991. Chemical evidence of kerogen formation in source rocks and oil shales via selective preservation of thin resistant out walls of microalgae: origin of ultralaminae. Geochim. Cosmochim. Acta 55, 1041–1050.

Derenne, S., Le Berre, F., Largeau, C., Hatcher, P., Connan, J., Raynaud, J.F., 1992. Formation of ultralaminae in marine kerogens via selective preservation of thin resistant out walls of microalgae. Org. Geochem. 19, 345–350.

Dickens, G.R., 2001. The potential volume of oceanic methane hydrates with variable external conditions. Org. Geochem. 32, 1179–1193.

Dudley, R., 1998. Atmospheric oxygen, giant Paleozoic insects and the evolution of aerial locomotor performance. J. Exp. Biol. 201, 1043–1050.

Dudley, R., 2000. The evolutionary physiology of animal flight: paleobiological and present perspectives. Annu. Rev. Physiol. 62, 135–155.

Flaviano, C., Le Berre, F., Derenne, S., Largeau, C., Connan, J., 1994. First indications of the formation of kerogen amorphous fractions by selective preservation; role of non-hydrolysable macromolecular constituents of eubacterial cell walls. Org. Geochem. 22, 711–759.

Goldhaber, M.B., Hatch, J.R., Callender, E.C., Irwin, E.R., Tuttle, M.L., Ayuso, R.A., Lee, L., Morrison, J.M., Grosz, A., Atkins, J.B., Black, D.D., Zappia, H., Pashin, J.C., Diehl, S.F., Sanzolone, R.F., Ruppert, L.F., Kolker, A., Finkelman, R.B., Bevins, H.E., 2002. Impact of elevated arsenic in coal on the geochemical landscape of the eastern U.S. Sixth International Symposium on the Geochemistry of the Earth's Surface (GES-6), Honolulu, Hawaii.

Goth, K., de Leeuw, J.W., Puttmann, W., Tegelaar, E.W., 1988. Origin of Messel oil shale kerogen. Nature 376, 7359–7361.

Graham, J.B., Dudley, R., Anguilar, N., Gans, C., 1995. Implications of the late Paleozoic oxygen pulse for physiology and evolution. Nature 375, 117–120.

Habicht, K.S., Gade, M., Thamdrup, B., Berg, P., Canfield, D.E., 2002. Calibration of sulfate levels in the Archean ocean. Science 298, 2372–2374.

Hatcher, P.G., Spiker, E.C., Szeverenyi, N., Maciel, G.E., 1983. Selective preservation and origin of petroleum-forming aquatic kerogen. Nature 305, 498–501.

Holland, H.D., 2002. Volcanic gases, black smokers and the great oxidation event. Geochim. Cosmochim. Acta 66, 3811–3826.

Jahren, A.H., Arens, N.C., Sarmiento, G., Guerrero, J., Amundson, R., 2001. Terrestrial record of methane hydrate dissociation in the Early Cretaceous. Geology 29, 159–162.

Katz, M.E., Pak, D.K., Dickens, G.R., Miller, K.G., 1999. The source and fate of massive carbon input during the latest paleocene thermal maximum. Science 286, 1–1533.

Leithold, E.L., Blair, N.E., 2001. Watershed control on the carbon loading of marine sedimentary particles. Geochim. Cosmochim. Acta 65, 2231–2240.

Lewan, M.D., 1980. Geochemistry of Vanadium and Nickel in Organic Matter in Sedimentary Rocks, Ph.D. dissertation, University of Cinncinnati.

Leythäuser, D., 1973. Effects of weathering on organic matter in shales. Geochim. Cosmochim. Acta 37, 120–133.

Lichtfouse, E., Chenu, C., Baudin, F., Leblond, C., DaSilva, M., Behar, F., Derenne, S., Largeau, C., Wehrung, P., Albrecht, P., 1998. A novel pathway of soil organic matter formation by selective preservation of resistant straight-chain biopolymers: chemical and isotopic evidence. Org. Geochem. 28, 411–415.

Littke, R., Klussman, U., Krooss, B., Leythäuser, D., 1991. Quantification of loss of calcite, pyrite and organic matter during weathering of Toarcian black shales and effects on kerogen and bitumen characteristics. Geochim. Cosmochim. Acta 55, 3369–3378.

Martini, A.M., Budai, J.M., Walter, L.M., Schoell, M., 1996. Microbial generation of economic accumulations of methane within a shallow organic-rich shale. Nature 383, 155–158.

Martini, A.M., Walter, L.M., Budai, J.M., Ku, T.C.W., Kaiser, C.J., Schoell, M., 1998. Genetic and temporal relations between formation waters and biogenic methane: upper Devonian antrim shale, Michigan Basin, USA. Geochim. Cosmochim. Acta 62, 1699–1720.

Masiello, C.A., Druffel, E.R.M., 2001. Carbon isotope geochemistry of the Santa Clara River. Glob. Biogeochem. Cycles 15, 407–416.

Mora, C.L., Driese, S.G., Colarusso, L.A., 1996. Middle to Late Paleozoic atmospheric CO_2 levels from soil carbonate and organic matter. Science 271, 1105–1107.

Pagani, M., Arthur, M.A., Freeman, K.H., 1999. Miocene evolution of atmospheric carbon dioxide. Paleoceanography 14, 273–292.

Perez-Jimenez, J.R., Young, L.Y., Kerkhof, L.J., 2001. Molecular characterization of sulfate-reducing bacteria in anaerobic hydrocarbon-degrading consortia and pure cultures using the dissimilatory sulfite reductase (dsrAB) genes. FEMS Microbiol. Ecol. 35, 145–150.

Petit, J.R., Jouzel, J., Raynaud, D., Barkov, N.I., Barnola, J.M., Basile, I., Bender, M., Chappellaz, J., Davis, M., Delaygue, F., Delmotte, M., Koktlyakov, V.M., Legrand, M., Lipenkov, V.Y., Lorius, C., Pepin, L., Ritz, C., Saltzmann, E., Stievenard, M., 1999. Climate and atmospheric history of the past 420,000 years from the Vostok ice core, Antarctica. Nature 399, 429–436.

Petsch, S.T., Berner, R.A., 1998. Coupling the geochemical cycles of C, P, Fe, and S: the effect on atmospheric O_2 and the isotopic records of carbon and sulfur. Am. J. Sci. 298, 246–262.

Petsch, S.T., Berner, R.A., Eglinton, T.I., 2000. A field study of the chemical weathering of ancient sedimentary organic matter. Org. Geochem. 31, 475–487.

Petsch, S.T., Smernik, R.J., Eglinton, T.I., Oades, J.M., 2001a. A solid state ^{13}C-NMR study of kerogen degradation during black shale weathering. Geochim. Cosmochim. Acta 65, 1867–1882.

Petsch, S.T., Eglinton, T.I., Edwards, K.J., 2001b. 14C-dead living biomass: evidence for microbial assimilation of ancient organic carbon during shale weathering. Science 292, 1127–1131.

Petsch, S.T., Edwards, K.J., Eglinton, T.I., 2003. Abundance, distribution and d13C analysis of microbial phospholipid-derived fatty acid in a black shale weathering profile. Org. Geochem. 34, 731–743.

Raymond, P.A., Bauer, J.E., 2000. Bacterial consumption of DOC during transport through a temperate estuary. Aquat. Microb. Ecol. 22, 1–12.

Raymond, P.A., Bauer, J.E., 2001a. DOC cycling in a temperate estuary: a mass balance approach using natural ^{14}C and ^{13}C isotopes. Limnol. Oceanogr. 46, 655–667.

Raymond, P.A., Bauer, J.E., 2001b. Riverine export of aged terrestrial organic matter to the North Atlantic Ocean. Nature 409, 497–500.

Raymond, P.A., Bauer, J.E., 2001c. Use of ^{14}C and ^{13}C natural abundances for evaluating riverine, estuarine and coastal DOC and POC sources and cycling: a review and synthesis. Org. Geochem. 32, 469–485.

Royer, D.L., Berner, R.A., Beerling, D.J., 2001a. Phanerozoic atmospheric CO_2 change: evaluating geochemical and paleobiological approaches. Earth-Sci. Rev. 54, 349–392.

Royer, D.L., Wing, S.L., Beerling, D.J., Jolley, D.W., Koch, P.L., Hickey, L.J., Berner, R.A., 2001b. Paleobotanical evidence for near present-day levels of atmospheric CO_2 during part of the Tertiary. Science 292, 2310–2313.

So, C.M., Young, L.Y., 1999. Isolation and characterization of a sulfate-reducing bacterium that anaerobically degrades alkanes. Appl. Environ. Microbiol. 65, 2969–2976.

So, C.M., Young, L.Y., 2001. Anaerobic biodegradation of alkanes by enriched consortia under four different reducing conditions. Environ. Toxicol. Chem. 20, 473–478.

Thompson, L.G., Mosley-Thomspon, E., David, M.E., Henderson, K.A., Brecher, H.H., Zagorodnov, V.S., Mashiotta, T.A., Lin,

P.N., Mikhalenko, V.N., Hardy, D.R., Beer, J., 2002. Kilimanjaro ice core records; evidence of Holocene climate change in tropical Africa. Science 298, 589–593.

Tuttle, M., Goldhaber, M., Ruppert, L., Hower, J., 2001. Arsenic in coal and stream sediments from the Appalachian Basin Kentucky (abs). USGS Conference on Arsenic in the Environment, February 21–22, 2001, Denver CO.

Tuttle, M.L.W., Goldhaber, M.B., Ruppert, L.F., Hower, J.C., 2002. Arsenic in rocks and stream sediments of the Central Appalachian Basin, Kentucky. Open File Report USGS, vol. 02-28.

Van Cappellen, P., Ingall, E.D., 1996. Redox stabilization of the atmosphere and oceans by phosphorus-limited marine productivity. Science 271, 493–496.

Warner, K.L., 2001. Arsenic in glacial drift aquifers and the implication for drinking water—Lower Illinois River Basin. Groundwater 39, 433–442.

Williamson, M.A., Rimstidt, J.D., 1994. The kinetics and electrochemical rate-determining step of aqueous pyrite oxidation. Geochim. Cosmochim. Acta 58, 5443–5454.

Yapp, C.J., Poths, H., 1992. Ancient atmospheric CO_2 pressures inferred from natural goethites. Nature 355, 342–344.

Available online at www.sciencedirect.com

ELSEVIER

Palaeogeography, Palaeoclimatology, Palaeoecology 219 (2005) 171–189

www.elsevier.com/locate/palaeo

Geo-biological aspects of coastal oil pollution

Luise Berthe-Corti*, Thomas Höpner

Institute for Chemistry and Biology of the Marine Environment (ICBM), University of Oldenburg, D-26111 Oldenburg, Germany

Received 14 May 2003; accepted 29 October 2004

Abstract

More than 10 years after the 1991 Gulf War oil spill on the Saudi-Arabian coast of the Arabian Gulf, natural remediation has only been partially successful. This fact demonstrates the importance of studying the preconditions for, and the process of, hydrocarbon degradation as well as the competing processes which prevent or slow them down. This paper deals with the preconditions of biodegradation: the presence of water, the availability of oxygen, the influence of temperature and the presence and type of degrading microorganisms. This paper discusses abiotic transformation as well as aerobic and anaerobic biodegradation. The importance of emulsification and hydrocarbon uptake into the degrading microorganisms is underlined. Competing processes include the conversion of liquid oil to viscous and finally solid material, the formation of solid sediment–oil mixtures and the clogging of sediment pores preventing oxygen from entering the pores. At first glance, the competing processes seem more "geological" while emulsification and hydrocarbon uptake seem more "biological". However, since the necessary energy input (e.g., waves and turbulence) is "geo"(physical), it becomes clear, on the one hand, that the biological process requires geophysical energy input and, on the other hand, is inhibited by geological competition. Oil pollution as well as remediation progress affect the impacted environment since they leave undegradable residues and intermediates as well as microbial biomass and its conversion products. Sediments are enriched with organic matter, whose properties and behaviour are altered. Even inorganic matter may be formed during hydrocarbon biodegradation.

The final part of this paper consists of case studies of the accident of the tanker BRAER (1993), the oil spill in the Arabian Gulf (1991), and the continual seeping of the wreck PRESTIGE (2002/03). These three cases demonstrate the importance of oil type, energy input, climatic conditions, as well as human interest in the use of the impacted coast, in determining speed and success of remediation and possible restoration measures.
© 2004 Elsevier B.V. All rights reserved.

Keywords: Oil pollution; Hydrocarbon; Biodegradation; Case studies; Aerobic degradation; Anaerobic degradation

1. Scope and focus

The conversion and the degradation of oil in a coastal environment are processes influenced by geological as well as biological factors. In approaching this situation, we assume that a formerly oil-free

---

* Corresponding author. Tel.: +49 441 798 3290; fax: +49 441 798 193290.
  E-mail address: luise.berthe.corti@uni-oldenburg.de (L. Berthe-Corti).

0031-0182/$ - see front matter © 2004 Elsevier B.V. All rights reserved.
doi:10.1016/j.palaeo.2004.10.020

environment has been contaminated by oil, either by spontaneous seeping from natural deposits or by man-made events. The resulting oil-contaminated spot is far from a state of equilibrium with the environment (in contrast to the situation within a natural deposit). This state of disequilibrium drives changes in the new geo-biological system, resulting in modifications and degradations of the oil and consequently of the environment as well.

This paper focuses solely on coastal oil contamination and on experiments aimed at controlling the conditions of oil/hydrocarbon degradation in nature. This limitation is appropriate, since world-wide experience has shown that a drifting oil slick is cause for great concern when it hits a coastline. As long as a drifting oil slick is at sea, it is subject to dissection, dispersion, degradation, and even anthropogenic oil defence measures. As soon as an oil slick hits the coast, however, a new geo-biological system develops, with its own processes and timescales. Put simply, oil at sea results in a system of two liquid phases consisting of nonmiscible components plus organic matter, detritus, organisms, gases. Oil on the coast results in a system with an additional, solid phase which replaces most of the aqueous phase. At sea, oil is subjected only to mild changes in the physical and chemical conditions such as temperature, irradiation, salinity, and nutrients. On the coast, however, it is exposed to sudden and strong changes such as between hot and cold, dry and inundated, sweet and saline, solid and viscous.

In general, this multiplicity of conditions allows oil to be degraded under a wide range of climatic and geographic conditions, as the following estimations demonstrate. The annual global anthropogenic marine oil discharge (tanker, platform, and pipeline accidents, land-born discharges, intentional discharges) is estimated to amount to between 2.8 and 5.5 million tons (Van Bernem and Lübbe, 1997). The global annual natural seepings of oil into the sea is estimated to range between 200,000 and 2 million tons ("best estimate" 600,000 tons) by the National Research Council (2002). Adding the average of the anthropogenic input and the best estimate of the natural input, the annual total input into the sea is within the range of 4.7 million tons. This amount is distributed over the seas very unevenly. The distribution follows the location of oil production platforms, of shipping lines, transport destinations, and surface-near deposits. "Hot spots" of oil spills and coastal pollution could be expected to occur in these areas, however, after more than 100 years of oil economy, no long-term accumulation of oil and degradation products has been observed there. Taking the coastlines of the Strait of Hormuz as an example, on the one hand, we find tar spots everywhere; on the other hand, their number is in no proportion to the local concentration of tankers passing the straits (Burns et al., 1982). The explanation is that oil degradation must occur at sea as well as on the coasts; and that it is apparently so effective that it is able to cope with the continuous arrival of new oil residues.

What is about crude oil and oil products that makes them such an environmental threat? Aside from the frequency and the amount of oil which regularly pollutes the environment, oil has many adverse properties: it is nonmiscible with water; it has the ability to coat dry particles; it contains numerous compounds, many of which are toxic; depending on its origin, it can have the properties of a fluid, a viscous or even solid matter; during weathering, it may change from fluid to solid; oil is sticky; oil layers slow down oxygen transport; and it has many other properties which are incompatible with many biological processes. A product of slow geo-biological processes, oil is organic and fossil at the same time; it constitutes an alien element in any environment.

2. Conditions of the hydrocarbon biodegradation

Scientific observations of oil spills and laboratory experiments have revealed the conditions necessary for biological hydrocarbon degradation. Some of the essential factors determining the hydrocarbon degradation in sediments are summarized in Table 1.

2.1. Presence of water

One key precondition is the presence of water (fresh or salty) and the emulsification of oil in water or of water in oil (the former is much more effective in terms of biological degradation). Emulsification requires an "energy input," in the case of a marine oil spill in form of waves and turbulence. In our case studies, we looked for the presence of such conditions

Table 1
Abiotic and biotic factors determining the hydrocarbon fate in sediment

Sediment characteristics
Grain size
Specific surfaces
Charge of surfaces
Water holding capacity
Hydrophobicity
Physicochemical factors/environment
Temperature
Moisture
Electron acceptors, e.g., oxygen, nitrate, sulfate, iron (III)
Nutrients and cometabolites
Organic carbon
Nitrogen
Phosphate
Biological activity
Abundance and metabolic activity of microorganisms
Bioturbation
Abundance and metabolic activity of other organisms living in sediment

and examined their effect on the biodegradation of oils (Chapter 5). We found that the energy input was present in the case of the oil spill caused by the tanker BRAER, while it was largely absent during the Arabian Gulf spill. Emulsification is supported by tensides irrespective of whether they are of biological origin or are added as chemicals. The role of biotensides/bioemulsifiers in the metabolism of hydrocarbons is discussed in Section 3.3. The questions of whether/or in which situation the storage and application of industrial tensides is helpful and environmentally compatible are not discussed here.

2.2. Availability of oxygen

Another necessary condition is the availability of oxygen. The amount of oxygen needed for biodegradation should not be underestimated. Aerobic degradation of an aliphatic hydrocarbon requires the 3.5-fold weight of oxygen! At sea or in coastal sediments, this oxygen is provided by the atmosphere by means of energy input or by photosynthetic activity, which only under very special conditions is strong enough to play an important role in the oxygen supply. For the biodegradation of emulsified oil floating on a rough water surface, oxygen supply is no problem; however, if sediments have been soaked with oil, this reduces the oxygen supply, which can then be a limiting factor

for biological hydrocarbon degradation. Contrary to the widespread assumption that aerobic hydrocarbon degradation needs an oxygen-saturated milieu, it is not necessary for the contaminated sites to show oxygen saturation to permit effective hydrocarbon degradation. It has been demonstrated that bacterial communities which originate from intertidal sediment or pure cultures of marine bacteria have a high capability for adapting to changes in the dissolved oxygen concentration, from oxygen saturation to suboxic conditions (Michaelsen et al., 1992; Berthe-Corti and Bruns, 1999, 2001; Bonin and Bertrand, 2000). The growth rate of the community and the biodegradation rate declined only at very low dissolved oxygen concentrations (<1% of saturation), and part of the hydrocarbon was converted into end-products other than CO_2 and biomass (Berthe-Corti and Ebenhöh, 1999).

2.3. Presence of nutrients

Nutrients are required mainly for biomass formation, so fixed ratios between hydrocarbons degraded and nutrients consumed could hardly to be expected. Nevertheless, in seawater mesocosm experiments, Gibbs (1975) found that nutrients not only limited the degradation, but that there was a narrow ratio: to degrade one gram of hydrocarbon, 60 mg of N (ammonia or nitrate) and 6 mg of P (phosphate) were needed. Harder et al. (1991) confirmed these data for a laboratory system in which hexadecane was degraded by autochthonous bacteria in a seawater sediment slurry. Roling et al. (2002) found that the addition of inorganic nutrients (N, P) was critical. However, the precise amount of the nutrients was not so important in the case of oil degradation in beach-sediment microcosms; and Delille et al. (2002) showed in a field study investigating crude oil degradation in an subantarctic intertidal sediment that the rate of oil degradation was improved by the addition of bioremediation agents such as a slow release fertilizer (EAP-22 from Els-Atochem or fish composts). Roy and Greer (2002) showed, although for soil rather than seawater, that the concentration and the type of nitrogen fertilizer (containing NO_3^- or NH_4^+, or a mixture thereof) added after a contamination were most important for the hydrocarbon degradation rate.

2.4. Temperature

For biological oil degradation, one would expect an increase in the reaction rate as the temperature rises, as is expected for every chemical and biological reaction. Dalyan et al. (1990) demonstrated that in a sediment slurry system, the hexadecane degradation followed the general temperature/reaction rate rule, according to which the rate doubles for every 10° rise in temperature. The rate reached zero at about 0 °C, and the increase in the degradation rate slowed at temperatures higher than 30 °C, not surprising for a system taken from a temperate intertidal zone. The influence of the temperature also depends on oil composition. Atlas (1975) investigated the influence of oil composition and temperature on the biodegradability of seven different crude oils. At 20 °C, lighter oils had greater abiotic losses and were more susceptible to biodegradation than heavier oils. These light crude oils, however, possessed toxic volatile compounds which evaporated only slowly at 10 °C, inhibiting the microbial degradation of these oils. Thus, oil seems to be degraded only extremely slowly, or not at all, under cold conditions. However, some early as well as some recent investigations in the Antarctic environment have demonstrated oil degradation in a cold marine environment or by microbial populations enriched from permanently cold sea water and sediment. Walker and Colwell (1974) isolated hydrocarbon-degrading species from the Chesapeake Bay at low temperature (0–10 °C), using a model petroleum containing 85.8% n-alkanes, 3.9% pristane and cyclic alkanes, as well as 6.4% mono- and polynuclear aromatics. The authors isolated the genera *Vibrio, Aeromonas, Pseudomonas*, and *Acinetobacter*, which are well-known alkane degraders under temperate conditions. Delille et al. (2002) showed degradation under subantarctic conditions in a field study. 2 l of oil were applied to sediment fields, degrading within six months under natural conditions.

2.5. Presence of microorganisms

A further precondition for oil degradation in marine environments is the presence of microorganisms able to degrade hydrocarbons and able to multiply under the given conditions. At sea, hydrocarbon-degrading microorganisms seem to be present everywhere, owing to three factors: (i) biogenic hydrocarbons are frequent, a fact revealed in an increasing number of chemical hydrocarbon analyses performed world-wide after a global increase in oil spills; (ii) natural seeps of hydrocarbons into the sea are frequent (see introduction); (iii) after more than a century of increased oil transports and oil spills in the world's oceans and seas, the marine environment has adapted to oil, especially in highly frequented areas such as the Arabian Gulf, the Campeche Bank in the Gulf of Mexico or the North Sea. Many studies were done to analyse the concentration and the spacial distribution of oil-degrading microorganisms in the sea or in sediments. Mostly, the authors analysed the total cell number and/or the concentration of hydrocarbon-degrading bacteria. Cell numbers in the water column range between 10^4 and 10^6 cells/ml. Data about cell numbers in sediments vary considerably, depending on the sediment and the environmental conditions and on whether the data are reported in milliliters or grams of sediment. Most data are in the range of 10^8 to 10^{11} cells per gram or per milliliter of sediment (Meyer-Reil, 1993). For antarctic sediment, Delille et al. (2002) reported 10^7 to 10^8 cells/ml. Hydrocarbon-containing foams floating on seawater for example can contain 10^7 to 10^9 cells/ml (Rambeloarisoa et al., 1984). When a marine site has been contaminated with crude oil, the community changes and hydrocarbon-degrading bacteria become enriched. Roling et al. (2002) showed that mainly cells from the *Alcanivorax/Fundibacter* group become enriched in beach sediment microcosms after oil contamination. Kasai et al. (2002) showed in beach-simulating tanks that the cell number of the *Cycloclasticus* species, which are able to degrade aromatic hydrocarbons, increased from 10^3 to 10^6 cells/g of sediment after the addition of crude oil.

2.6. Hydrocarbon-degrading microorganisms

For many decades, it has been known that hydrocarbon-degrading microorganisms are widespread in marine habitats. The isolated organisms comprise Procaryotes and Eucaryotes, such as Eubacteria, yeasts, and filamentous fungi, but also some cyanobacteria and algae (Cerniglia, 1992; Frohne et al., 1991; Schauer, 2001). There seems to be no exclusive affiliation of hydrocarbon-degrading bacteria with a special phylogenetic group, although many of the known aerobic hydrocarbon-degrading bacteria belong to the gamma subgroup of the Proteobacteria. The capacity to degrade hydrocarbons has been demonstrated within Gram-negative as well as Gram-positive species. According to Schauer (2001), 0.1–3% of the known Eubacteria, 15–20% among the yeasts, and 2–5% among filamentous fungi are able to degrade hydrocarbons.

In contrast to the extensive information about hydrocarbon-degrading aerobic species, less is known about species which are able to degrade hydrocarbons under anoxic conditions. However, an increasing number of Procaryotes which are able to degrade hydrocarbons by an anaerobic metabolism have been described within the past 10 years. Among these are sulfate-reducing, denitrifying, dissimilatory iron(III)-reducing, and methanogenic bacteria (Widdel and Rabus, 2001). Degradation is not restricted to distinct classes of hydrocarbons, e.g., alkanes, but also comprises aromatics, including polyaromatic and heteroaromatic molecules (Spormann and Widdel, 2000). Most species which are known to anaerobically degrade aromatic hydrocarbons are of fresh-water or terrestrial origin. The species of the *Azoarcus–Thauera* branch of the beta-subclass of Proteobacteria have been especially well investigated. However, toluene-degrading denitrifiers belonging to the beta- or gamma subclass of Proteobacteria were also isolated from marine habitats.

The spectrum of hydrocarbons which can be degraded seems to be species-specific, and the species seem to have a substrate-specific degradation capacity. The ability of a species to degrade aromatic hydrocarbons does not automatically imply that this species is able to degrade aliphatic substances. Alkane-oxidizing bacteria, for example, can be divided into three groups: the methane-oxidizing bacteria, the bacteria utilizing other gaseous alkanes, and those growing on liquid or solid alkanes (Ashraf et al., 1994; Gibson (Chapter 4), 1984; Perry, 1980; Ratledge, 1978).

Most liquid and solid alkanes with a chain length $>C_{12}$ are degraded fairly well. Benzene and substituted aromatics as well as polyaromatic hydrocarbons up to four rings are readily degraded. But under laboratory conditions, in pure cultures, polyaromatics

with more than four rings are considered to be recalcitrant (Mahro and Kästner, 1993). Their elimination from the environment seems to be mainly due to cooperative effects within microbial communities, combined with geo-biological processes resulting in a ring cleavage of molecules with a high number of aromatic rings. This results in incorporations of phenolic residues into the humic matrix.

2.7. Hydrocarbon-degrading microbial communities

Earlier investigations analysed the total cell number, or the concentration, of hydrocarbon-degrading cells in marine sediment and/or seawater (Atlas, 1984; Rambeloarisoa et al., 1984; Sorkhoh et al., 1990). However, due to analytical restrictions, it was difficult to analyse the structure of microbial communities. With the development of biochemical and molecular investigative methods, the colonization of marine sediments increasingly has become the subject of extensive analyses (Moyer et al., 1994; Gray and Herwig, 1996; Llobet-Brossa et al., 1998; Knoblauch et al., 1999).

Microbial communities are often associated with surfaces (e.g., hydrocarbon droplets, tar balls or sediment particles), where they form multispecies biofilms. These biofilms have high cell densities and form specific microenvironments in respect to substrate concentration (hydrocarbon, other nutrients such as N and P) and physicochemical conditions (pH, oxygen concentration, etc.). The structure and the activity of the biofilm communities are dependent on these microenvironments.

In the simplest case, the metabolic capacity of a community is merely an additive effect of the capability of each community member to degrade different compounds of the hydrocarbon mixtures (e.g., from crude oil). This is an unrealistic assumption and would not enable a complete degradation of hydrocarbon mixtures (such as oil), since some oil components seem to be degraded only cometabolically. Biofilm communities, however, seem to perform cometabolic and cooperative transformations. Berthe-Corti and Fetzner (2002), for example, discussed a possible cooperative effect of two enzyme systems involved in *n*-alkane degradation (ammonia monooxygenase and alkane hydroxylase) at very low dissolved oxygen concentrations.

Microbial communities also perform consecutive cometabolic and metabolic transformations. This would enable hydrocarbon compounds to be completely degraded, even those compounds that are regarded as recalcitrant when their degradation is investigated using pure cultures under laboratory conditions. Biodegradation of cyclic alkanes and condensed cyclic alkanes has only been demonstrated for microbial communities and not for single species. Anaerobic degradation of toluene by pure cultures of sulfate-reducing or denitrifying bacteria leads to the accumulation of the dead-end products benzylsuccinate and benzylfumarate (Evans et al., 1992). Degradation of m-, or o-cresol by denitrifying bacteria results in the accumulation of 4-OH-3-CH$_3$-benzoate and 4-OH-2-CH$_3$-benzoate, respectively (Rudolphi et al., 1991). But these metabolites are not found under natural conditions. In microbial communities, these dead-end products can serve as substrates for further species, thus leading to the complete mineralization of the original substrate.

Microbial communities often show an additional effect which is a specific "answer" to the environmental situation, resulting in the changes of the metabolic capacity of single community members, e.g., variations in the growth rate or variations in the K_M-value of enzymes (Gibson, 1984 (Chapter 15), Berthe-Corti and Ebenhöh, 1999).

Regarding the fate of oils in coastal zones, biodegradation at extremely low dissolved oxygen concentrations in the transition zone between oxic and anoxic conditions is of special importance. Coastal marine sediments are characterized by steep oxygen and redox gradients. These gradients result in microbial communities which use different electron acceptors (oxygen, nitrate, sulfate, iron(III), mangane (IV), etc.) but live in very close proximity. Berthe-Corti and Bruns (1999, 2001) demonstrated that alkane-degrading microbial communities which had been taken from marine intertidal sediment and which were grown in suboxic and quasi-anoxic chemostat cultures (dissolved oxygen concentration 1.6 to ≤0.8 µmol O$_2$/l) contained populations utilizing oxygen, nitrate, or sulfate as an electron acceptor. The composition and the metabolic activity of the communities were stable at dissolved oxygen concentrations between oxygen saturation and suboxic conditions, and they were only affected when the conditions changed from quasi-

anoxic (<0.8 µmol O$_2$/l) to completely anoxic. The result was a reduction in growth rate and rate of hexadecane mineralization.

3. Steps and reactions

3.1. Fundamental reactions of aerobic degradation

The fundamental reactions of the aerobic hydrocarbon decomposition have been well known for several decades. Suitable surveys are contained in the books of Atlas (1984) and Gibson (1984). Even though many details have been published since, such as the degradation of aliphatic alkenes (Ensign, 2001), the fundamental steps are still valid and enable us to understand the dependence of the processes on environmental conditions (Fig. 1a,b). Experiments on the laboratory scale as well as observation of polluted sites have made it possible to estimate the impact of oil degradation on sediment.

The key step of hydrocarbon degradation is the addition of one oxygen atom, in some cases, two oxygen atoms, to the hydrocarbon molecule, which is then converted to an alkanol (in the case of aliphatic hydrocarbons) or to a phenol (in the case of aromatic molecules). In some species, an epoxide is the first intermediate. This activation makes the hydrocarbon more soluble in water, marks a reactive site, and introduces a reactive site for the next reactions. The reaction requires energy, which is typically generated via the oxidation of a reduced biological intermediate such as NADH, which itself is reoxidized by an electron acceptor.

For the degradation of alkanes, different enzyme systems are known which carry out the primary attack. An omega-hydroxylase system consisting of three proteins (the rubredoxin reductase, a rubredoxin and an omega-hydroxylase) was isolated and characterized from *Pseudomonas* (Van Beilen et al., 1994). In some bacterial or fungal species as well as in mammalian cells, there are enzyme systems which depend on cytochrome *P*450 acting as a terminal oxidase. The main intermediates of the alkane degradation are fatty acids, which are produced from the alkanols via aldehydes. These acids can be further decomposed by the pathway typical of physiological carboxylic acid degradation, in which the molecule is shortened stepwise. However, fatty acids can also be excreted by the cells and accumulate in the environment. Once released, they can produce ambiguous effects. On the one hand, fatty acids can serve as a carbon source for bacteria of a community, thus enhancing the hydrocarbon degradation. On the other hand, fatty acids (chain length <14 C) can inhibit growth and hydrocarbon metabolism because they interfere with the cell membrane (Atlas and Bartha,

1973). This provokes a toxic effect and reduces growth.

Different degradative pathways have been demonstrated for aromatic substrates. The choice of the pathway depends on the type of the organism and/or on the type of the aromatic molecule, especially on its substituents and (in the case of polyaromatic molecules, PAH) on the number of rings (Cerniglia, 1992; Gibson, 1984). For an overview of the fundamental possibilities of PAH biodegradation, three different

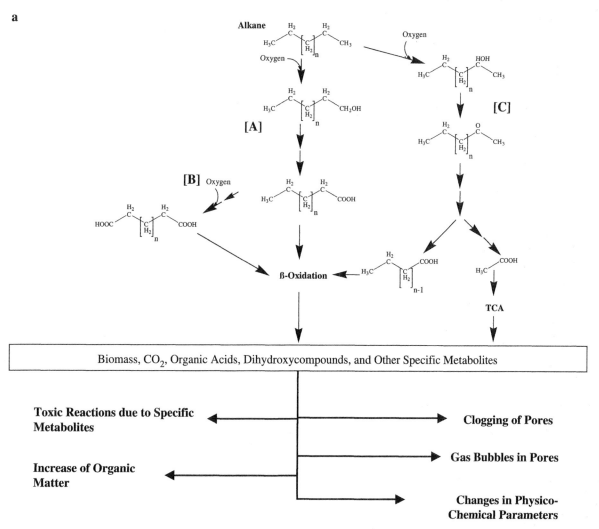

Fig. 1. Aerobic degradation of mineral oil hydrocarbons with its environmental impact. (a) Biodegradation of *n*-alkanes: metabolism begins with the activity of a monooxygenase which introduces a hydroxyl group into the aliphatic chain. [A]—monoterminal oxidation, [B]—biterminal oxidation, [C]—subterminal oxidation); TCA—tricarboxylic acid cycle. (b) Biodegradation of aromatic hydrocarbons: metabolism begins with the activity of a monooxygenase [1] or a dioxygenase [2] which introduce one or two atoms of oxygen; it can also begin with unspecific reactions [3].

b

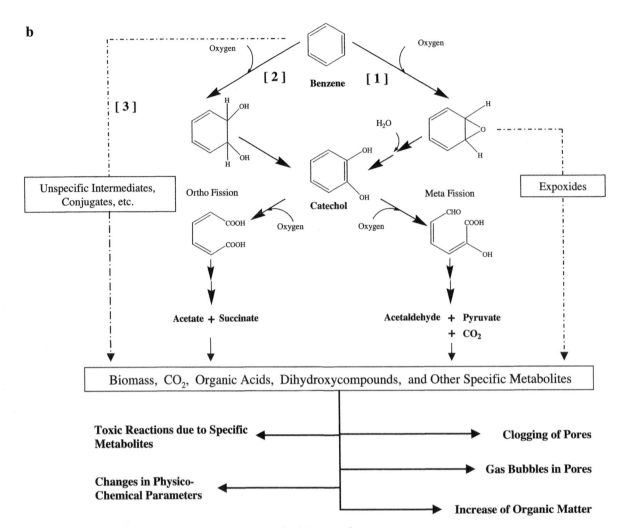

Fig. 1 (*continued*).

metabolic routes considered to be the main pathways are summarized here. Fig. 2 shows the biotic and abiotic elimination of PAH and its impact on sediments. Some of the details shown are explained later on in the Sections 3.4 and 4.

3.1.1. Complete mineralization or the dioxygenase pathway

This pathway is taken mainly by bacteria. The monoaromatic molecule or one ring of the polyaromatic system is attacked by a dioxygenase, and the molecule is oxidized stepwise via formation of a diol and subsequent ring cleavage. Pyruvate is one of the main intermediates of the pathway. The main

products are biomass and carbon dioxide. An accumulation of dead-end products is rare and occurs mostly when cells are deficient in their degradation pathway. The "disadvantage" of this pathway is that only ring systems of up to four rings are mineralized. Systems with a higher number of rings seem to be recalcitrant.

3.1.2. Cometabolic transformation or the monooxygenase pathway

This pathway has been mainly demonstrated for yeasts and fungi, but it also occurs in bacteria and in some algae. The respective PAH-degrading species can only perform the degradation if a

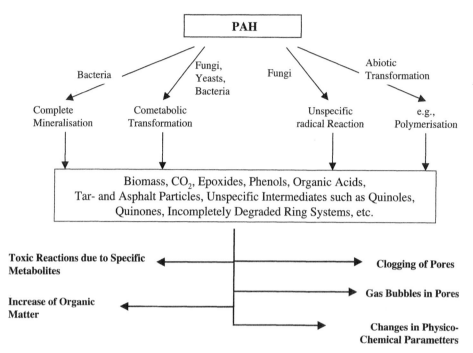

Fig. 2. Schematic overview of the main metabolic routes of the degradation of monoaromatic and polyaromatic hydrocarbons (PAH).

compound is available which can serve as a source of carbon and energy. The characteristic enzymes which perform ring cleavage are monooxygenases (e.g., Cyt *P*450). The monooxygenase activity results in the formation of an epoxide which is highly reactive, resulting in toxic or mutagenic activity. Epoxides may also be transformed to trans-dihydrodiols. The latter have not been metabolized further in pure cultures in the laboratory and have to be regarded as dead-end products. However, no such metabolites have been detected in soil or in sediment.

3.1.3. Unspecific oxidation via radical reactions

The wood-destroying white rot fungi, e.g., have been shown to destroy the structure of lignin via the activity of extracellular peroxidases and phenol oxidases. They attack the phenolic molecule structure by a nonspecific action, thus also attacking other aromatic structures such as PAH. The type of cleavage product is not predictable.

Frequent metabolites of PAHs are quinones, quinoles, and ring systems with a ring number lower than that of the original substance. These compounds

may be incorporated into sediments and alter the sediment structure (compare Fig. 2).

3.2. Remarks about anaerobic hydrocarbon degradation

For many decades, it was assumed that hydrocarbons undergo biodegradation only in the presence of molecular oxygen. However, in 1988 Evans and Fuchs published a review paper on the anaerobic degradation of aromatic compounds, and Aeckersberg et al. (1991) reported on a sulphate-reducing bacterium able to anaerobically mineralize hexadecane. Since that time, a great deal of work has been done on the anaerobic degradation of aliphatic and aromatic hydrocarbons. It has been demonstrated that anaerobic hydrocarbon degradation is not uncommon in nature although, in most cases, it is considerably slower than aerobic degradation. Denitrifying, sulfate-reducing, and iron(III)-reducing strains collected at different sites (terrestrial, aquifers, fresh-water and marine systems) are able to anaerobically metabolize hydrocarbons. The same has been demonstrated for the phototrophic bacterium *Blastochloris sulfoviridis* strain ToP1, which uses light as an energy source (Spormann and Widdel, 2000). Even methanogenic consortia have been shown to degrade hydrocarbons (Anderson and Lovley, 2000; Edwards and Grbic-Galic, 1994). The metabolic routes of alkane degradation seem to function differently and are not completely understood yet. Several authors have discussed a terminal or subterminal addition of a one-carbon moiety or a fumarate molecule to the alkane as an activation mechanism (So et al., 2003; Wilkes et al., 2002; Fig. 3). For aromatic molecules, it has been demonstrated that alkyl benzenes which have a methyl group as a side chain undergo an enzymic addition of fumarate, most likely via a radical mechanism. This was demonstrated for toluene. Alkyl benzenes with side chains of two or more carbon atoms are activated by dehydrogenation of the side chain. This has been shown for ethyl- and propylbenzene

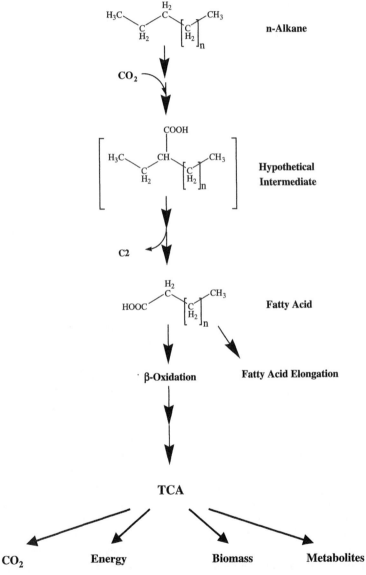

Fig. 3. Proposed pathway for anaerobic degradation of *n*-alkanes; activation via addition of a C1-moiety (subterminal carboxylation at C3). Pathway according to So et al. (2003); TCA—tricarboxylic acid cycle.

(Spormann and Widdel, 2000). A scheme of the anaerobic degradation is shown in Fig. 4. Its impact on the sediments is discussed in the Sections 3.4 and 4.

3.3. Emulsification and uptake of hydrocarbons

Released into an aqueous environment, hydrocarbons form droplets and tend to adsorb to suspended organic or inorganic particles. Reaching intertidal or beach regions, they adsorb to sediment particles. The degree of adsorption depends on the physicochemical characteristics (e.g., grain size, chemical composition, and charge) of the particles as well as on the type of oil. Depending on the different conditions of the distribution in the sediment, the bioavailability of the oil varies considerably. Microorganisms have to cope with these factors and to make the nonpolar hydrocarbon phase bioavailable. This involves specific processes which include (i) uptake of hydrocarbons dissolved in the aqueous phase by the cells; (ii) mechanical interaction of cells with hydrocarbon droplets; (iii) production and excretion of bioemulsifiers. Bioemulsifiers improve the bioavailability of hydrocarbons by the stabilization of the water–hydrocarbon emulsions. They enhance the solubilization of hydrocarbons and consequently the uptake of hydrocarbons into the cells. In addition, bioemulsifiers facilitate the adherence of the cells to the droplet surfaces. Production and function of bioemulsifiers have been subject to intense investigations, mainly with n-alkanes as a carbon source (Haferburg et al., 1986).

Several microorganisms produce bioemulsifiers which are high molecular mass molecules of a polar and a nonpolar moiety. The polar moiety is often a sugar such as rhamnose or a trehalose, whereas the nonpolar moiety is often a long-chain hydroxy fatty acid (Kosaric et al., 1987). Much of the work on the production and function of bioemulsifiers was done with *Acinetobacter* strains and with yeasts. But also other hydrocarbon-degrading genera, such as *Pseudomonas*, *Achromobacter*, *Arthrobacter*, *Brevibacterium*, *Corynebacterium*, *Rhodococcus*, and the marine genus *Alcanivorax* are known to comprise strains which produce bioemulsifiers. Bioemulsifiers are not only of physiological importance for the microorganisms producing the emulsifier but, once excreted to the environment, they also facilitate oil availability for other microorganisms incapable of

producing emulsifiers. In addition, they can serve as a carbon source for other bacteria of a community, thus increasing cometabolic effects in oil-degrading communities. Bioemulsifiers provoke a change of heavy metal solubility in sediments, thus contributing to a change in the sediment composition.

Production of bioemulsifiers is accompanied by the formation of biofilms to which microorganisms adhere. This biofilm formation has different effects. On the one hand, it improves the contact between cells and oil and helps the cells to remain fixed on the oil droplet, facilitating oil degradation especially in the water column. On the other hand, cell growth in the biofilm increases its thickness, which, especially in sediments, may cause problems in the bioavailability of gases and nutrients such as molecular oxygen, N-, and P-compounds. This effect slows down oil degradation.

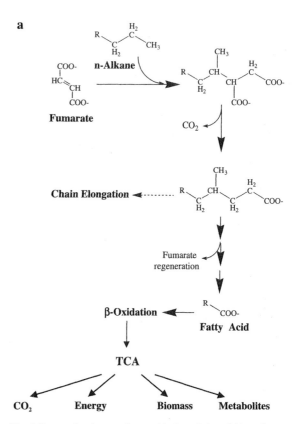

Fig. 4. Proposed pathways of anaerobic degradation of (a) n-alkanes and (b) aromatic hydrocarbons; activation via addition of fumarate, [1]—succinate. Pathways according to Spormann and Widdel (2000), and Wilkes et al. (2002); TCA—tricarboxylic acid cycle.

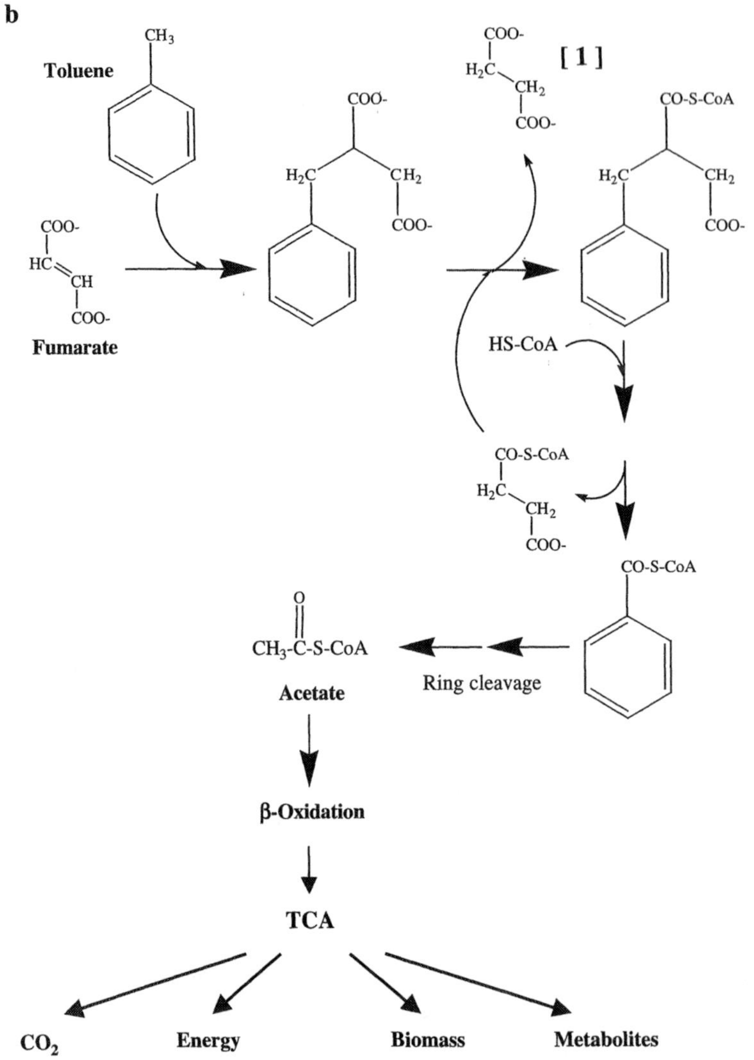

b

Fig. 4 (*continued*).

3.4. Competing processes

The ideal preconditions for biodegradation cited above occur only rarely, e.g., in the case of a rough and nutrient-rich sea or on energy-rich tidal flats. Mostly, however, the reality of oil spills is very different. The ideal steps are rendered difficult, slowed down, or made impossible by competing processes. Such influences are exemplified by the case studies.

Heavy oils or heavy oil products such as heavy fuel oil or bunker oil C behave very differently from the light oils described above. Heavy oils incorporate suspended matter, debris, biomass, and even garbage, which increases their viscosity and decreases their biodegradability. Due to their viscosity, the energy needed to emulsify heavy oils is very great. Solar irradiation causes the evaporation of the light components and photodecomposition, resulting in unpredictable compounds. Oil carpets are formed. Where they meet the coastline, beaches are covered. Their removal by natural forces is very slow or even impossible, and technical purification is expensive and troublesome. Biological degradation is extremely slow because the low oil surface to volume ratio limits the bioavailability of the oil.

Oil biodegradation works well on the open sea but proceeds differently on beaches. Vast areas of tidal beaches can be covered by oil when there is wind onshore during the ebb tide. If this oil cover is subjected to strong sun irradiation, the oil does not float up during the next flood because the light components have evaporated. The sediment is soaked with the sticky oil. Tides and wind add further sediment, and the initially liquid, later viscous, pollutant becomes more and more solidified (Tannenbaum et al., 1987). This solidified material is only slowly attacked by waves, hampering biodegradation because the available surface is too small. Irradiation and the catalysing capacity of particle surfaces help to convert a part of the original mixture of small molecules into high molecular mass material of low solubility, forming tar and finally asphalt. Such products appear as geological rather than organic matter.

Experience has shown that it is difficult for organisms to settle on oil layers. The Arabian Gulf spill presented a new experience in so far as thick and vital cyanobacterial mats developed on oil covers within a few months, introducing biomass as well as aeolian and hydrodynamic sediments fixed by the growing mats. This observation was welcomed initially (Sorkhoh et al., 1992) but then turned out to be disappointing because biodegradation was not favoured (Höpner et al., 1996). In some cases, colonization opened the oil crusts; in other cases, it formed stable covers which prevented the access of oxygen to deeper layers, helping to preserve the pollution. The latter clearly transformed the polluted system, resulting in geo-biological matter that had never been present before. This "new" geo-biological matter dominated the sites after the spill. In the upper eulittoral and the lower supratidal zone, calcareous incrustation and solid salt supported the conversion of the oil into rock-like matter with a life span of 10 or more years (Fig. 5).

4. Effects on the environment

The assumption that a hydrocarbon mixture which is released into the environment can be completely converted to carbon dioxide and will disappear over time, given the necessary conditions, is wrong. There

Fig. 5. Eight years after the 1991 Arabian Gulf spill (Saudi Arabia, 30 km north of Jubail Industrial City): A solidified and calcified oil crust, partially dissolution of which is caused primaril by eolic forces. The number "14" on the photograph denotes Saudi Arabian observation station no. 14 (24 in total), "98" denotes the year when the photograph was taken. Photograph: Höpner.

are, among others, at least three effects which prevent such an optimistic expectation:

- A bacterial or fungal population propagates and forms biomass while decomposing a substrate. The weight ratios of hydrocarbons degraded to the formed biomass may show values near 1:1. This conversion is energetically driven by the fact that the formal oxidation number of the hydrocarbon–carbon is in the range of −2, while the typical oxidation number of a biomass–carbon is zero (being now a carbohydrate carbon). This makes it clear that the conversion of a weight unit of hydrocarbon into the corresponding weight of biomass is an energy-forming process which can proceed easily under environmental conditions.
- Bacteria form and excrete products while degrading hydrocarbons. Some of these products act as bioemulsifiers, which enhance the bioavailability of hydrocarbons. The amount of products as compared with the degraded hydrocarbon can be considerable and is highly dependent on the conditions. As described above, bioemulsifiers can serve as a carbon and energy source, so that the effect is transient (e. g., Zajic and Mahomedy, 1984).
- Crude oil and some industrial hydrocarbon products contain compounds which are not degradable, not completely degradable, or extremely slowly degradable. Some examples: (i) material containing hetero-atoms (such as sulfur); (ii) asphaltenes and tars dissolved or suspended in a liquid phase, which exhibit high molecular mass properties; (iii) highly branched or condensed compounds. This reduced biodegradability is relevant if the material in question has previously undergone weathering or aging (Tannenbaum et al., 1987). This means that nondegradable residues will remain; how much and what kind of residues depends on the decomposing hydrocarbon mixture, the organisms, and the conditions.

If biodegradation is the core of the remediation strategy, the expected effect is not a complete removal of the pollutant, which would result from complete mineralization, but rather the partial replacement by biomass and other biogenic matter

and their conversion and degradation to nontoxic products. Generally, the latter cause less severe environmental impact than the original pollutant. But if the geo-biological implications are uncertain, it is important to keep in mind that the polluting hydrocarbons do not only suffer a direct geo-biological transformation but that they additionally begin a geo-biological development indirectly via the biomass path and via metabolic products (compare Figs. 1 and 2).

To illustrate environmental effects, we have included a field observation and the results of a laboratory investigation which lead to the conclusion that hydrocarbon biodegradation may alter a sedimentary or soil environment by forming or transforming inorganic matter. The field observation was made during the long-term study of the coastal oil pollution of 1991 in Saudi Arabia. It was found that the tar balls within a sandy beach sediment were wrapped in a calcium carbonate cover. In a similar way, the outer walls of oil-filled crab burrows sometimes had been converted to calcified tubes (Höpner unpublished). A possible explanation is that the oil biodegradation had led to a local carbon dioxide formation trapped by the calcium, which had a high concentration because of high local salinities.

The laboratory investigation was published by Lovley and Lonergan (1990) who investigated the anaerobic mineralization of toluene, phenol and p-cresol by the bacterium GS-15 which uses iron (III) as an electron acceptor. They observed the nearly stoichiometric formation of solid magnetite (an inorganic iron(II)oxide). Its formation is the direct consequence of the physiological activity of the hydrocarbon degrading organisms. Lovley and Lonergan (1990) pointed to the environmental implications: the natural accumulation of the inorganic mineralic magnetite around hydrocarbon seeps must be attributed to microbial oxidation of hydrocarbon compounds coupled to Fe(III) reduction.

5. Case studies

The material for case studies was provided by the International Tanker Owners Pollution Federation (ITOPF, 2003). Their list of marine oil spills over

700 tons contains 410 cases between 1973 and 2001. For our case studies, we decided not to select the AMOCO CADIZ spill (1978, 223,000 m^3) because the very thorough scientific documentation comprises only the first year after the spill. The data were discussed in a meeting in 1979, and the corresponding documentation was published in 1981 (Centre Océanologique de Bretagne, 1981). This time frame would be too short to allow a geo-biological study. We did not select the EXXON VALDEZ spill (1989, 37,000 m^3, Schnoor, 1991; Stoker et al., 1992) because the restoration expense was exceptional ("... employment of more than 11,000 people, utilization of essentially the entire world's supply of containment booms and skimmers, and an expenditure of more than two billion dollars, Maki, 1991). In addition, the Exxon company maintained an incomparable public relations program which evoked the unfortunate impression that an oil spill of such magnitude can be eradicated within 3 years or less. Such a program will probably never be repeated, regardless of the severity of potential spills.

The tanker BRAER grounded off Shetland on Jan 5, 1993 and lost 85,000 tons of oil. "The ecological disaster widely predicted by the media was apparently largely averted" (Glegg and Rowland, 1996), and the case disappeared from public discussion within a few days. Why this case developed so favourably and why the environmental impact was so limited make it a suitable case study:

- The BRAER load consisted of "an unusual partly biodegraded North Sea crude oil with a high natural dispersability."
- "The (very severe) weather played a very important part in controlling the fate of the oil such that oil-in-water, rather than water-in-oil, emulsions were produced."
- Due to the weather conditions, "the beaching of oil was minimized."
- "The subsequent transport of much of the oil appears to have been determined mainly by the water movement than by wind direction."

This is not the complete truth, but the selected facts show that of two possible extreme developments (rapid degradation or long-term preservation), the first one occurred. This shows that the nature of

the oil is one of the factors which govern the severity of a spill. Special attention should be given to a high-energy input, which is a more decisive factor than high temperature. Oil-in-water emulsion is much more effective in respect to biological degradation than water-in-oil emulsion. Beaching of oil has a more serious impact than when it floats on water. At the same time, beaching is the origin of most geo-biological implications. All of these negative consequences were prevented during the BRAER event.

The largest oil spill in history, the Arabian Gulf spill of 1991, is not listed in the ITOPF statistics (which contains only tanker accidents). In Jan/Feb 1991, an estimated amount of 1 million m^3 of light Kuwait oil were released from various points of Kuwait into the Gulf. It is not clear whether this amount includes the proportion which evaporated between release and stranding (about 40%; Watt, 1996).

The spill separated into two parts which behaved very differently. Up to 700,000 m^3 drifted out onto the open sea, and about 700,000 stranded. About 200,000 m^3 were skimmed off and recovered (Tawfiq and Olsen, 1993). The drifting part had disappeared by July of the same year—apparently biodegraded. The stranded part spoiled about 600 km of the Saudi Arabian coastline with very different consequences. There were both narrow and wide strips (up to 1 km), as well as both small and large amounts (up to 20 kg/m^2; Höpner et al., 1993). Mostly the pollution has remained until today. The very different effects were not only due to the variable coastline but also to extremely high spring tides shifting the oil far into the upper intertidal zone. The clean-up measures were of a rather experimental character. To remove the estimated 130 million m^3 of polluted intertidal sediments would have been not possible and would have caused unacceptable secondary damage.

The most severely polluted coastal strip was observed annually, the last time in October 2001 (Höpner, 2001). The results summarized here led to the conclusions given above under the heading "Competing processes."

Oil which hits a shoreline spreads out horizontally, is soaked up by the sediment vertically, and loses its volatile part by evaporation. This process is relatively fast. Afterwards, the slower processes start.

In the case of soft coasts, which prevail in the studied area, we observed a number of alternative scenarios:

- Oil cannot penetrate into the ground, mainly because the sediment is fine grained. On the surface, it is subjected to heat and irradiation. It is slowly converted to tar and finally to asphalt, which resists biological attacks. Only strong mechanical forces are able to break this cover up into tar pieces.
- Asphalt, calcareous incrustation, and salt crystallisation converted oil-soaked sediments into rock-hard matter, which remains that way for many years.
- Oil is soaked into a dry sediment. An oil sediment mixture is formed in which the oil ages and is partially oxidized, since oxygen access is possible. The initially moist mixture solidifies because macromolecular products are formed. They form brown mixtures, which are found often in the area. It is so persistent that it is found almost unchanged even after 10 years. It can be decomposed to tar pieces by wave action.
- Tar pieces can be transported and decomposed mechanically. The smaller the pieces are, the better the biological degradation. Tar balls do not have a long life span.
- Oil can be covered by wet finely grained sediment and by cyanobacterial mats. This means that it is enclosed in an oxygen-free environment and cannot be biodegraded aerobically. It persists in the same form as when it was buried; it has not changed its consistency, not even after 10 years.

In contrast to those conditions which favour persistence, there are distinct conditions which render biological degradation possible:

- Bioturbation provides oxygen. Microbiological degradation starts if nutrients and some moisture are also present, in addition. Where these conditions are present, oil has no longer been found after 10 years.
- Oil was deposited on areas exposed to wave action and tidal currents, allowing mechanical degradation to take place. The resulting particles and pieces were biologically degraded in the seawater. This process is still going on in many places.

- Oxygen had access to the surface of oil soaked sediments. In this case, biodegradation starts again. In many places, oiled sediments were found to be covered by a layer of clean sediment. The thickness of this upper layer increased in the course of time (this has to be distinguished from cases where wind or current deposit clean sediments on the top of oiled ones).

These three biodegradation processes are so slow that we found only three (of 24) places where they have completed their work within 10 years.

The Gulf spill has taught us that degrading biological processes compete with geological ones. Among the first processes, bioturbation (caused almost exclusively by digging crabs) is dominant, while the latter tend to preserve layered soil structures which were formed during or immediately after the spill. The result of the competition is directed by the frequency of the flooding rather than by the oil load.

The case of the tanker PRESTIGE was still going on at the time this article was being written (Feb 2003); this fact opens the area for speculation. The PRESTIGE was carrying 77,000 tons of heavy fuel oil. It grounded on Nov 13, 2002 off Cape Finisterre, Northern Spain, where it lost estimated 5000 tons. It was towed to the southwest for about 150 sm, broke in two on Nov 19, and the bow and stern sank into water about 2 miles deep. At the end of January, the wreck was still releasing approximately 5 tons oil per day. Beach contamination now extends from the northern Portuguese over the northern Spanish to the southern French coast. From aerial views, contamination is "intermittent," and the "extremely viscous" contaminant requires manual rather than mechanised work. By the end of January, about 41,000 of solid and 21,000 tons of liquid waste has been collected, and the costs were estimated at 215–320 million € plus 80–250 million € losses in fisheries and aquaculture (ITOPF, 2003). Since the loss of birds is considered "moderate," the main damage estimates refer to human coastal use, such as the tourist and fishing industries.

Viscous oil disperses only very slowly and approaches the beaches mainly as isolated slicks and tar balls. This causes patchy pollution, leaving the main parts of the beaches clean—as long as it is cold, and the oil remains viscous. As soon as the temperature rises, the viscous matter will liquefy and will cause secondary contamination, so that rapid removal is necessary.

We have seen again that the severity of an oil spill is dependent on the type of the oil spilled, heavy fuel oil being especially serious for several reasons (consider the very similar accident of the tanker ERIKA, which spilled 20,000 tons of heavy fuel oil north of the coast of Brittany). An oil which is rather persistent at sea as compared with a light crude oil is also suspected to relatively persist as a coastal contaminant, all the more because parts of the coastline are rather inaccessible and prevent the oil's removal. This means that some of the oil will enter the very slow geo-biological processes.

6. Conclusions

Remediation–the healing of oil spill damage with the aim of restoring the original ecological state using only natural means–is an interplay of geological and biological factors. "Geological" includes the physical influences. "Biological" includes the production of new matter as well as biodegradation. To emphasise the requirements of the geological as well as of the biological contribution, it would be useful to imagine the following: What would be the fate of a coastal oil spill in a sterile environment, and what changes can biological activity provoke in the absence of physical forces?

In a sterile environment, bioemulsifiers and biofilms would not be formed. The surface of the spilled oil would be exposed to heat, irradiation, airborne dust, and water-born sediments. Even if mechanical forces, such as waves and turbulence, distributed and dispersed the oil, the particles formed would be unstable because they are sticky, and there are no biofilms and bioemulsifiers to prevent reclotting. Evaporation would increase the viscosity until solidification. This would be supported by light-mediated chemical polymerisations and by inclusion of dust and sediment particles. Dry sediments would be soaked with oil, and finally, the formation of solid matter would prevail. This might stay at the surface or might be buried in the sediment, but it would not be altered or would only be altered by extremely slow chemical reactions. Transport processes would not be excluded, and the

geological and physical forces might conceal the spilled matter but could not really remove it.

The second scenario: The best biological potentials fail if the forces needed for the emulsification are absent, if particle surfaces adsorbing oil are lacking, if oxygen is not supplied, and degradation products are not removed. To transfer this situation to the laboratory means to inoculate a fermentor containing water and oil but not to stir nor aerate it. Oil degradation will proceed extremely slowly; it hardly takes place at all. From this, we learn that the remediation of an oil spill is an interplay of physicochemical/geological and biological effects, a fact which is supported by the scientific studies as well as the case studies.

The ITOPF oil spill statistics (ITOPF, 2003) list 1593 ship-born oil spills (1183 between 7 and 700 tons each, 410 with more than 700 tons). All of these cases are very different. The differences begin with the type and amount of oil spilled and the way in which it is distributed. They continue with the precautionary measures taken in the affected area and the prevailing environmental conditions (climate, weather, etc.). The ecosystem concerned may be dependent on self-remediation, on technical measures, or on both. In any case, the result is a more or less damaged ecosystem which will be left to a further development. How it will develop further is a question of the existing conditions. There are so many variables that every case is different; it is hardly possible to extrapolate from one to the next. What is clear is that, to estimate the geo-biological development of oil spills, three main questions have to be asked: What is the composition of the oil spilled? How active are the geological and geophysical forces? How can the potential of biodegradation be assessed?

References

Aeckersberg, F., Bak, F., Widdel, F., 1991. Anaerobic oxidation of saturated hydrocarbons to carbon dioxide by a new type of sulfate-reducing bacterium. Arch. Microbiol. 156, 5–14.

Anderson, R.T., Lovley, D.R., 2000. Hexadecane decay by methanogenesis. Nature 404, 722–723.

Ashraf, W., Mihdhir, A., Murrell, J.C., 1994. Bacterial oxidation of propane. FEMS Microbiol. Lett. 122, 1–6.

Atlas, R.M., 1975. Effects of temperature and crude oil composition on petroleum biodegradation. Appl. Microbiol. 30, 396–403.

Atlas, R.M., 1984. Petroleum Microbiology. Macmillan Publishing, New York, NY.

Atlas, R.M., Bartha, R., 1973. Inhibition by fatty acids of the biodegradation of petroleum. Antonie van Leeuwenhoek 39, 257–271.

Berthe-Corti, L., Bruns, A., 1999. The impact of oxygen tension on cell density and metabolic diversity of microbial communities in alkane degrading continuous-flow cultures. Microb. Ecol. 37, 70–77.

Berthe-Corti, L., Bruns, A., 2001. Composition and activity of marine alkane-degrading bacterial communities in the transition from suboxic to anoxic conditions. Microb. Ecol. 42, 46–55.

Berthe-Corti, L., Ebenhöh, W., 1999. A mathematical model of cell growth and alkane degradation in Wadden Sea sediment suspensions. Biosystems 49, 161–189.

Berthe-Corti, L., Fetzner, S., 2002. Bacterial metabolism of n-alkanes and ammonia under oxic, suboxic and anoxic conditions. Acta Biotechnol. 22, 299–336.

Bonin, P., Bertrand, J.C., 2000. Influence of oxygen supply on heptadecane mineralization by Pseudomonas nautica. Chemosphere 41, 1321–1326.

Burns, K.A., Villeneuve, J.P., Anderlin, V.C., Fowler, S.W., 1982. Survey of tar, hydrocarbon and metal pollution in the coastal waters of Oman. Mar. Pollut. Bull. 13, 240–247.

Centre Océanologique de Bretagne, 1981. Amoco Cadiz: fates and effects of the oil spill. Consequences d'une pollution accidentelle par les hydrocarbures, Cent. Natl. Exploit. Oceans, Paris (Fr.) p. 881.

Cerniglia, C.E., 1992. Biodegradation of polycyclic aromatic hydrocarbons. Biodegradation 3, 351–368.

Dalyan, U., Harder, H., Höpner, T., 1990. Hydrocarbon biodegradation in sediments and soils. A systematic examination of physical and chemical conditions: Part III. Temperature. Erdöl Kohle Erdgas Petrochem. 43, 435–437.

Delille, D., Delille, B., Pelletier, E., 2002. Effectiveness of bioremediation of crude oil contaminated subantarctic intertidal sediment: the microbial response. Microb. Ecol. 44, 118–126.

Edwards, E.A., Grbic-Galic, D., 1994. Anaerobic degradation of toluene and o-xylene by a methanogenic consortium. Appl. Environ. Microbiol. 60, 313–322.

Ensign, S.A., 2001. Microbial metabolism of aliphatic alkenes. Biochemistry (Mosc.) 40, 5845–5853.

Evans, W.C., Fuchs, G., 1988. Anaerobic degradation of aromatic compounds. Annu. Rev. Microbiol. 42, 289–317.

Evans, P.J., Ling, W., Goldschmidt, B., Ritter, E.R., Young, L.Y., 1992. Metabolites formed during anaerobic transformation of toluene and o-xylene and their proposed relationship to the initial steps of toluene mineralization. Appl. Environ. Microbiol. 58, 496–501.

Frohne, J., Kettrup, A., Riepe, W., Schulz-Berendt, V., 1991. Mikrobieller Abbau von Kohlenwasserstoffen und Kohlenwasserstoffverbindungen. DGMK Forschungsbericht. DGMK Deutsche Wissenschaftliche Gesellschaft für Erdöl, Erdgas und Kohle e.V., Hamburg, pp. 461–501.

Gibbs, C.F., 1975. Quantitative studies on marine biodegradation of oil: I. Nutrient limitation at 14 degrees C. Proc. R. Soc. Lond., B Biol. Sci. 188, 61–82.

Gibson, D.T., 1984. Microbial Degradation of Organic Compounds. Dekker, New York.

Glegg, G.A., Rowland, S.J., 1996. The Braer oil spill-hydrocarbon concentrations in intertidal organisms. Mar. Pollut. Bull. 32, 486–492.

Gray, J.P., Herwig, R.P., 1996. Phylogenetic analysis of the bacterial communities in marine sediments. Appl. Environ. Microbiol. 62, 4049–4059.

Haferburg, D., Hommel, R., Claus, R., Kleber, H.P., 1986. Extracellular microbial lipids as biosurfactants. Adv. Biochem. Eng. Biotechnol. 33, 53–93.

Harder, H., Kuerzel-Seidel, B., Höpner, T., 1991. Hydrocarbon biodegradation in sediments and soils. A systematic examination of physical and chemical conditions: Part 4. Special aspects of nutrient demand. Erdöl Kohle Erdgas Petrochem. 44, 59–62.

Höpner, T., 2001. Jubail marine wildlife sanctuary: oil pollution status report October 2001. Unpublished.

Höpner, T., Felzmann, H., Struck, H., van Bernem, K.H., 1993. The nature and extent of oil contamination on Saudi Arabian Gulf beaches: examinations of beaches of Dawhat ad Dafi and Dawhat ad Musallamiya in summer 1991 and winter 1991/92. Arab. J. Sci. Eng. 18, 243–255.

Höpner, T., Yousef, M., Berthe-Corti, L., Felzmann, H., Struck, H., Al-Thukair, A., 1996. Cyanobacterial mats on oil polluted sediments - start of a promising self remediation process? In: Krupp, F., Abuzinada, A.H., Nader, J.A. (Eds.), A Marine Wildlife Sanctuary of the Arabian Gulf: Environmental Research and Conservation Following the 1991 Gulf War Oil Spill. National Commission for Wildlife Conservation and Developmen (Riyadh), Forschungsinstitut Senckenberg (Frankfurt), Frankfurt a.M./Jubail, pp. 85-95.

ITOPF, 2003. International Tanker Owners Pollution Federation Limited. http://www.itopf.com.

Kasai, Y., Kishira, H., Harayama, S., 2002. Bacteria belonging to the genus cycloclasticus play a primary role in the degradation of aromatic hydrocarbons released in a marine environment. Appl. Environ. Microbiol. 68, 5625–5633.

Knoblauch, C., Jørgensen, B.B., Harder, J., 1999. Community size and metabolic rates of psychrophilic sulfate-reducing bacteria in Arctic marine sediments. Appl. Environ. Microbiol. 65, 4230–4233.

Kosaric, N., Cairns, W.L., Gray, N.C.C., 1987. Biosurfactants and Biotechnology. Marcel Dekker, New York.

Llobet-Brossa, E., Rosselló-Mora, R., Amann, R., 1998. Microbial community composition of Wadden sea sediments as revealed by fluorescence in situ hybridization. Appl. Environ. Microbiol. 64, 2691–2696.

Lovley, D.R., Lonergan, D.J., 1990. Anaerobic oxidation of toluene, phenol, and p-cresol by the dissimilatory iron-reducing organism, GS-15. Appl. Environ. Microbiol. 56, 1858–1864.

Mahro, B., Kästner, M., 1993. Der mikrobielle Abbau polyzyklischer aromatischer Kohlenwasserstoffe (PAK) in Böden und Sedimenten: mineralisierung metabolitenbildung und entstehung gebundener Rückstände. Bioengineering 9, 50–58.

Maki, A.W., 1991. The Exxon Valdez oil spill: initial environmental impact assessment: Part 2. Environ. Sci. Technol. 25, 24–29.

Meyer-Reil, L., 1993. Mikrobielle Besiedlung und Produktion. In: Meyer-Reil, L., Köster, M. (Eds.), Mikrobiologie des Meeresbodens. Gustav Fischer, Jena, pp. 38–81.

Michaelsen, M., Hulsch, R., Höpner, T., Berthe-Corti, L., 1992. Hexadecane mineralization in oxygen-controlled sediment-seawater cultivations with autochthonous microorganisms. Appl. Environ. Microbiol. 58, 3072–3077.

Moyer, C.L., Dobbs, F.C., Karl, D.M., 1994. Estimation of diversity and community structure through restriction fragment length polymorphism distribution analysis of bacterial 16S rRNA genes from a microbial mat at an active, hydrothermal vent system, Loihi Seamount, Hawaii. Appl. Environ. Microbiol. 60, 871–879.

National Research Council, 2002. Committee on Oil in the Sea: Oil in the Sea: III. Inputs, Fates, and Effects. National Academy Press, Washington, DC.

Perry, J.J., 1980. Propane utilization by microorganisms. Adv. Appl. Microbiol. 26, 89–115.

Rambeloarisoa, E., Rontani, J.F., Giusti, G., Duvnjak, Z., Bertrand, J.C., 1984. Degradation of crude oil by a mixed population of bacteria isolated from sea-surface foams. Mar. Biol. 83, 69–81.

Ratledge, C., 1978. Degradation of aliphatic hydrocarbons. In: Watkinson, R.F. (Ed.), Developments in Biodegradation of Hydrocarbons: 1. Applied Science Publishers, London, pp. 1–46.

Roling, W.F.M., Milner, M.G., Jones, D.M., Lee, K., Daniel, F., Swannell, R.J.P., Head, I.M., 2002. Robust hydrocarbon degradation and dynamics of bacterial communities during nutrient-enhanced oil spill bioremediation. Appl. Environ. Microbiol. 68, 5537–5548.

Roy, R., Greer, C.W., 2002. Hexadecane mineralization and denitrification in two diesel fuel-contaminated soils. FEMS Microbiol. Ecol. 32, 17–23.

Rudolphi, A., Tschech, A., Fuchs, G., 1991. Anaerobic degradation of cresols by denitrifying bacteria. Arch. Microbiol. 155, 238–248.

Schauer, F., 2001. Abbau und Verwertung von Mineralölbestandteilen durch Mikroorganismen. Bodden 11, 3–31.

Schnoor, J.L., 1991. The Alaska oil spill: its effects and lessons. Environ. Sci. Technol. 25, 14.

So, C.M., Phelps, C.D., Young, L.Y., 2003. Anaerobic transformation of alkanes to fatty acids by a sulfate-reducing bacterium, strain Hxd3. Appl. Environ. Microbiol. 69, 3892–3900.

Sorkhoh, N.A., Ghannoum, M.A., Ibrahim, A.S., Stretton, R.J., Radwan, S., 1990. Crude oil and hydrocarbon-degrading strains of *Rhodococcus rhodochrous* isolated from soil and marine environments in Kuwait. Environ. Pollut. 65, 1–17.

Sorkhoh, N., Al-Hasan, R., Radwan, S., Höpner, T., 1992. Self-cleaning of the Gulf. Nature 359, 109.

Spormann, A.M., Widdel, F., 2000. Metabolism of alkylbenzenes, alkanes, and other hydrocarbons in anaerobic bacteria. Biodegradation 11, 85–105.

Stoker, S.W., Neff, J.M., Schroeder, T.R., McCormick, D.M., 1992. Biological Conditions in Prince William Sound, Alaska. Following the Valdez Oil Spill: 1989–1992. Beringian Resources, Arthur D. Little, Cook Inlet Fisheries Consultants, and Woodward-Clyde Consultants, Prince William Sound.

Tannenbaum, E., Starinsky, A., Aizenshtat, Z., 1987. Light-oils transformation to heavy oils and asphalts—assessment of the amounts of hydrocarbons removed and the hydrological-geological control of the process. Exploration for Heavy Crude Oil and Natural Bitumen. The American Association of Petroleum Geologists, Tulsa, Oklahoma, pp. 221–231.

Tawfiq, N.I., Olsen, D.A., 1993. Saudi Arabia's response to the 1991 Gulf oil spill. Mar. Pollut. Bull. 27, 335–345.

Van Beilen, J.B., Wubbolts, M.G., Witholt, B., 1994. Genetics of alkane oxidation by *Pseudomonas oleovorans*. Biodegradation 5, 161–174.

Van Bernem, C., Lübbe, T., 1997. Öl im Meer-Katastrophen und langfristige Belastungen. Wissenschaftliche Buchgesellschaft, Darmstadt.

Walker, J.D., Colwell, R.R., 1974. Microbial degradation of model petroleum at low temperatures. Microb. Ecol. 1, 63–95.

Watt, I., 1996. A summary of the clean-up techniques used in the Jubail Marine Wildlife Sanctuary after the Gulf War oil spill and an assessment of their benefit to intertidal recovery. In: Krupp, F., Abuzinada, A.H., Nader, J.A. (Eds.), A Marine Wildlife Sanctuary of the Arabian Gulf: Environmental Research and Conservation Following the 1991 Gulf War Oil Spill. National Commission for Wildlife Conservation and Developmen (Riyadh), Forschungsinstitut Senckenberg (Frankfurt), Frankfurt a.M./Jubail, pp. 116-127.

Widdel, F., Rabus, R., 2001. Anaerobic biodegradation of saturated and aromatic hydrocarbons. Curr. Opin. Biotechnol. 12, 259–276.

Wilkes, H., Rabus, R., Fischer, T., Armstroff, A., Behrends, A., Widdel, F., 2002. Anaerobic degradation of *n*-hexane in a denitrifying bacterium: further degradation of the initial intermediate (1-methylpentyl)succinate via C-skeleton rearrangement. Arch. Microbiol. 177, 235–243.

Zajic, J.E., Mahomedy, A.Y., 1984. Biosurfactants: Intermediates in the Biosynthesis of Amphipatic Molecules in Microbes. Macmillan Publishing, New York, NY, pp. 221–297.

Available online at www.sciencedirect.com

Palaeogeography, Palaeoclimatology, Palaeoecology 219 (2005) 191–198

ELSEVIER

www.elsevier.com/locate/palaeo

Biogeochemistry: now and into the future

Gregory A. Cutter*

Department of Ocean, Earth, and Atmospheric Sciences, Old Dominion University, Norfolk, VA 23529-0276, USA

Received 20 September 2003; accepted 29 October 2004

Abstract

Biogeochemistry can be defined as the mutual interactions (two-way) between the biology and chemistry of the Earth system, and as such is clearly an important component of the broader discipline of geobiology. It is a well-developed field, having a dedicated journal and textbook, and many hundreds of publications appearing in the scientific literature each year. Perhaps the best example of biogeochemistry is the balance between photosynthesis and respiration, autotrophy and heterotrophy, on Earth, the so-called Redfield equation. The modern practice of biogeochemical research has been focused on present-day processes such as those in the Redfield equation, but new tools and approaches are beginning to explore the biogeochemistry of the ancient Earth; this is the challenge for biogeochemists in the future.
© 2004 Elsevier B.V. All rights reserved.

Keywords: Biogeochemistry; Autotrophy; Heterotrophy; Interdisciplinary; Earth system

Of the disciplines under the rubric of geobiology, biogeochemistry is perhaps one of the more mature in name, if not in practice. Indeed, a biogeochemistry journal was established 20 years ago (Biogeochemistry; Kluwer) and a textbook that is in its second edition was first published in 1991 (Biogeochemistry: An analysis of global change by W.H. Schlesinger; Academic Press). A simple search of the Earth sciences literature shows over 600 peer-reviewed articles in 2003 alone having "biogeochemistry" or "biogeochemical" in their titles or abstracts. This tremendous use of the term suggests a well-established definition of biogeochemistry, presumably a balanced study of the mutual interactions (feedback) between the biology and chemistry of the Earth system. However, closer examinations of its practice shows the studies are largely BIOgeochemistry, bioGEOchemistry, or biogeoCHEMISTRY; a balanced approach is seldom part of these studies. However, I would argue that studying the interactions between the chemistry and biology of the Earth system is a crucial metric of calling one's science "biogeochemistry". Studying the individual components is certainly important, but interactions and feedback make it biogeochemistry in my opinion. Therefore, this article is not a thorough review of biogeochemistry, but rather an illustration from my perspective of what this sub-discipline of Earth science is and what it can be. In the context of

* Tel.: +1 757 683 4929; fax: +1 757 683 5303.
  E-mail address: gcutter@odu.edu.

0031-0182/$ - see front matter © 2004 Elsevier B.V. All rights reserved.
doi:10.1016/j.palaeo.2004.10.021

geobiology, biogeochemistry certainly has prominent roles, for example: biogeochemical processes physically link biological and geological components of the Earth system through chemical intermediates (e.g., organic acids or even CO_2 in weathering); portions of the Earth's climate system can be biogeochemically controlled (e.g., Charlson et al., 1987); and biogeochemical research can provide chemical and biochemical markers/proxies for biological processes when no fossil records exist.

Considering the definition above, what would be a good example of biogeochemistry, or to use the colloquial question, if it walks like a duck, quacks like a duck, etc., is it a duck? As an aquatic scientist, I will focus on aquatic biogeochemistry, although most of my discussions have parallels in terrestrial biogeochemistry. I believe that the classic example of biogeochemistry is the work started in the 1930s by Alfred C. Redfield and the so-called Redfield equation for the marine system (Redfield et al., 1963):

$$106CO_2 + 16NO_3 + H_3PO_4 + 122H_2O$$

$$\Leftrightarrow [(CH_2O)_{106}(NH3)_{16}H_3PO_4] + 138O_2$$

Most recognize that the forward reaction is photosynthesis and the reverse is respiration, and the wonderful part of this equation is that it involves interchanges between dissolved ions and gases (i.e., the hydrosphere and atmosphere), and particulate organic matter, all of it mediated by organisms (the biosphere). It should also be clearly understood that these relations are not based on first principles, but are totally empirical. Not surprisingly, one can find many "violations" of the equation (the 106:16:1 is highly dependent upon the organisms performing the reactions, their growth stage, etc.), but the principle that it represents, chemistry/geochemistry affecting biology *and* the reverse, is the essence of biogeochemistry.

While the Redfield equation is given on a molecular/organism basis, in a closed system such as here on Earth, the balance between autotrophy and heterotrophy that it represents controls the geochemistry of most minor (e.g., nutrient) and many trace elements in the aquatic system and vice versa. Furthermore, the autotrophic–heterotrophic balance on Earth (represented by more than just the Redfield

equation since it does not include biogenically linked, reduced compounds such as pyrite) also controls the amount of oxygen in the atmosphere on long time scales (e.g., Berner and Canfield, 1989; Berner, 2003). This demonstrates that the study of biogeochemistry is of global significance (or should be). And finally, the idea of balancing autotrophy and heterotrophy embodied in the Redfield equation has a lesson for today: net heterotrophy (e.g., including additional heterotrophy in the form of internal combustion engines) that is due to the respiration of "out of phase" (i.e., ancient) net autotrophy (fossil fuels) leads to excess CO_2 in the present-day atmosphere.

The Redfield equation shows us that the study of biogeochemistry, perhaps not in name, but in discipline, has been around for quite a while. So, are there other, more recent, examples of biogeochemistry in practice? Actually, extension of the Redfield approach to the biologically required trace elements (e.g., Fe, Zn, Mo, Se) also provides excellent examples. Liebig's Law of the Minimum suggests that when elements and compounds are biologically required in a fixed proportion (e.g., 106:16:1 for the Redfield C/N/P), the lack of any one of them will limit growth. This seems rather obvious and when one examines the nutrient concentrations of the world's oceans, they appear to be largely in Redfield proportions and the productivity of surface waters (i.e., the fixation rate of CO_2-carbon into organic matter) parallels the nutrient concentrations. However, there are certain regions of the ocean, particularly those surrounding the Antarctic continent where the nutrient concentrations are elevated, yet phytoplankton productivity is anomalously low (so-called "high nutrient, low chlorophyll" or HNLC waters). There must be another limiting nutrient/factor, and the work of the late John Martin and coworkers (Martin and Fitzwater, 1988; Martin and Gordon, 1988; Martin et al., 1989, 1991) using analyses of water column concentration profiles and experimental manipulations of phytoplankton incubations suggest that it is dissolved iron. In other words, initial experiments revealed that adding dissolved iron to HNLC waters contained in trace metal clean, polycarbonate flasks, triggers the growth of phytoplankton. (Note: Some species did grow while others were unaffected, and zooplankton grazers were excluded, but the message is still, iron additions result in more growth of autotrophs).

The tools for studying iron biogeochemistry (water column sampling, lab or field manipulations) have expanded to a much larger scale when researchers following Martin's lead (e.g., Coale et al., 1998, 2004; Boyd et al., 2000; Gervais et al., 2002; Tsuda et al., 2003) performed mesoscale iron fertilization experiments (add dissolved Fe directly into HNLC surface waters, not flasks, and measure the changes in nutrients, phytoplankton biomass and growth, particulate organic carbon export, etc.). Perturbing the chemical makeup of a water mass via iron additions, and the subsequent monitoring of the biological and geochemical responses (interactions and feedbacks) are biogeochemical in nature. Furthermore, unlike the early Redfield work that focused on the first trophic level (phytoplankton), the Fe-addition experiments examined the biogeochemical interactions up to the zooplankton (i.e., grazer) and microbial (grazers and recyclers) levels (e.g., Cochlan, 2001; Landry et al., 1997). Although one can argue about the "legitimacy" (or even morality) of such ecosystem manipulations, these experiments have gone well beyond laboratory (controlled) studies or field collections, and developed new methods for studying Earth system processes (e.g., Langrangian studies using inert tracers like SF_6 to mark the water mass and therefore allowing in situ processes and rates to be quantified; Law et al., 2001).

These mesoscale experiments largely paralleled the development of the Redfield approach (lab experiments to field sampling on ocean basin scales), but Martin also expanded his notion of iron limitation into the temporal domain by considering its paleo-implications on a global scale. The so-called "iron hypothesis" (Martin, 1990) suggested that the glacial low CO_2, interglacial higher CO_2 could be explained by enhanced atmospheric Fe inputs to HNLC surface waters, increased productivity and organic C export, and therefore atmospheric CO_2 drawdown during glacial periods. The biogeochemical signals to test this hypothesis lie in glacial ice (for CO_2 and aerosol Fe data), sediments (for org C fluxes directly or via proxies; indicators of enhanced aerosol transport and deposition), and integration of the disparate data sets using simulation models (e.g., Moore et al., 2001). While the iron hypothesis built a temporal and tracer/proxy aspect to modern biogeochemical research, it also included the importance of transport (e.g., upwelling of iron from deeper, higher iron concen-

tration waters in the thermocline, atmospheric transport pathways and deposition to the surface ocean) in biogeochemical studies. Indeed, many current biogeochemical investigations lack integration of the physical transport aspects into their studies (should we coin another name, "Physical Biogeochemistry" or "Biogeochemical Physics"?).

The study of trace element biogeochemistry has also brought to light the importance of what was previously an interest of aquatic chemists only—chemical speciation. At the risk of enraging those who did not enjoy college chemistry, the concentration of a dissolved ion is not important in a chemical reaction that is in anything but extremely dilute solutions, it is the activity of that ion that matters (where the activity of the ion, $a_i = \gamma_i C_i$, with C_i being the concentration and γ_i the activity coefficient). In the high ionic strength of seawater, the activity coefficients of most ions drop well below 1.0 (where activity=concentration). This effect was compiled in classic papers by Turner et al. (1981) and Byrne et al. (1988) for the inorganic constituents of fresh and sea waters. In this regard, one can think of the activity of an ion as the concentration of the ion that is free to participate in a given reaction, not complexed/interacting with other ions. The biological consequences of this chemical speciation were shown by Brand et al. (1983, 1986), where growth (positive for essential elements and negative for toxic ones) of a phytoplankton species was affected by the concentration of the free trace metal ion, and not its total concentration. These biological studies were complimented by the discoveries that many trace metal ions in seawater are complexed with organic ligands (still largely unidentified) that drop their free ion concentrations (or activities) to less than 1% of the total ion (e.g., Cu, Moffett, 1995; Zn, Bruland, 1989; Fe, Rue and Bruland, 1995). Of course, none of this is really biogeochemistry without demonstrating the interactions of trace metals, their chemical speciation, and biological processes. For a single metal, copper, this was nicely demonstrated by Croot et al. (2000), where the production of organic ligands by several phytoplankton species varied directly with the activity of free copper. Perhaps one of the most eloquent examples of the role of trace metal speciation on biogeochemical cycling is the work of Bruland et al. (1991). These investigators examined the biogeochemistries of Co, Cu, Fe, Mn, Ni, and Zn using a combination of lab and field measurements to illustrate antagonistic and synergistic interactions between these metals, their speciation, and phytoplankton growth. The significance of this work is that it included multiple metals (more realistic), some of which were essential while others were toxic, and that it had implications for not only the productivity of surface waters, but also the trophic structure of the biotic community (e.g., selection for phytoplankton species more tolerant of toxic trace metals or that make stronger ligands to complex the metals).

It should be noted that although the inclusion of chemical speciation into the study of trace metal biogeochemistry might seem state of the art, the role of chemical speciation in biogeochemical cycles was appreciated by Redfield and coworkers during their studies. It has long been known that the Redfield equation changes with the chemical form of nitrogen being taken up by autotrophs:

$$106CO_2 + 16NH_3 + H_3PO_4 + 106H_2O$$

$$\Leftrightarrow [(CH_2O)_{106}(NH3)_{16}H_3PO_4] + 106O_2$$

In this case, the uptake of ammonia is preferred over nitrate due to the energetic advantage of not having to reduce nitrogen (from +5 to −3 in the first Redfield equation above). And with respect to trace metals, as early as 1969, Barber and Ryther suggested that the production of "organic chelators" (ligands) could affect primary production. Thus, chemical speciation has a biogeochemical influence at many levels in the ocean's biogeochemistry. This fact has necessitated the development of many analytical techniques, and field and laboratory approaches to exploit them.

Overall, the present status of biogeochemical studies is one in which a suite of excellent analytical methods exist to determine the concentration, chemical form, and even isotopic composition (ratios) of many inorganic and organic compounds in the aquatic environment. In addition, lab incubations to carefully control environmental variables have been developed to study specific biota–compound interactions. These are complimented by field incubations and manipulations, with the end member being the mesoscale ecosystem perturbations done in the ocean (e.g., Coale et al., 2004) and even lakes (e.g., Elser et

al., 2000). Most importantly, for any examination of an integrated system (implicit in biogeochemistry), we have learned to assemble interdisciplinary teams of scientists who can pool their talents and communicate with their colleagues—work as a system; this is a crucial, and even limiting, factor for interdisciplinary research (Duarte and Piro, 2001 present an excellent summary of this interdisciplinary problem).

Excellent examples of studying the interactions between the chemistry and biology of the Earth system exist, but what can/should we expect in the future? Certainly, new analytical and molecular methods are needed for detailed studies of many biogeochemical processes, and more kinetic data will take biogeochemistry from a descriptive, qualitative to a more quantitative and widely applicable science. It was noted above that physical transport has been largely overlooked in many biogeochemical studies, but more significantly, trophic interactions in an ecosystem have also been inadequately addressed. The Redfield equation only includes the two major trophic levels, autotrophs and heterotrophs, but the Earth has many trophic levels and interactions that affect the whole system. The mesoscale iron enrichment experiments (e.g., Coale et al., 2004) have crudely examined trophic interactions in a biogeochemical cycle, but much more is needed. These are some of the problems to be addressed by biogeochemical researchers in the near future, but perhaps the greatest challenge may be to look back through geologic time to reveal the biogeochemical processes of the past and therefore describe how the Earth system evolved to its present state. This is the direct connection between biogeochemistry and geobiology.

The fact that the Earth's present-day atmosphere is 21% oxygen is clear evidence of the biogeochemical linking of photosynthesis, respiration, and nutrients. Using a variety of geological and isotopic tracers (e.g., Holland and Beukes, 1990; Farquhar et al., 2000), the dramatic increase in atmospheric oxygen levels appears to have begun over 2 billion years ago. However, these studies are very broad and not focused on specific biogeochemical processes. More detailed studies of biogeochemical cycling have emerged for the Proterozoic ocean, some 2.5–0.54 billion years before present. This eon is significant for the marked changes that occurred in the oceans and atmosphere due to photosynthesis and types of respiration (oxic and anoxic) consuming organic matter. For example, Bjerrum and Canfield (2002) argued that oceanic primary production and therefore oxygenation of the atmosphere between 3.2 and 1.9 billion years before present was limited by the removal of phosphate onto iron oxides, which themselves were a result of photosynthesis (i.e., the Banded Iron Formations that are globally distributed). This study was far more biogeochemical than many others examining the Proterozoic since it included the feedback between geochemical (adsorption to iron oxides) and biological (photosynthesis) processes.

Using S isotopes in pyrite found just after the Banded Iron Formations disappear from the geologic record, Canfield (1998) postulated that while the surface of the planet (and surface oceans) were becoming oxic, the deep ocean remained anoxic and sulfidic (as opposed to anoxic without high concentrations of hydrogen sulfide). He explained the rise in hydrogen sulfide concentrations as being due to an increase in sulfate from oxic weathering and the consequent rise in bacterial sulfate reduction as the primary form of anoxic respiration. While his sulfur isotope findings may seem largely geochemical, they have important ramifications for the ocean's biogeochemical processes. Indeed, Anbar and Knoll (2002) made the argument that sulfidic waters in the deep ocean effectively removed dissolved iron and molybdenum from solution as insoluble sulfides (as well as Cd, Cu, and Zn), thus limiting their availability for

enzymes required in the biochemical processes of nitrogen fixation (nitrogenase) and assimilation (nitrate reductase). This would then mean that autotrophic productivity in the surface ocean was nitrogen-limited (rather than phosphate-limited earlier in the eon). This would also have repercussions on the evolution of eukaryotes that are far more sensitive to nitrogen limitation than the older autotrophs (bacteria and archaea).

Recently, Arnold et al. (2004) obtained molybdenum isotope data for the mid-Proterozoic to confirm the removal of this element as predicted, establishing a link, albeit indirect, to the biogeochemistry of the Proterozoic. Besides Mo and S, there are other useful isotopic tracers such as ^{15}N in sedimentary organic matter that allow surface water nitrogen cycling processes to be revealed (e.g., Francois et al., 1992; Ganeshram et al., 2000) in the Pleistocene, although extrapolating this proxy to earlier periods may suffer from diagenetic effects (Sigman et al., 1999). There are also many paleotracers of seawater chemistry (e.g., Cd/Ca in forams as a proxy of dissolved phosphate concentrations; Boyle, 1981, 1988), but deriving the complete biogeochemical setting of the paleo-ocean may require the linkage of many data sets (multiple proxies), something that is only beginning to be employed effectively for biogeochemistry.

Tracing the status of the Redfield equation can serve as an example of effective paleobiogeochemistry. High organic carbon in black shales deposited

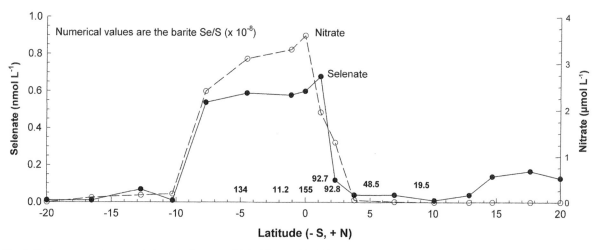

Fig. 1. Surface water selenate and nitrate in the equatorial Pacific at 160°W (Cutter and Bruland, 1984) and core top Se/S in barites at 140°W.

during the Cretaceous are due to net autotrophy, but were they a result of an increase in autotrophy itself (large organic carbon flux to the sediments) through abundant nutrients and light, or are these shales the result of slow heterotrophic respiration, perhaps mediated by water column anoxia? Most of the periods of black shale formation have been called "ocean anoxic events" (OAEs), suggesting the later formation process. Barite ($BaSO_4$) is a diagenetically stable mineral and its accumulation rate in oxic sediments shows an excellent (but empirical) correlation with new production (carbon export) in surface waters (Dymond et al., 1992, Francois et al., 1995; Paytan et al., 1996). The accumulation rates of barite during many of these OAEs show marked increases (and preceded the OAEs), arguing for a high flux mechanism, although an anoxic contribution cannot be completely eliminated (Paytan et al., 2004).

In the OEA studies, the mechanisms driving these higher fluxes (i.e., the forward Redfield equation), presumably relieving nutrient (N, P) or micronutrient (e.g., Fe) limitation, were not examined. Towards this end, and as an example, the dissolved trace element selenium exists in multiple oxidation states in seawater, the most abundant of which, selenate (Se+6), has depth and horizontal surface distributions that closely follow nitrate (Fig. 1). Thus, selenate may act as a proxy for nitrate in much the same way that dissolved Cd is a proxy for phosphate (Boyle, 1981). More significantly, selenate is incorporated into the barite crystal structure (Gaines et al., 2000; $BaSeO_4$ has a low solubility product like $BaSO_4$), and determining the Se/S ratio in marine barites may simply and accurately record upper ocean nitrate concentrations over long periods of time. To assess whether this new proxy may work, some preliminary data from core top barite Se/S across the equatorial Pacific (where there are also surface water selenate data) are shown in Fig. 1. The average barite Se/S in the equatorial upwelling region is $92 \pm 64 \times 10^{-8}$, while core top barite Se/S ratios underlying oligotrophic waters that were not part of this Pacific N–S transect average $11.5 \pm 5.8 \times 10^{-8}$, indicating that the values from the equatorial Pacific upwelling region are significantly enriched like the mixed layer concentrations of selenate. Although limited in coverage, these data make a case that the Se/S ratio in barite is recording mixed layer selenate. Because of the strong correlation between selenate and nitrate ($r=0.92$ for the

data in Fig. 1), marine barite is also recording the surface water nitrate concentrations. Thus, with barite accumulation tracking C export, both surface water nitrate concentration and the new production (C flux) that depends on this nitrate may be simultaneously measured with the same sediment phase. The determination of barite Se/S has yet to be done for the OAEs, but this single phase/multiproxy approach may prove useful for many environments, and the addition of Se isotope determinations in barite may even provide some redox information on the OAEs (Se isotope systematics appear to be similar to that of S; Johnson et al., 1999). Other proxies in other sediment phases (e.g., biomarkers for specific phytoplankton; Cd/Ca in forams) could further delineate the processes affecting the Redfield status of past surface waters. And this multiproxy approach, I suggest, is the challenge for future biogeochemists looking back in time.

References

Anbar, A.D., Knoll, A.H., 2002. Proterozoic ocean chemistry and evolution: a bioinorganic bridge? Science 297, 1137–1142.

Arnold, G.L., Anbar, A.D., Barling, J., Lyons, T.W., 2004. Molybdenum isotope evidence for widespread anoxia in mid-Proterozoic oceans. Science 304, 87–90.

Barber, R.T., Ryther, J.H., 1969. Organic chelators: factors affecting primary production in the Cromwell Current upwelling. J. Exp. Mar. Biol. Ecol. 3, 191–199.

Berner, R.A., 2003. The long-term carbon cycle, fossil fuels and atmospheric composition. Nature 426, 323–326.

Berner, R.A., Canfield, D.E., 1989. A new model for atmospheric oxygen over Phanerozoic time. Am. J. Sci. 289, 333–361.

Bjerrum, C.J., Canfield, D.E., 2002. Ocean productivity before about 1.9 Gyr ago limited by phosphorus adsorption onto iron oxides. Nature 417, 159–162.

Boyd, P.W., et al., 2000. A mesoscale phytoplankton bloom in the polar Southern Ocean stimulated by iron fertilization. Nature 407, 695–702.

Boyle, E.A., 1981. Cadmium, zinc, copper, and barium in foraminifera tests. Earth Planet. Sci. Lett. 53, 11–35.

Boyle, E.A., 1988. Cadmium: chemical tracer of deepwater paleoceanography. Paleoceanography 3, 471–489.

Brand, L.E., Sunda, W.G., Guillard, R.R.L., 1983. Limitation of marine phytoplankton reproductive rates by zinc, manganese, and iron. Limnol. Oceanogr. 28, 1182–1195.

Brand, L.E., Sunda, W.G., Guillard, R.R.L., 1986. Reduction of marine phytoplankton reproduction rates by copper and cadmium. J. Exp. Mar. Biol. Ecol. 96, 225–250.

Bruland, K.W., 1989. Oceanic zinc speciation: complexation of zinc by natural organic ligands in the central North Pacific. Limnol. Oceanogr. 34, 267–283.

Bruland, K.W., Donat, J.R., Hutchins, D.A., 1991. Interactive influences of bioactive trace metals on biological production in oceanic waters. Limnol. Oceanogr. 36, 1555–1577.

Byrne, R.H., Kump, L.R., Kantrell, K.J., 1988. The influence of temperature and pH on trace metal speciation in seawater. Mar. Chem. 25, 163–181.

Canfield, D.E., 1998. A new model for Proterozoic ocean chemistry. Nature 396, 450–453.

Charlson, R.J., Lovelock, J.E., Andreae, M.O., Warren, S.G., 1987. Oceanic phytoplankton, atmospheric sulphur, cloud albedo and climate. Nature 326, 655–661.

Coale, K.H., Johnson, K.S., Fitzwater, S.E., Blain, S.P.G., Stanton, T.P., Coley, T.L., 1998. IronEx-I, an in situ iron-enrichment experiment: experimental design, implementation and results. Deep-Sea Res., II, Top. Stud. Oceanogr. 45, 919–945.

Coale, K.H., et al., 2004. Southern Ocean iron enrichment experiment: carbon cycling in high- and low-Si waters. Science 304, 408–414.

Cochlan, W.P., 2001. The heterotrophic bacterial processes in response during a mesoscale iron enrichment (IronEx II) in the eastern equatorial Pacific. Limnol. Oceanogr. 46, 428–435.

Croot, P.L., Moffett, J.W., Brand, L.E., 2000. Production of extracellular Cu complexing ligands by eucaryotic phytoplankton in response to Cu stress. Limnol. Oceanogr. 45, 619–627.

Cutter, G.A., Bruland, K.W., 1984. The marine biogeochemistry of selenium: a re-evaluation. Limnol. Oceanogr. 29, 1179–1192.

Duarte, C.M., Piro, O., 2001. Interdisciplinary challenges and bottlenecks in the aquatic sciences. Limnol. Oceanogr. Bull. 10, 57–61.

Dymond, J., Suess, E., Lyle, M., 1992. Barium in deep-sea sediment: a geochemical proxy for paleoproductivity. Paleoceanography 7, 163–181.

Elser, J.J., Sterner, R.W., Galford, A.E., Chrzanowski, T.H., Findlay, D.L., Mills, K.H., Paterson, M.J., Stainton, M.P., Schindler, D.W., 2000. Pelagic C:N:P stoichiometry in a eutrophied lake: responses to a whole-lake food-web manipulation. Ecosystems 3, 293–307.

Farquhar, J., Bao, H., Thiemens, M., 2000. Atmospheric influence of Earth's earliest sulfur cycle. Science 289, 756–758.

Francois, R., Altabet, M.A., Burckle, L.H., 1992. Glacial to interglacial changes in surface nitrate utilization in the Indian sector of the southern ocean as recorded by sediment d^{15}N. Paleoceanography 7, 589–606.

Francois, R., Honjo, S., Manganini, S., Ravizza, G., 1995. Biogenic barium fluxes to the deep-sea: implications for paleoproductivity reconstruction. Glob. Biogeochem. Cycles 9, 289–303.

Gaines, R.V., Skinner, H.C.W., Foord, E.E., Mason, B., Rosenzwei, A., King, V.T., 2000. Dana's new mineralogy. J. Wiley.

Ganeshram, R., Pedersen, T., Calvert, S., McNeill, G., Fontugen, M., 2000. Glacial–interglacial variability in denitrification in the world's oceans: causes and consequences. Paleoceanography 15, 361–376.

Gervais, F., Riebesell, U., Gorbunov, M.Y., 2002. Changes in primary productivity and chlorophyll a in response to iron fertilization in the Southern Polar Frontal Zone. Limnol. Oceanogr. 47, 1324–1335.

Holland, H.D., Beukes, N., 1990. A paleoweathering profile from Griqualand West South Africa: evidence for a dramatic rise in atmospheric oxygen between 2.2 and 1.9 bybp. Am. J. Sci. 290A, 1–34.

Johnson, T.M., Herbel, M.J., Bullen, T.D., Zawislanski, P.T., 1999. Selenium isotope ratios as indicators of selenium sources and oxyanion reduction. Geochim. Cosmochim. Acta 63, 2775–2783.

Landry, M.R., Barber, R.T., Bidigare, R.R., Chai, F., Coale, K.H., Dam, H.G., Lewis, M.R., Lindley, S.T., McCarthy, J.J., Roman, M.R., Stoecker, D.K., Verity, P.G., White, J.R., 1997. Iron and grazing constraints on primary production in the central equatorial Pacific: an eqPac synthesis. Limnol. Oceanogr. 42, 405–418.

Law, C.S., Watson, A.J., Liddicoat, M.I., Stanton, T., 2001. Sulphur hexafluoride as a tracer of biogeochemical and physical processes in an open-ocean iron fertilisation experiment. Deep-Sea Res., II, Top. Stud. Oceanogr. 45, 977–994.

Martin, J.H., 1990. Glacial–interglacial CO_2 change: the iron hypothesis. Paleoceanography 5, 1–13.

Martin, J.H., Fitzwater, S., 1988. Iron deficiency limits phytoplankton growth in the north-east Pacific subarctic. Nature 331, 341–343.

Martin, J.H., Gordon, R.M., 1988. Northeast Pacific iron distributions in relation to phytoplankton productivity. Deep-Sea Res. 35, 177–196.

Martin, J.H., Gordon, R.M., Fitzwater, S.E., Broenkow, W.W., 1989. VERTEX: phytoplankton/iron studies in the Gulf of Alaska. Deep-Sea Res. 36, 649–680.

Martin, J.H., Gordon, R.M., Fitzwater, S., 1991. The case for iron. Limnol. Oceanogr. 36, 1793–1802.

Moffett, J.W., 1995. Temporal and spatial variability of copper complexation by strong chelators in the Sargasso Sea. Deep-Sea Res., I, Oceanogr. Res. Pap. 42, 1273–1295.

Moore, J.K., Doney, S.C., Glover, D.M., Fung, I.Y., 2001. Iron cycling and nutrient-limitation patterns in surface waters of the World Ocean. Deep-Sea Res., II, Top. Stud. Oceanogr. 49, 463–507.

Paytan, A., Kastner, M., Chavez, F., 1996. Glacial to interglacial fluctuations in productivity in the equatorial Pacific as indicated by marine barite. Science 274, 1355–1357.

Paytan, A., Martinez-Ruiz, F., Engle, M., Ivy, A., Wankel, S.D., 2004. Using sulfur isotopes to elucidate the origin of barite associated with high organic matter accumulation events in marine sediments. In: Amend, J.P., Edwards, K.J., Lyons, T.W. (Eds.), Sulfur Biogeochemistry—Past and Present, Special Paper-Geological Society of America, vol. 379, pp. 151–160.

Redfield, A.C., Ketchum, B.H., Richards, F.A., 1963. The influence of organisms on the composition of sea-water. In: Hill, M.N. (Ed.), Interscience, The Sea, vol. 2, pp. 26–77.

Rue, E.L., Bruland, K.W., 1995. Complexation of iron(III) by natural organic ligands in the Central North Pacific as determined by a new competitive ligand equilibration/adsorptive cathodic stripping voltammetric method. Mar. Chem. 50, 117–138.

Sigman, D.M., Altabet, M.A., Francois, R., McCorkle, D.C., Gaillard, J.-F., 1999. The isotopic composition of diatom-bound nitrogen in Southern Ocean sediments. Paleoceanography 14, 118–134.

Tsuda, A., et al., 2003. A mesoscale iron enrichment in the Western Subarctic Pacific induces a large centric diatom bloom. Science 300, 958–961.

Turner, D.R., Whitfield, M., Dickson, A.G., 1981. The equilibrium speciation of dissolved components in freshwater and seawater at 25 °C and 1 atm pressure. Geochim. Cosmochim. Acta 45, 855–881.

Printed and bound by CPI Group (UK) Ltd, Croydon, CR0 4YY

08/05/2025

01864933-0004